D0207502

A C S   S Y M P O S I U M   S E R I E S   **553**

Nitrosamines and Related N-Nitroso Compounds

Chemistry and Biochemistry

Richard N. Loeppky, EDITOR

University of Missouri—Columbia

Christopher J. Michejda, EDITOR

National Cancer Institute—Frederick Cancer Research and Development Center

Developed from a symposium sponsored by the Division of Agricultural and Food Chemistry at the 204th National Meeting of the American Chemical Society, Washington, D.C., August 23–28, 1992

American Chemical Society, Washington, DC 1994

Library of Congress Cataloging-in-Publication Data

Nitrosamines and related N-nitroso compounds: chemistry and biochemistry / Richard N. Loeppky, Christopher J. Michejda.

p.   cm.—(ACS symposium series, ISSN 0097–6156; 553)

"Developed from a symposium sponsored by the Division of Agricultural and Food Chemistry at the 204th National Meeting of the American Chemical Society, Washington, D.C., August 23–28, 1992."

Includes bibliographical references and index.

ISBN 0–8412–2856–6

1. Nitrosamines—Congresses.

I. Loeppky, Richard N.   II. Michejda, Christopher J., 1937–   .
III. American Chemical Society. Division of Agricultural and Food Chemistry.   IV. American Chemical Society. National Meeting (204th: 1992: Washington, D.C.)   V. Series.

QD305.N8N62   1994
615.95′142—dc20   94–2014
                                                                      CIP

Foreword

THE ACS SYMPOSIUM SERIES was first published in 1974 to provide a mechanism for publishing symposia quickly in book form. The purpose of this series is to publish comprehensive books developed from symposia, which are usually "snapshots in time" of the current research being done on a topic, plus some review material on the topic. For this reason, it is necessary that the papers be published as quickly as possible.

Before a symposium-based book is put under contract, the proposed table of contents is reviewed for appropriateness to the topic and for comprehensiveness of the collection. Some papers are excluded at this point, and others are added to round out the scope of the volume. In addition, a draft of each paper is peer-reviewed prior to final acceptance or rejection. This anonymous review process is supervised by the organizer(s) of the symposium, who become the editor(s) of the book. The authors then revise their papers according to the recommendations of both the reviewers and the editors, prepare camera-ready copy, and submit the final papers to the editors, who check that all necessary revisions have been made.

As a rule, only original research papers and original review papers are included in the volumes. Verbatim reproductions of previously published papers are not accepted.

M. Joan Comstock
Series Editor

Contents

NITROSAMINE OCCURRENCE:
BRIEF DISCUSSIONS OF RESEARCH

Preface

N-NITROSO COMPOUNDS MAY HAVE A SIGNIFICANT ROLE in human carcinogenesis. The ubiquity of their precursors leads to the ready formation of nitrosamines and other N-nitroso compounds in the human micro- and macroenvironment. Therefore, there is a continuing need to understand the extent and mechanisms of N-nitroso compound formation, as well as to understand the mechanisms of their bioactivation and detoxification. Recent discoveries have demonstrated several pathways for the endogenous formation of N-nitroso compounds. Ingested or endogenous nitrogenous substrates can react with nitrous acid in the stomach or be nitrosated either there or elsewhere by nitrosating agents arising from the endogenous formation of NO or the bacterial reduction of nitrate.

This volume focuses on the chemistry and biochemistry of N-nitroso compounds. It ranges from topics covering fundamental chemical mechanistic and analytical studies, biochemical models, and DNA adduct formation, to the use of molecular biological methods and other techniques to investigate biological effects of N-nitroso compounds.

The symposium from which this volume grew was intended to refocus attention on the chemical and biochemical research of N-nitroso compounds, rather than the epidemiological efforts that have been stressed in many international meetings in recent years. This volume presents aspects of the most important current work on N-nitroso compounds. In addition to the research papers presented in the first 22 chapters, we have included 23 brief summaries of other current research. The authors are the major investigators in the field. This book is important to those chemists who actively work to reduce human exposure to carcinogenic agents and to those many chemists who seek through their work to achieve a better understanding of the mechanisms of the carcinogenic process. The book will also be meaningful to those who want to enhance their knowledge of the chemistry and biochemistry of N-nitroso compounds.

Acknowledgments

We are grateful to the American Chemical Society Division of Agricultural and Food Chemistry, which provided financial support for this symposium. The cooperation and suggestions of Richard A. Scanlan of Oregon State University are especially appreciated. We also appreciate the technical assistance of ACS editors Rhonda Bitterli and Barbara Tansill. We are particularly indebted to Christina A. Wells, who served most cooperatively between appointments, as our technical editor. Her organizational talents, high standards, and excellent editorial skills have improved this volume significantly.

RICHARD N. LOEPPKY
Department of Chemistry
University of Missouri
Columbia, MO 65211

CHRISTOPHER J. MICHEJDA
ABL–Basic Research Program
Frederick Cancer Research and Development Center
National Cancer Institute
P.O. Box B
Frederick, MD 21702

Received November 15, 1993

Chapter 1

Nitrosamine and *N*-Nitroso Compound Chemistry and Biochemistry

Advances and Perspectives

Richard N. Loeppky

Department of Chemistry, University of Missouri, Columbia, MO 65211

A review of the significant developments, and changes in focus and perspective in the chemistry and biochemistry of nitrosamines and other N-nitroso compounds over the past twelve years is presented. Research in this field has moved away from identifying nitrosamine contaminants of commercial mixtures. Significant effort has been directed at understanding the possible role of endogenous nitrosation in man and development of analytical chemical markers to be used in molecular epidemiology. The successes and failures of this approach are reviewed. A highly significant sidelight to this research was the monumental discovery that NO is produced by many cells. Macrophages produce quantities sufficient to lead to the nitrosation of amines and the deamination of DNA. Fragments of nitrosamines derived from nicotine have been detected in human hemoglobin and DNA. Significant advances have been made in understanding the formation of nitrosamines from a variety of nitrogen containing compounds and new methods for blocking nitrosamine formation have been developed. The biochemistry of the α-hydroxylation of simple nitrosamines and other biochemical activation pathways applicable to nitrosamines containing OH, C=O, and other groups is much better understood and is reviewed. Possible future research needs and directions are presented.

Nitrosamines constitute a family of potent carcinogens which are formed readily from a diverse set of nitrogen compounds, and nitrite or its derivatives (nitrogen in a formal +3 oxidation state). The ubiquity of the precursor compounds and the relative ease of formation of nitrosamines has led to the statement "like uninvited guests at a party, nitrosamines just won't go away" (*1*). In this chapter, we review the research imperatives and developments which have directed the field over the course of the last twelve years (2) and comment briefly on the future.

Perspective. The carcinogenic properties of nitrosamines and other N-nitroso compounds have been known for over thirty years (3). It would seem that, over the course of this period, we would have a nearly complete knowledge of the chemistry and biochemistry of these compounds, and that little further research in this area would be necessary. Both the biochemistry and chemistry of these compounds, however, is sufficiently complex, that many of the answers sought by researchers have been slow to emerge. Research emphasis has also been impacted by governmental policy decisions and funding priorities.

Over three hundred nitrosamines have been tested for carcinogenicity and 90% of these compounds show activity (3). Careful dose response studies have shown that many of these compounds have a high degree of potency, and di-methylnitrosamine has been found to be carcinogenic in more than 20 species of animals (3). More importantly, no species has been found to be exempt from carcinogenicity of this compound. Nitrosamines have the interesting characteristic of being relatively organ specific carcinogens.

Background. N-nitroso compounds can be placed in two broad categories. Of these the nitrosamines, being amides of nitrous acid, are the most stable and are formally derived from the reaction of the secondary amine with nitrous acid. The second class of the N-nitroso compounds (N-nitrosoamide type) are those substances which have a carbonyl group attached to the nitrogen bearing the NO group. Members of this class include the reactive N-nitrosoamides, N-nitrosocarbamates, and N-nitrosoureas. The instability of compounds of the N-nitroso amide type is due to the joining of two very electropositive functional groups (NNO and CO). N-Nitrosoamides have been proposed to rearrange as shown in equation 1. The hypothetical intermediate decomposes to a diazonium ion. These compounds also undergo base catalyzed decomposition to give diazoalkanes. While the mechanism of catalyzed and solvolytic decomposition of compounds of the N-nitro-soamide type is still an active area of study, the N-nitrosoureas (equation 2) and N-nitrosoguanidines appear to decompose by at least two pathways (4,5). One reaction path involves the attack of a nucleophile at the C=O (or

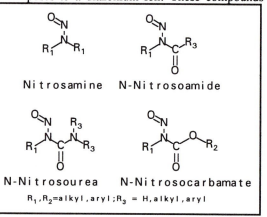

C=N-) to generate a tetrahedral intermediate which decomposes with the generation of a diazotate ion (R-N=N-O⁻). Biologically prevalent thiols can catalyze the decomposition of N-methyl-N-nitroso-N'-nitroguanindine (MNNG) by

this pathway (*4*). The other pathway generates the diazotate ion by means of a β-elimination involving a H on the attached heteroatom. This pathway also generates reactive isocyanates. Regardless of mechanism, potent alkylating agents are readily generated from compounds of the N-nitrosoamide type.

While compounds of the N-nitrosoamide type are known to be direct mutagens (require no biochemical activation) the same is not true for nitrosamines. Nitrosamines must be metabolized in order to elicit their mutagenic or carcinogenic properties (*3*). While the mode of carcinogenic biochemical activation is not known for all nitrosamines, many of them are activated through the process of α-hydroxylation. As shown in equation 3, the α-hydroxy nitrosamines resulting from this process are chemically unstable and decompose readily to diazonium ions, which are aggressive alkylating agents. In biological organisms, the result is the alkylation of DNA, RNA, and protein. There is good evidence that these processes are related to the adverse biological effects of these compounds (*3*).

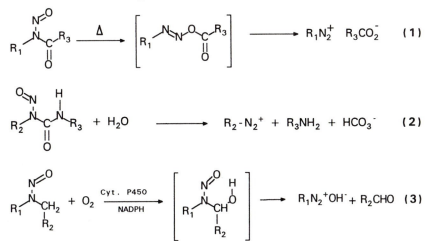

N-Nitroso Compounds and Human Carcinogenesis. Since the potent animal carcinogenicity of these compounds is the driving force for much of the research in the field, it is important to ask, are these compounds human carcinogens? The answer to this question is important because of the ease of which nitrosamines are formed both exogenously and endogenously. All humans excrete the non-carcinogenic amino acid derivative N-nitrosoproline in their urine (*6*). This finding and other studies clearly demonstrate that N-nitroso compounds are formed within the human body (*7*). Over the past twelve year period a great deal of effort has been expended in an attempt to establish an epidemiological connection between human cancer and endogenous N-nitroso compound formation and exogenous exposure to preformed N-nitroso compounds. While highly definitive conclusions are difficult to make, the evidence connecting nitrosamines and more reactive N-nitroso compounds to human cancer can be summarized as follows: Tobacco and tobacco smoke contain significant concentrations of nitrosamines. Two of the more carcinogenic are N-nitrosonornicotine (NNN) and 4-methylnitrosamino-1-(3-pyridyl)-1-butanone (NNK), both derived from the nitrosation of nicotine (*8*). Careful dose response studies have shown NNK to be a powerful lung carcinogen in three species of rodents, despite the mode of administration (*8*). Molecular fragments from NNK and NNN have been found in the hemoglobin and DNA of humans exposed to high concentrations of these nitrosamines (*11*). Epidemiological studies have linked oral cavity cancer to snuff usage (*10*). To date, the nitrosamines derived from the nicotine alkaloids are the only significant carcinogenic agents found in smokeless tobacco (*11*). On the other hand, because of the many carcinogenic agents in tobacco, a precise epidemiological connection between tobacco related cancers and nitrosamines is difficult to make.

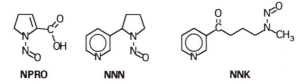

NPRO   **NNN**   **NNK**

There is also good evidence linking N-nitroso compounds to gastric cancer in areas where there is a high dietary nitrate and nitrite intake. The collaborative work of Correa and Tannenbaum (*12*) is notable and discussed below. This evidence, coupled with the fact that nitrosamines are extremely potent carcinogens in a wide variety of animals, regardless of species, suggests that nitrosamines are also human carcinogens.

N-Nitroso Compounds as Drugs. While the deleterious effects of many N-nitroso compounds has been the focal point of much research and practical work over the last several decades, several new types of compounds are receiving clinical attention and new efforts continue to be directed at developing new efficacious antitumor drugs from N-nitrosoureas. Two N-nitrosoureas, BCNU (bischloroethyl-nitrosourea) and CCNU (1-chloroethyl-3-cyclohexyl-1-nitrosourea), have been used as anti cancer agents in the clinic for a number of years. Yet, the chemistry and mechanism of undesirable toxicity of these substances

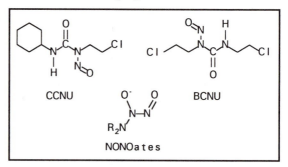

is not completely understood. In addition to attempts to reduce their toxicity through mechanism studies and structure change, other research has sought to develop N-nitrosoureas which are covalently linked to molecules which bind to the receptors of specific tumors.

Following the discovery of the biological significance of NO, considerable research has been directed at the development of "NO releasing" drugs. Among the promising drugs in this field are the "NONOates", a set of new substances, chemically related to N-nitroso compounds. "NONOates" (an acronym for the compound family) are capable of releasing NO and have the promise of treating a number of diseases (see Keefer, this volume). This discovery is a direct outgrowth of research in the nitrosamine field.

Changes in Research Orientation

The 1970's witnessed the development of a powerful and highly selective device for detecting and quantitating nitrosamines and related N-nitroso compounds (*13*). This device, the TEA Analyzer (a trade name), is connected as a detector to either a GC, or a HPLC using a non-aqueous system. The development of the detector led to the discovery of the occurrence of trace quantities of nitrosamines in a variety of commercial products. A great deal of activity was spent on the development of methods for preventing the formation of these compounds and other analytical methodology for their detection in a wide variety of matrices. At the end of this period, more than a decade ago, many researchers believed that the next step involved the development of a sensitive analytical methodology for the detection of non-volatile nitrosamines using reversed phase HPLC to provide separation. The expectation was, that this development would, through detection, give rise to a new wave of research on nitrosamine formation and blocking of non-volatile N-nitroso compounds. These expectations have largely not been realized.

Beginning in 1980 significant policy and regulatory philosophy changes in the U.S. toward hazardous contaminants, such as nitrosamines, has resulted in very little information on nitrosamine occurrence during the intervening period. A great deal of work on the chemistry and biochemistry of nitrosamines as it relates to their occurrence and carcinogenic properties has been done in Germany, however. The German government sponsored a *Schwerpunkt* program under the direction of Rudolph Preussmann of the German Federal Cancer Research Institute. This program was very successful in revealing and eliminating many of the hazards due to volatile nitrosamines in that country and had a world-wide impact (*14*). The program, however, ended in 1982 and the general level of research in this area has also diminished considerably.

For a number of years, the International Agency for Research in Cancer (IARC), a subdivision of the World Health Organization (WHO), sponsored a number of conferences and publications in their scientific series on the chemistry, biochemistry, and other properties of nitrosamines and N-nitroso compounds (*15-17*). During the period of the last decade, much of the research sponsored and featured at these meetings has been directed away from the chemistry and biochemistry of these compounds and toward epidemiology. There has been a great thrust by this agency, through its programs, to determine whether nitrosamines are human carcinogens. This research was stimulated by the finding that humans synthesize and excrete N-nitrosoproline in their urine. Numerous studies directed at correlating this possible marker of *in vivo* N-nitroso compound formation have met with mixed success but promise significant findings (see below).

While not much information has been forthcoming from the private sector regarding the occurrence of nitrosamines and methods for blocking their prevention, significant advances have been made through publicly supported research. German researchers (Pruessmann and Spiegelhalder (*18*), Eisenbrand, (*19-20*)) have focused a great deal of attention on nitrosamine formation in the occupational setting and have developed specific strategies for reducing nitrosamine formation in the rubber industry (see Speigelhalder et al., this volume) and in metal working fluids. Other work in our laboratory has resulted in the generation of both monomeric (*21*) and polymeric blocking agents (*22,23*) which are very effective at blocking nitrosamine formation under real world conditions. Research has led to recommendations which will lessen endogenous formation of N-nitroso compounds (*24*). There are probably numerous other unpublicized examples of effective measures which have been taken to reduce nitrosamine exposure. However, the lack of published surveillance information from government laboratories, particularly in the United States, does not permit anything but an anecdotal assessment in this area.

Endogenous Nitrosation

Studies which showed that simultaneous feeding of nitrite and secondary amines to animals generates tumors led to the hypothesis that nitrosamines could form endogenously (*25-27*). The possible deleterious effects of dietary nitrite were recognized quite early in the history of nitrosamine carcinogenesis and measures were taken to limit nitrite concentrations in foods. Because of both its natural occurrence and its widespread use in intensive agriculture, nitrate exposure and dietary intake is much more common than that of nitrite. Careful studies showed that nitrate is partially reduced in the oral cavity to nitrite which then reacts with stomach acid to produce nitrous acid and actively nitrosates nitrogen compounds (*28,29*). Nitrite does not circulate in the bloodstream because it is oxidized by oxyhemoglobin to nitrate. Some of the nitrate which circulates in the bloodstream is reduced in the saliva to nitrite. The quantity generated by this latter pathway is often larger than that produced from the immediate reduction of nitrate following consumption (*28*).

The relatively facile and rapid metabolism of nitrosamines has confounded research on the extent and nature of endogenous nitrosation. Metabolic oxidation of nitrosamines often results in the formation of reactive α-hydroxynitrosamines which, in addition to the production of deleterious electrophiles, generate smaller molecules which easily enter existing metabolic pathways and are not easily traced. Nevertheless, Preussmann and coworkers demonstrated the urinary excretion of dimethylnitrosamine (DMN) in humans following the administration of the readily nitrosated anlagesic drug aminopyrine. DMN is produced by gastric nitrosation of aminopyrine but its metabolism was inhibited by the simultaneous administration of ethanol (*30*). Nitrosamino acids such as nitrosoproline are not readily metabolized and their use as markers of endogenous nitrosation has been sought.

Prior to the discovery of endogenous NPRO synthesis, Tannenbaum's group at MIT was involved in an extensive investigation of the nature and extent of endogenous nitrosation (*31,32*). Key to this investigation was an understanding of nitrate balance and disposition in humans. Reports dating back to the early part of this century suggested that nitrate was synthesized by humans. Through a set of careful and elaborate experiments, Tannenbaum and colleagues used $^{15}NO_3^-$ to determine that humans are net synthesizers of nitrate, but only 60% of the labeled nitrate could be accounted for *(31)*. These data also suggested reductive metabolism for nitrate. During the course of experiments designed to determine the source of the biosynthesized nitrate, serendipity played a very important role in leading the investigators to a monumental discovery. Unusually high rates of nitrate production were found in an individual who was suffering from gastroenteritis and had a high fever. This observation led the Tannenbaum group to speculate that excess nitrate synthesis was related to an immune response to infection, a hypothesis which they subsequently proved in animal experiments *(32)*.

Marletta, a colleague of Tannenbaum's, sought the source and mechanism of nitrate synthesis in response to infectious toxic stimulus of the immune system. His research soon led to the discovery that macrophages were responsible for synthesizing nitrite and nitrate in response to challenge by toxic lipopolysaccharide *(33,34)*. Marletta then showed that the source of the nitrogen in these molecules comes from the guanidino nitrogens of L-arginine *(33)*. Subsequent collaborative work between Tannenbaum's and Marletta's groups demonstrated that stimulated macrophages were capable of amine nitrosation and that the source of the nitrosating agents (eventually being analyzed as nitrite and nitrate) was nitric oxide, NO *(35)*.

These discoveries had importance far beyond the field of nitrosamine chemistry and carcinogenesis. The discovery of Marletta and Tannenbaum resulted in finding a key puzzle piece in biochemistry and physiology *(36)*. Nitric oxide is now recognized to be a very important short lived signaling agent and effector of guanylate cyclase. It is also known as the principal constituent of what has been referred to for years as EDRF, the endothelial-derived relaxing factor, which is responsible for the relaxation and dilation of the blood vessels. Many different types of cells have now been shown to be capable of synthesizing NO from L-arginine. Macrophages, which are recruited as foreign cell killers to the sites of infection have an extremely high capacity for NO production. (See Tannenbaum and Wishnok, this volume, for a more complete discussion.)

Bacteria are also capable of mediating the nitrosation of amines *(37-38)*. This supposition had remained in a speculative status for many years because many of the studies in this field were experimentally flawed, not properly controlled, or plagued with other difficulties *(39)*. It is well known that there are many bacteria capable of reducing nitrate to nitrite. If the resulting nitrite is released into an acidic environment then chemical nitrosation processes can occur. It has been shown, however, that bacteria are capable of mediating the nitrosation of amines under the conditions where the pH is too high to permit HNO_2 mediated nitrosation *(37)*. This finding is of significance not only to the area of endogenous nitrosation, where bacterial colonization may be a problem, but also in the many processes which involve fermentation. The presence of amines or other nitrogen compounds in complex fermentation mixtures containing nitrate can clearly lead to amine nitrosation through bacterial mediated nitrate reduction and nitrosation.

The IARC and several other organizations sponsored large scale studies aimed at correlating the level in N-nitrosoproline excretion in humans with cancer risk in high risk areas *(40)*. Studies were done in China, Japan, Colombia, and several European countries. Careful analysis of the data from many of these studies failed to reveal a significant correlation between N-nitrosoproline excretion and the risk (*40*). To some extent this is not surprising since common dietary constituents such as vitamin C, phenolic compounds, and other substances are capable of scavenging nitrosating agents. Moreover, the effect of bacterial colonization on endogenous nitrosation through synthesis of nitrosating agents further confounds these studies. There has been one notable success, however, in the work of Correa and colleagues (*12*). A careful study done in the country of Colombia has demonstrated a positive correlation between stomach cancer and nitrate and nitrosoproline excretion in the urine. This study gives evidence that nitrosamines are human carcinogens.

Nitrosamines and Molecular Epidemiology

An important development in the area of carcinogenesis research has been the emergence of what is called "molecular epidemiology." While this new subfield has many different aspects, fundamentally it involves an attempt to correlate various molecular markers with cancer risk. In the nitrosamine area this research is best exemplified by the work of Hecht and his colleagues at the American Health Foundation (see pertinent chapters in this volume). Success has depended upon a knowledge of the metabolism and DNA and/or protein binding characteristics of fragments produced from a given nitrosamine. Hecht's group has shown that molecular fragments derived from either N-nitrosonornicotine or the related potent carcinogen NNK can be found in both the DNA and the hemoglobin of humans exposed to relatively high concentrations of the parent nitrosamines (*9,41*). This research requires not only an elucidation of the metabolic transformations of the nitrosamine and the nature of the electrophilic intermediates generated from them, but an establishment of the structure of the specific DNA or protein adducts produced from these reactive species. Very sensitive and precise analytical methodology is required. Animal studies have shown that the extent of adduct formation in either DNA or hemoglobin is related to dose and metabolic characteristics (*8*). These methods give the promise of being able to correlate adduct levels with cancer risk.

Analytical Methodology

As was discussed in the introduction of this paper, the field of nitrosamine chemistry has benefited significantly from the development of a highly specialized GC and HPLC detector called a TEA (*13*). The activity of this detector depends upon the thermal decomposition of nitrosamines to give NO which is then reacted with ozone to produce chemiluminescent nitrogen dioxide. The detector works very well with both GC and HPLC systems which use nonaqueous eluents. The TEA detector does not give reproducible data when interfaced with HPLC systems using reversed phase chromatography (aqueous eluents). This limitation has been overcome through the development of a new detector (*42,43*). This detector relies upon the photochemical dissociation of the nitroso compound and detects the resulting NO as nitrite, after oxidation and hydrolysis of the N_2O_3 produced. It works very effectively with reversed phase HPLC systems. The methodology has recently been incorporated into a commercially available system called a "Nitrolite" detector. The instrumentation has not been available long enough to have had significant impact, however.

Considerable effort has been put into the detection of nitrosamines and other nitroso compounds as a group (*44*). "Group" determination of the total quantity of N-nitroso compounds present in a sample, however, has some pitfalls and is particularly problematic with samples which contain large quantities of nitrite. A degree of reliability has been achieved, however, through some relatively recent refinements. Pignatelli and coworkers have published a number of papers on this methodology and the reader is referred to her paper in this volume and references contained therein. The continuing problem with this methodology is the observation of false positives. Nevertheless, the methodology has been used to estimate the N-nitroso content of various commercial mixtures as well as biological samples such as gastric juice. The "Nitrolite" detector can be used for group detection of N-nitroso compounds, as well, but too little published material is available at this time to permit a critical evaluation.

Occurrence

Although a number of reviews on nitrosamine occurrence have appeared recently (*45-52*), changes in focus and government policies have significantly reduced research on the occurrence of nitrosamines and other nitroso compounds in materials to which humans are exposed. Nitrosamine formation in the rubber industry continues to be a problem despite considerable progress (see Speigelhalder et al., this volume). A survey of the patent literature reveals that there has been a good deal of activity in the development of methods for removing dipropylnitrosamine from triflaurin, a common herbicidal agent.

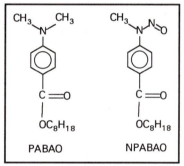

Relatively little other research has been published on the occurrence of nitrosamines in various pesticide preparations in recent years. Several surveys have indicated that the quantities of various nitrosamines found in cured meat have continued to remain at low levels through reduced nitrite concentrations (*50*). Similarly, following the discovery of the mechanism of dimethylnitrosamine formation in beer, changes in malting procedures have produced very low levels of nitrosamines in many of the world's beers and other malted beverages (*51*). On the other hand, relatively large concentrations of several nitrosamines have been found in various food samples from China (*52*). Several new nitrosamines have been found in tobacco products, but the principal carcinogenic nitrosamines, NNN and NNK, are always present (11).

Many cosmetic samples and other personal care items continue to contain N-nitrosodiethanolamine and some other nitrosamines (see Havery, this volume). A new nitrosamine, 2-ethylhexyl 4-N-methyl-N-nitrosaminobenzoate (NPABAO), which is derived from a common sunscreen ingredient (2-ethylhexyl 4-N,N-dimethylaminobenzoate, PABAO) was found in a number of sunscreens and other items such as lip gloss and lipstick (*53*). The details of the formation of this nitrosamine under the conditions of commercial formulation are not available, if known. However, we have demonstrated that it forms readily from minor amounts of the contaminating secondary amine and, most interestingly, more slowly but still rapidly from the corresponding tertiary amine, PABAO (*54*). A detailed report has concluded that this nitrosamine is not mutagenic in a number of different assays (*55*). On the other hand, our Ames tests demonstrated dose response mutagenicity for this substance (*54*). Its structural characteristics suggest low carcinogenicity, but as is the case with most nitrosamines, it is beneficial to find ways to prevent its trace formation in products containing materials such as a sunscreens.

Nitrosamine Formation

Considerable progress has been made during the last ten to twelve year period in understanding different mechanisms of nitrosamine formation which may be important in the human environment. Some of this work, particularly that done in Europe, has been stimulated by governmental adoption of what is known as the WHO NAP nitrosation test (*56,57*). This test determines whether N-nitroso compounds are formed when a nitrogen compound is incubated with relatively low concentrations of nitrous acid. Application of this test has resulted in the finding that several nitrosamines are formed rapidly from several amine containing drugs and other materials. A number of other compounds have been found to rapidly react with nitrous acid to produce nitrosamines (*58*) (see Loeppky et al., this volume). Several other studies have demonstrated how nitrosamines can be formed during cooking processes. For example, at cooking temperatures ionic nitrite can react with glycerol esters (fat) by means of an SN_2 displacement mechanism to produce glycerol nitrite esters. These esters readily react with amines to produce nitrosamines (*59*). This process, called "ester mediated nitrosation" (Scheme 1), can also

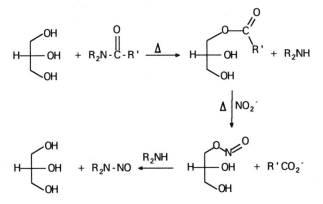

Scheme 1
Ester Mediated Nitrosation

result in nitrosamine formation when amides, ionic nitrite and nonvolatile alcohols (such as glycerol) are heated together under cooking temperatures. Nitrosamines derived from the amides can form under these conditions. This process can also occur during the formation of commercial preparations containing amides of diethanolamines. Hotchkiss' group has demonstrated that oxides of nitrogen can add to double bonds in unsaturated fats to generate compounds capable of facile nitrosation (Scheme 2) (*60*). Recent work by Challis' group has revealed mechanisms for rapid nitrosamine formation from bronopol (2-bromo-2-nitro-1,3-propanediol) (*61,62*). Bronopol is used as an antimicrobial agent in certain cosmetic and other personal care formulations. It liberates formaldehyde and nitrite which can catalyze nitrosamine formation when heated in the presence of amines.

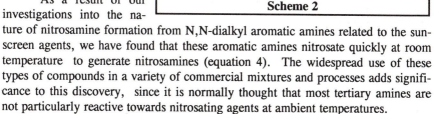

Scheme 2

As a result of our investigations into the nature of nitrosamine formation from N,N-dialkyl aromatic amines related to the sunscreen agents, we have found that these aromatic amines nitrosate quickly at room temperature to generate nitrosamines (equation 4). The widespread use of these types of compounds in a variety of commercial mixtures and processes adds significance to this discovery, since it is normally thought that most tertiary amines are not particularly reactive towards nitrosating agents at ambient temperatures.

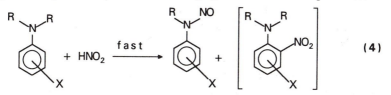

Pignatelli and others have carefully examined the formation of nitrosamines from carbohydrates in the presence of amino acids. The products of these reactions are N-nitroso Amadori compounds, substances which are quite polar and in some cases unstable (*63*). Chinese workers have detected the formation of nitroso peptides and other N-nitrosoamides in the stomach contents of humans at high risk for gastric cancer in that country (*64*). Other work has demonstrated that these substances are relatively reactive and are direct acting mutagens.

Chemistry and Biochemistry of Activation

A great deal of progress has been made in understanding the biochemistry of the enzymatic α-hydroxylation process (*65*) (equation 2.), which appears to be very important in the carcinogenic activation of many nitrosamines (*66*). The principal enzymes involved in the catalysis of this transformation are various isozymes of cytochrome P-450. While some of these isozymes have an overlapping substrate specificity, the finding of localized high concentrations of specific isozymes in particular tissues or parts of tissue has helped explain one facet of the organ specificity associated with nitrosamine carcinogenesis. Rat esophageal tissue has localized cytochrome P-450 isozymes which have the capacity to selectively α-hydroxylate methylalkylnitrosamines on the alkyl chain and produce DNA methylation from the resulting methyldiazonium ion (67). These compounds are selective esophageal carcinogens in these animals.

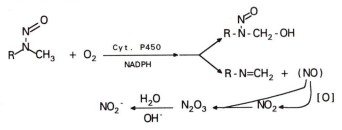

Scheme 3

Collaborative work between Keefer's and Yang's groups showed that the cyt. P450 mediated metabolism of nitrosamines can lead to competitive α-hydroxylation or denitrosation (Scheme 3) (68,69). The latter process can result in the detoxification of a nitrosamine while the former leads to highly reactive electrophilic intermediates. These competitive pathways follow the abstraction of an α-hydrogen atom from the nitrosamine to generate a radical whose fate depends upon its structure and the environment (P-450 isozyme). The radical has been independently generated (Scheme 4) and found to lose NO at rates slightly greater than the rates of hydrogen abstraction from the best H atom donors (70,71) (e.g. thiols). In the case of the cytochrome P-450 2E1 catalyzed reaction of dimethylnitrosamine with oxygen, only 20% of the radical loses NO. The remainder is trapped by the oxygenated enzyme bound heme to yield the α-hydroxynitrosamine.

α-Hydroxynitrosamines have a half life of less than 1 min. at 37°, but are more stable in non aqueous media (72). Wiessler et. al have synthesized glucuronide (73) and phosphate (74) conjugates of α-hydroxynitrosamines. The former derivatives have been shown to be produced *in vivo* and could account for some of the organ specific characteristics of nitrosamine carcinogenesis.

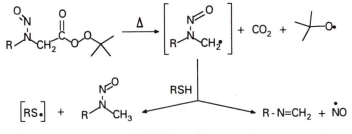

Scheme 4

While there has been a general recognition that the first step in the carcinogenic activation of many nitrosamines involves α-hydroxylation this does not appear to be true for many commonly occurring nitrosamines such as N-nitrosodiethanolamine (NDELA) and related compounds. The side chains of nitrosamines bearing β-hydroxy groups do not appear to be α-hydroxylated (*75*). The puzzle posed by the carcinogenic activation of these compounds has generated a number of interesting hypotheses and model chemical and biochemical studies to test them (Scheme 5). NDELA, as well as other β-hydroxynitrosamines, are substrates for mammalian liver alcohol dehydrogenase (*76-80*). The resulting α-nitrosamino aldehydes are unusually reactive, direct acting mutagens capable of transferring their N-nitroso group to other amines as well as the primary amino groups in DNA (*76,77,79*). This latter process leads to deamination and a change in base-pairing characteristics. α-Nitrosamino aldehydes also generate glyoxal fragments which specifically bind to guanine in DNA (*81,82*).

Other pathways proposed for the carcinogenic activation of these substances involve chain shortening (*83-85*), followed by α-hydroxylation, or sulfate conjugation of the hydroxyl group (*86,87*). The latter process is proposed to be followed by and intramolecular nucleophilic displacement by the nitrosamine oxygen to generate a more reactive 3-alkyl-1,2,3-oxadiazolinium ion capable of alkylating DNA. The reader is directed to the chapters (this volume) by Michejda and Eisenbrand for a more detailed discussions.

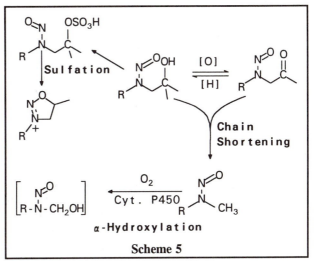

Scheme 5

Mitochondrial enzymes, normally involved in fatty acid metabolism, have been shown to catalyze the decarboxylation of certain β-oxidized-nitrosamino acids (*88*). These data help to explain the potent bladder carcinogenicity of dibutylnitrosamine and its metabolites. Other chemistry employing β-oxopropyl-N-nitrosocarbamates has provided a model for the chain shortening and DNA methylation observed for the important β-oxopropyl pancreatic carcinogens (*84,89*). The diazotates produced from these transformations undergo and intramolecular cyclization to the carbonyl followed by scission to generate diazomethane and acetate.

Metabolism studies on the tobacco specific nitrosamines and other related cyclic nitrosamines have advanced significantly (*8,90*). In many cases, the nature of the electrophilic intermediates which attack DNA have been elucidated and the structure of specific DNA adducts has been proven. This research is of considerable value in the newly developing field of molecular epidemiology which is based on the isolation of specific DNA or hemoglobin adducts of carcinogens and relate human metabolic capacity and exposure to possible disease states. A good understanding of the metabolism of nitrosamines, as well as other carcinogens, is important not only for a better understanding of the carcinogenic process and as a test of various hypotheses dealing with the first steps in tumorogenesis, but is essential to the analytical chemistry associated with molecular epidemiology. Nevertheless, much of the chemistry and biochemistry of these processes is very complex and requires further elucidation.

Future Directions

A vexing problem associated with the prevention of cancer which may be associated with nitrosamines or other N-nitroso compounds, is the ease with which these compounds form from common precursors. Formation occurs *in vivo* and under conditions of common chemical manufacturing, processing, and storage. Despite their ready metabolism, highly selective chemistry suitable for the destruction of nitrosamines in foods or commercial preparations to which humans are exposed is unknown. (Nitrosamines can be easily destroyed through the use of HBr/acetic acid or catalytic reduction.) The adage, "An ounce of prevention is worth a pound of cure," is certainly appropriate. Numerous successful measures have been taken to eliminate or significantly reduce the formation of nitrosamines in commercial products. Prevention of exposure to carcinogenic contaminants such as nitrosamines requires a good knowledge of the reactions by which they form, excellent yet simple analytical methodology for their detection, an array of effective blocking agents and strategies, the development of chemistry for their selective removal, a knowledge of their acute and chronic toxicity and, where appropriate, an understanding of their metabolism and mode of bioaction. The finding that N-nitrosation reactions can occur *in vivo* places additional emphasis on the importance of understanding the reactive susceptibility of many common drugs and other substances toward N-nitrosation. Progress in these areas, like all scientific endeavors, requires effective idea and information exchange between all sectors of the scientific community.

Now that commercial instrumentation appears to be available for the detection of N-nitroso compounds eluting from aqueous HPLC systems, it would seem appropriate to determine whether deleterious substances bearing the N-nitroso functionality constitute a significant exposure problem for humans. Regular and unbiased publication of surveillance data on the level of various N-nitroso compounds in specific classes of commercial products would seem to be welcome by many. The instrumentation to support such an effort seems to be available, but more simple and automated analytical procedures require development.

It is also important to learn more about the biochemistry and biological properties of N-nitroso compounds which are prevalent in the human environment. An understanding of the metabolism of these compounds and the nature of the chemically reactive products generated from them facilitates a knowledge of their protein and DNA related chemistry. This information permits the structural elucidation of DNA and protein adducts and the development of sensitive analytical chemical methods for their detection. Exposure patterns and metabolic activation capacities can then be assessed in humans through the methods of molecular epidemiology. All of this information is very important in the construction of accurate risk assessment.

Because of the linkage between the chemistry and biochemistry of nitrosamines and other N-nitroso compounds and cancer, and the ability of chemists to actively participate in the reduction of human exposure to these compounds, vigorous chemical and biochemical research and cooperation in this field is imperative. The "field" is again at a turning point, and the opportunity for significant accomplishment is at hand.

Acknowledgments

The author is grateful for research support from the National Cancer Institute (R37 CA24916) and from the National Institute of Environmental Health Sciences (R01 ES03953), as well as the helpful suggestions of Drs. Stephen Hecht, Christopher J. Michejda, and Steven R. Tannenbaum during the preparation of this manuscript.

References

(1)   Allen, W. "Carcinogens that won't go away" *St. Louis Post Dispatch*, Aug. 30, **1992**, p. B4.
(2)   Scanlan, R. A.; Tannenbaum, S. R.; Editors. *American Chemical Society Symposium Series, No. 174: N-Nitroso Compounds* American Chemical Society: Washington, D. C., **1981**.
(3)   Preussmann, R.; Stewart, B. W. *N-Nitroso carcinogens*; *ACS Monogr.*, **1984**, *182* (Chem. Carcinog., 2nd Ed., Vol. 2), 643-828.
(4)   Galtress, C. L.; Morrow, P. R.; Nag, S.; Smalley, T. L.; Tschantz, M. F.; Vaughn, J. S.; Wichems, D. N.; Ziglar, S. K.; Fishbein, J. C. *J. Am. Chem. Soc.* **1992**, *114*, 1406-1411.
(5)   Santala, T.; Fishbein, J. C. *J. Am. Chem. Soc.* **1992**, *114*, 8852-8857.
(6)   Ohshima, H.; Bartsch, H. *Cancer Res.* **1981**, *41*, 3658-3662.
(7)   Bartsch, H.; Montessano, R. *Carcinogenesis (London)* **1984**, *5*, 1381-1393.
(8)   Hoffmann, D.; Rivenson, A.; Chung, F. L.; Hecht, S. S. *Crit. Rev. Toxicol.* **1991**, *21*, 305-311.
(9)   Carmella, S. G.; Kagan, S. S.; Kagan, M.; Foiles, P. G.; Palladino, G.; Quart, A. M.; Quart, E.; Hecht, S. S. *Cancer Res.* **1990**, *50*, 5438-5445.
(10) Winn, D. M.; Blot, W. J.; Shy, C. M.; Pickle, L. M.; Toledo, A.; Fraumeni, J. F., Jr. *New Engl. J. Med.* **1981**, *304*, 745-749.
(11) Brunnemann, K. D.; Hoffmann, D. *Crit. Rev. Toxicol.* **1991**, *21*, 235-240.
(12) Stillwell, W. G.; Glogowski, J.; Xu, H. X.; Wishnok, J. S.; Zavala, D.; Montes, G.; Correa, P.; Tannenbaum, S. R. *Cancer Res.* **1991**, *51*, 190-194.
(13) Fine, D. H. *IARC Sci. Publ.*, **1983**, *45* (Environ. Carcinog.: Sel. Methods Anal., v6), 443-8, 473-90.
(14) Preussmann, R.; Editor. *The Nitrosamine Problem (Das Nitrosamin-Problem)*; Verlag Chemie: Weinheim, Fed. Rep. Ger., **1983**.
(15) O'Neill, I. K.; Von Borstel, R. C.; Miller, C. T.; Long, J.; Bartsch, H., Eds. *IARC Scientific Publication No. 57, N-Nitroso compounds: Occurrence, biological effects and relevance to human cancer*; International Agency for Research on Cancer: Lyon, Fr, **1984**.
(16) Bartsch, H.; O'Neill, I. K.; Schulte-Hermann, R.; Editors. *IARC Scientific Publications, No. 84: Relevance of N-Nitroso Compounds to Human Cancer: Expo-

sures and Mechanisms; International Agency for Research on Cancer: Lyon, Fr., **1987**.
(17) O'Neill, I. K.; Chen, J.; Bartsch, H.; Editors. *IARC Scientific Publications, No. 105: Relevance to Human Cancer of N-Nitroso Compounds, Tobacco and Mycotoxins*; International Agency for Research on Cancer: Lyon, Fr., **1991**.
(18) Preussmann, R.; Spiegelhalder, B. Ger. Offen. DE 4012797, **1991**.
(19) Eisenbrand, G. Ger. Offen. DE 3939475 A1 6 Jun, **1991** Addn. to Ger. Offen. 3,818,495.
(19) Eisenbrand, G. Ger. Offen. DE 3939474 A1 6 Jun, **1991**.
(21) Wilcox, A. L.; Bao, Y. T.; Loeppky, R. N. *Chem. Res. Toxicol.* **1991**, *4*, 373-381.
(22) Bao, Y. T.; Loeppky, R. N. *Chem. Res. Toxicol.* **1991**, *4*, 382-389.
(23) Loeppky, R. N.; Bao, Y. T. U. S. Patent 5, 087,671 **1991**
(24) Tannenbaum, S. R.; Wishnok, J. S.; Leaf, C. D. *Am. J. Clin. Nutr.* **1991**, *53* (1 suppl.), 247S-250S.
(25) Sander, J.; Schweinsberg, F.; Menz, H. P. *Hoppe Seyler's Z. Physiol. Chem.* **1968**, *349*, 1691-1697.
(26) Mirvish, S. S. *Toxicol. Appl. Pharm.* **1975**, *31*, 325-351.
(27) Eisenbrand, G; *Drug Dev. Eval.* **1990**, *16* (Signif. N-Nitrosation Drugs), 47-69.
(28) Tannenbaum, S. R.; Sinskey, A. J.; Weisman, M.; Bishop, W. *J. Nat. Cancer Inst.* **1974**, *53*, 79-84.
(29) Tannenbaum, S. R.; Archer, M. C.; Wishnok, J. S.; Bishop, W. W. *J. Natl. Cancer Inst.* **1978**, *60*, 251-253.
(30) Spiegelhalder, B.; Preussmann, R. *Carcinogenesis (London)* **1985**, *6*, 545-548.
(31) Green, L. C.; De Luzuriaga, K. R.; Wagner, D. A.; Rand, W.; Istfan, N.; Young, V. R.; Tannenbaum, S. R. *Proc. Natl. Acad. Sci. U. S. A.* **1981**, *78*, 7764-7768.
(32) Wagner, D. A.; Young, V. R.; Tannenbaum, S. R. *Proc. Natl. Acad. Sci. U. S. A.* **1983**, *80*, 4518-4521.
(33) Marletta, M. A.; Yoon, P. S.; Iyengar, R.; Leaf, C. D.; Wishnok, J. S. *Biochemistry* **1988**, *27*, 8706-8711.
(34) Marletta, M. A. *Chem. Res. Toxicol.* **1988**, *1*, 249-257.
(35) Miwa, M.; Stuehr, D. J.; Marletta, M. A.; Wishnok, J. S.; Tannenbaum, S. R. *Carcinogenesis (London)* **1987**, *8*, 955-958.
(36) Marletta, M. A.; Tayeh, M. A.; Hevel, J. M. *BioFactors* **1990**, *2*, 219-225.
(37) Calmels, S.; Ohshima, H.; Bartsch, H. *J. Gen. Microbiol.* **1988**, *134*, 221-226.
(38) Suzuki, K.; Mitsuoka, T. *IARC Sci. Publ.*, **1984**, *57* (N-Nitroso Compd: Occurrence, Biol. Eff. Relevance Hum. Cancer), 275-81.
(39) Ralt, D.; Tannenbaum, S. R.; ACS Symp. Ser. **1981**, *174* (N-Nitroso Compd.), 157-64.
(40) Bartsch, H.; Ohshima, H.; Pignatelli, B.; Calmels, S. *Cancer Surv.* **1989**, *8*, 335-362.
(41) Foiles, P. G.; Murphy, S. E.; Peterson, L. A.; Carmella, S. G.; Hecht, S. S. *Cancer Res.* **1992**, *52* (9 Suppl.), 2698s-2701s.
(42) Shuker, D. E. G.; Tannenbaum, S. R. *Anal. Chem.* **1983**, *55*, 2152-2155.
(43) Conboy, J. J.; Hotchkiss, J. H. *Analyst (London)* **1989**, *114*, 155-159.
(44) Pignatelli, B.; Chen, C. S.; Thuillier, P.; Bartsch, H. *Analyst (London)* **1989**, *114*, 1103-1108.
(45) Ellen, G. *Drug Dev. Eval.* **1990** *16* (Signif. N-Nitrosation Drugs), 19-46.
(46) Walker, R. *Food Addit. Contam.* **1990**, *7*, 717-768.
(47) Tricker, A. R.; Preussmann, R. *Mutat. Res.* **1991**, *259*, 277-289.

(48) Lijinsky, W. *Adv. Mod. Environ. Toxicol.*, **1990** *17* (Environ. Occup. Cancer: Sci. Update), 189-207.
(49) Loeppky, R. N. In *The Nitrosamine Prob..*, Preussmann, R., Ed.; Verlag Chem.: Weinheim, Fed. Rep. Ger., **1983**; pp. 305-317.
(50) Liu, R. H.; Conboy, J. J.; Hotchkiss, J. H. *J. Agric. Food Chem.* **1988**, *36*, 984-987.
(51) Scanlan, R. A.; Barbour, J. F.; Chappel, C. I. *J. Agric. Food Chem.* **1990**, *38*, 442-443.
(52) Gao, J.; Hotchkiss, J. H.; Chen, J. , Relavance Hum. Cancer N-Nitroso Compd. ed.; International Agency for Research on Cancer: IARC Sci. Publ., **1991**; Vol. 105. pp 219-222.
(53) Meyer, T. A.; Powell, J. B. *J. Assoc. Off. Anal. Chem.* **1991**, *74*, 766-771.
(54) Loeppky, R. N.; Hastings, R.; Sandbothe, J.; Heller, D.; Bao, Y.; Nagel, D. *IARC Sci. Publ.,***1991** *105* (Relevance Hum. Cancer N-Nitroso Compd., Tob. Mycotoxins), 244-252
(55) Dunkel, V. C.; San, R. H. C.; Harbell, J. W.; Seifried, H. E.; Cameron, T. P. *Environmental and Molecular Mutagenesis* **1992**, *20*, 188-198.
(56) Coulston, F.; Dunne, J. F., Eds. *The Potential Carcinogenicity of Nitrosatable Drugs. WHO Symposium*; Ablex Publ. Corp.: Norwood, NJ, **1978**.
(57) Eisenbrand, G.; Bozler, G.; Nicolai, H. v, Eds. *The Significance of N-Nitrosation of Drugs*, Drug Development and Evaluation; Gustav Fischer Verlag: Stuttgart, **1990**; Vol. 16.
(58) Loeppky, R. N.; Shevlin, G.; Yu, L. *Drug Dev. Eval.*, **1990**, *16* (Signif. N-Nitrosation Drugs), 253-66.
(59) Loeppky, R. N.; Tomasik, W.; Millard, T. G. *IARC Sci. Publ.* **1984,** *57* (N-Nitroso Compd: Occurrence, Biol. Eff. Relevance Hum. Cancer), 353-63.
(60) Liu, R. H.; Conboy, J. J.; Hotchkiss, J. H. *J. Agric. Food Chem.* **1988**, *36*, 984-987.
(61) Challis, B. C.; Yousaf, T. I. *J. Chem. Soc., Chem. Commun.* **1990**, 1598-1599.
(62) Challis, B. C.; Yousaf, T. I. *J. Chem. Soc., Perkin Trans. 2*, **1992**, 283-6.
(63) Pignatelli, B.; Malaveille, C.; Friesen, M.; Hautefeuille, A.; Bartsch, H.; Piskorska, D.; Descotes, G. *Food Chem. Toxicol.* **1987**, *25*, 669-680.
(64) Zhang, R. F.; Deng, D. J.; Chen, Y.; Wu, H. Y.; Chen, C. S. *IARC Sci. Publ.,***1991** *105* (Relevance Hum. Cancer N-Nitroso Compd., Tob. Mycotoxins),152-157.
(65) Yang, C. S.; Yoo, J. S. H.; Ishizaki, H.; Hong, J. *Drug Metab. Rev.* **1990**, *22*, 147-159.
(66) Michejda, C. J.; Kroeger-Koepke, M. B.; Koepke, S. R.; Magee, P. N.; Chu, C. *Banbury Rep.*, **1982** *12* (Nitrosamine Hum. Cancer), 69-85.
(67) Ludeke, B.; Meier, T.; Kleihues, P. ; *IARC Sci. Publ.,***1991** *105* (Relevance Hum. Cancer N-Nitroso Compd., Tob. Mycotoxins), 286-93.
(68) Wade, D.; Yang, C. S.; Metral, C. J.; Roman, J. M.; Hrabie, J. A.; Riggs, C. W.; Anjo, T.; Keefer, L. K.; Mico, B. A. *Cancer Res.* **1987**, *47*, 3373-3377.
(69) Keefer, L. K.; Anjo, T.; Wade, D.; Wang, T.; Yang, C. S. *Cancer Res.* **1987**, *47*, 447-452.
(70) Loeppky, R. N.; Li, E. *Chem. Res. Toxicol.* **1988**, *1*, 334-336.
(71) Loeppky, R. N.; Li, Y. E. *IARC Sci. Publ.*, **1991** *105* (Relevance Hum. Cancer N-Nitroso Compd., Tob. Mycotoxins), 375-82.
(72) Mochizuki, M.; Anjo, T.; Okada, M. *Tetrahedron Lett.* **1980**, *21*, 3693-3696.
(73) Wiench, K.; Frei, E.; Schroth, P.; Wiessler, M. *Carcinogenesis (London)* **1992**, *13*, 867-872.

(74) Frei, E.; Frank, N.; Wiessler, M. In *N-Nitroso Compd. Pap. Int. Symp.*, Bhide, S. V.; Rao, K. V. K., Eds.; Omega Sci.: New Delhi, India., **1990**; pp. 197-206.
(75) Farrelly, J. G.; Stewart, M. L.; Lijinsky, W. *Carcinogenesis (London)* **1984**, *5*, 1015-1019.
(76) Loeppky, R. N.; Tomasik, W.; Kovacs, D. A.; Outram, J. R.; Byington, K. H. *IARC Sci. Publ.*, **1984**, 57 (N-Nitroso Compd: Occurrence, Biol. Eff. Relevance Hum. Cancer), 429-36.
(77) Airoldi, L.; Bonfanti, M.; Fanelli, R.; Bove, B.; Benfenati, E.; Gariboldi, P. *Chem. Biol. Interact.* **1984**, *51*, 103-113.
(78) Eisenbrand, G.; Denkel, E.; Pool, B. *J. Cancer Res. Clin. Oncol.* **1984**, *108*, 76-80.
(79) Loeppky, R. N.; Tomasik, W.; Kerrick, B., E. *Carcinogenesis (London)* **1987**, *8*, 941-946.
(80) Denkel, E.; Pool, B. L.; Schlehofer, J. R.; Eisenbrand, G. *J. Cancer Res. Clin. Oncol.* **1986**, *111*, 149-153.
(81) Chung, F. L.; Hecht, S. S. *Carcinogenesis (London)* **1985**, *6*, 1671-1673.
(82) Loeppky, R. N.; Tomasik, W.; Eisenbrand, G.; Denkel, E. *IARC Scientific Pub. No. 84* **1987**, 94-99.
(83) Loeppky, R. N.; Outram, J. R. *IARC Sci. Publ.*, **1982**, *41* (N-Nitroso Compd: Occurrence Biol. Eff.), 459-72.
(84) Leung, K. H.; Archer, M. C. *Chem. Biol. Interact.* **1984**, *48*, 169-179.
(85) Leung, K. H.; Archer, M. C. *Carcinogenesis (London)* **1985**, *6*, 189-191.
(86) Michejda, C. J.; Kroeger-Koepke, M. B.; Kovatch, R. M. *Cancer Res.* **1986**, *46*, 2252-2256.
(87) Kroeger-Koepke, M. B.; Koepke, S. R.; Hernandez, L.; Michejda, C. J. *Cancer Res.* **1992**, *52*, 3300-3305.
(88) Janzowski, C.; Jacob, D.; Henn, I.; Zankl, H.; Pool-Zobel, B. L.; Wiessler, M.; Eisenbrand, G. *IARC Sci. Publ.*, **1991**, *105* (Relevance Hum. Cancer N-Nitroso Compd., Tob. Mycotoxins), 332-8.
(89) Bartzatt, R.; Nagel, D. *Physiol. Chem. Phys. Med. NMR* **1991**, *23*, 29-34.
(90) Chung, F. L.; Hecht, S. S.; Palladino, G.; *IARC Sci. Publ.*, **1986** *70* (Role Cyclic Nucleic Acid Adducts Carcinog. Mutagen.), 207-25.

RECEIVED December 6, 1993

N-Nitroso Compound Exposure, Formation, and Blocking

Chapter 2

Nitrosamines in Sunscreens and Cosmetic Products

Occurrence, Formation, and Trends

Donald C. Havery and Hardy J. Chou

U.S. Food and Drug Administration, Mail Stop HFS 127, 200 C Street, SW, Washington, DC 20204

Several nitrosamines, most of which have been shown to be carcinogenic in laboratory animals, have been identified in cosmetic products over the last fifteen years. These compounds are primarily formed from amine precursors and nitrosating agents in the cosmetic product but are also introduced into the product through cosmetic raw materials. Market surveys of sunscreens and cosmetics have revealed levels of nitrosodiethanolamine up to 45 ppm and levels of 2-ethylhexyl 4-(N-methyl-N-nitrosamino) benzoate up to 21 ppm. Results from market surveys of cosmetic products over the past fifteen years do not suggest any trend in the nitrosamine levels in these products. A summary of the occurrence and levels of nitrosamines in cosmetic products along with strategies for the reduction or elimination of these compounds is presented.

The presence of nitrosamines has been reported in a number of consumer products including bacon, beer, rubber nipples and pacifiers, pesticides, drugs, tobacco and cutting fluids (*1-10*). Nitrosamines were first reported in cosmetic products in 1977 when Fan et al. (*11*) described the presence of nitrosodiethanolamine in products such as lotions, shampoos, and cosmetics.

Most nitrosamines that have been tested have been found to be carcinogenic in laboratory animals (*12-16*). However, the potential health significance of the presence of nitrosamines in cosmetic products depends on several factors. These include the specific nitrosamine involved (carcinogenic potency), the level of contamination, the type of product (i.e., creams and lotions versus wash-off products), the frequency of use, the degree of absorption through the skin, and nitrosamine stability (exposure to UV light). This paper reviews the current knowledge about the formation and occurrence of nitrosamines in sunscreens and cosmetics in addition to presenting strategies for the reduction or elimination of these compounds.

Formation of Nitrosamines in Cosmetic Products

Nitrosamines are generally formed by the reaction of secondary amines, and to a lesser extent tertiary amines, with a nitrosating agent such as nitrite at an optimum pH of 3.4 in aqueous solutions (*17,18*). Nitrosation can also occur at alkaline pH by the action of oxides of nitrogen on amines (*19,20*); this reaction is enhanced by the presence of formaldehyde (*21*).

The nitrosation potential for several cosmetic raw materials has been demonstrated (*22-24*). In some cases, nitrosamines are formed by the nitrosation of the primary component of the raw material while, in others, nitrosamines are formed from the nitrosation of other amines present in the raw material as contaminants. The tertiary amine triethanolamine (TEA) nitrosates rather slowly (*25*). However, the cosmetic raw material TEA typically contains the secondary amine diethanolamine (DEA) as a contaminant. The higher the levels of DEA in TEA used in a cosmetic product, the more nitrosamine may be formed in the resulting product.

Studies have been conducted to determine if nitrosamines are formed in cosmetic products from precursor amines and nitrosating agents, or are present in raw materials used to formulate these products. Limited surveys of cosmetic raw materials have shown that nitrosamines are present in some but not all raw materials (Table I). Based on these findings and other unpublished results obtained in our laboratory, it appears that while some degree of nitrosamine contamination of cosmetic products may be due to the raw materials, formation in the product is the principal means by which nitrosamines find their way into cosmetics.

Source of Amines. More than 8,000 raw materials are used in the formulation of cosmetic products. Many of these are potential sources of nitrosatable compounds and include amides, alkanolamines, fatty acid alkanolamides, amine oxides and secondary, tertiary and quaternary amines. These raw materials function as thickeners, emulsifiers, emollients, conditioners, foam stabilizers, cleansing and wetting agents, lubricants, skin protectants and anti-irritants.

When considering potential sources of nitrosatable amines in cosmetics, it is necessary to keep in mind that cosmetic raw materials vary in their degree of purity. Some grades of triethanolamine, for example, are intended to contain 15% of the secondary amine diethanolamine. Thus, the name of the ingredient that appears on the product ingredient label may not always adequately identify potential sources of nitrosatable amines in the product.

Source of Nitrosating Agents. Potential nitrosating agents in cosmetics include nitrite, oxides of nitrogen, alkyl nitrites and C-nitro compounds. Nitrite may be present as a contaminant in cosmetic raw materials as a result of storage in nitrite-treated drums or as an impurity from nitration reactions during synthesis. Results from a survey of organic cosmetic raw materials, however, suggests that these may not be a significant source of nitrite (*27*). In this survey of 14 organic raw materials, nitrite was found at >50 ppb in only 14% of the 114 samples.

Inorganic raw materials and pigments used as ingredients in cosmetics may also be a source of nitrite in finished products. In a study of 63 inorganic raw materials, nitrite was found in 70% of the samples at levels of 1 to >5 ppm (27). However, a study of nitrite in inorganic pigments showed that nitrite levels in these materials did not contribute significantly to nitrosamine formation in cosmetic products (30).

The C-nitro compound 2-bromo,2-nitro-1,3-propanediol (BNPD), which is used as a cosmetic preservative, releases nitrite slowly over time, and may be a source of nitrite in cosmetic products containing this preservative. C-Nitro compounds can also transnitrosate amines to produce nitrosamines (31).

Oxides of nitrogen, normal constituents of air, particularly polluted air, are also a potential nitrosating agent in cosmetics. N_2O_3 and N_2O_4 nitrosate amines under alkaline conditions (19,20), and NO_2 is readily absorbed from the air by triethanolamine (32). Once applied to the skin, water from a cosmetic emulsion evaporates, essentially leaving a nonaqueous layer. Oxides of nitrogen readily absorb into a nonpolar matrix and nitrosate amines to produce nitrosamines (33).

Further studies suggest that other nitrosating agents may form in cosmetic products from the reaction of raw materials like alkanolamines with nitrite or oxides of nitrogen to produce alkyl nitrites (33). Alkyl nitrites are effective nitrosating agents, even under nonacidic conditions (34).

Nitrosation in a Cosmetic Matrix. Nitrosamine formation in a cosmetic matrix is unique in that many products are emulsions consisting of both a water and an oil phase. Simple amines like DEA, TEA, and nitrite would be soluble in the aqueous phase while long chain fatty acid alkanolamides and oxides of nitrogen would be found primarily in the oil phase. This suggests that there is likely to be more than one mechanism of nitrosamine formation operating in cosmetic matrices, depending on which precursors are present and their relative solubilities. In a study by Powell et al. (33) TEA containing approximately 15% DEA was not nitrosated by nitrite to form nitrosodiethanolamine (NDELA) in a nonaqueous system. However, following the addition of stearic acid, DEA in the TEA raw material was nitrosated to give NDELA. In this case, stearic acid apparently combined with nitrite to form a nitrite ester which "carried" the nitrosating agent into the nonaqueous phase. For the same reasons, nitrosation of oil soluble TEA-stearate occurred most rapidly in nonpolar solvents. The presence of charged surfactants also affects nitrosation of amines in emulsion systems by affecting the distribution of the precursors in the two phases (35).

Though nitrosation in cosmetic matrices is often slow, cosmetic products may remain on store shelves and in consumer's cabinets for extended periods of time. Provided there are sufficient precursors present, nitrosamines can continue to form in the product. Studies in our laboratory have shown that if a cosmetic contains nitrosamine precursors, nitrosamine levels can increase in the product with time, especially at increased temperatures. A sunscreen product containing the sunscreen 2-ethylhexyl 4-(N,N-dimethylamino) benzoate (Padimate O) and BNPD was purchased and analyzed in 1987 and found to be nitrosamine free. In 1990 that same product, stored at room temperature since 1987, contained 8 ppm of the nitrosamine derivative of the secondary amine typically found in Padimate O. Some products that are allowed to remain in the sun, and thereby become elevated in temperature, may form higher levels of nitrosamines.

Seasonal products such as sunscreens that are not sold by the end of the summer may be retained until the following year and the conditions of storage may affect the nitrosamine levels in these products.

Formaldehyde Catalysis. The ability of formaldehyde to catalyze nitrosation reactions is well known (*21,36*). At the alkaline pH of cutting fluids for example, NDELA readily forms from the reaction of DEA with nitrite in the presence of formaldehyde (*10*).

Cosmetic products are typically preserved against bacterial and fungal contamination, often by the addition of formaldehyde-releasing preservatives and generally have a pH ranging from 6 to 8. The cosmetic preservative BNPD has undergone considerable scrutiny for its role in nitrosamine formation (*37*) since, at alkaline pH, BNPD decomposes to yield formaldehyde and nitrite (*38*). In the presence of BNPD, NDELA is readily formed from DEA present in the cosmetic raw material TEA at pH 6 (*39*). This has significant implications for cosmetics containing BNPD and other formaldehyde-releasing preservatives.

Inhibition of Nitrosamine Formation in Cosmetic Products. Nitrosamine formation in consumer products such as bacon has been effectively reduced by the addition of nitrosating agent scavengers. Studies have identified effective nitrosation inhibitors for cosmetic products. Cosmetic emulsions would require both hydrophobic and hydrophilic nitrosation inhibitors to be effective in both phases. Several water-soluble inhibitors have been evaluated, including ascorbic acid, sodium ascorbate, potassium sorbate, sodium bisulfite and propyl gallate (*25,27,35,39-41*). Oil-soluble inhibitors that have been examined include butylated hydroxytoluene (BHT), butylated hydroxyanisole, α-tocopherol, and ascorbyl palmitate (*27,41,42*). Ascorbic acid and sodium bisulfite were the most effective, inhibiting nitrosation at >99%. BHT, α-tocopherol, sodium ascorbate, and ascorbyl palmitate were effective inhibitors in some but not all systems studied.

Occurrence of Nitrosamines in Cosmetic Products

Total Nitrosamines. Because a wide variety of nitrogen-containing raw materials are used in cosmetic products, the potential exists for a diversity of nitrosamines to form if nitrosating agents are present. Two nitrosamines, nitrosodiethanolamine (NDELA) and 2-ethylhexyl 4-(N-nitroso-N-methyl) benzoate (NMPABAO) have been found frequently in cosmetic or sunscreen products. A number of other nitrosamines have also been found in a variety of cosmetic products (Table II).

In an attempt to identify cosmetic products potentially containing nitrosamines, a rapid screening procedure for total nitrosamines in cosmetics was developed (*43*). Results of a small market survey of cosmetic products conducted in 1986 indicated higher levels of apparent total nitrosamines than the levels of NDELA found in the product (Table III). Other positive peaks have been observed during analysis of cosmetics with the Thermal Energy Analyzer, a selective detector for nitrosamines. The implication of these observations is that there may be other unidentified nitrosamines in the product that contribute to the total amount.

Table I. Nitrosamines in Cosmetic Raw Materials

Year	No. of Positive Samples	Total No. Samples	N-Nitrosamine Identity[a]	Levels (ppb)	Ref.
1980	0	several	[b]	[b]	26
1981	30	99	NDELA	50->1,000	27
1984	[c]	[c]	NDiPLA	20-1,300	28
1984/87	2	4	NMOR	11-120	[d]
1988	6	9	NDELA	<5-1,460	29
1988	3	9	NDiPLA	13-465	29

[a] NDELA: N-nitrosodiethanolamine; NMOR: N-nitrosomorpholine; NDiPLA: N-nitrosodiisopropanolamine.
[b] None found.
[c] Not given.
[d] FDA unpublished data.

Table II. Nitrosamines in Cosmetic Products

Date Purchased[a]	No. Positive Products	Total No. Products	N-Nitrosamine Identity[b]	Levels (ppb)	Ref.
1980	6	7	NMDDA	20-600	44
1980/81	8	11	NMODA	28-969	25
1980/81	3	38	NMODA	42-261	25
1981	12	13	NMDDA	11-873	44
1981	11	13	NMTDA	8-254	44
1981/82	26	145	NMOR	up to 640	45
1981/82	50	145	NDMA	up to 24	45
1981/82	1	145	NDEA	15	45
1981/82	1	145	NDPA	35	45
1981/82	1	145	NDBA	15	45
1984/87	12	18	NDiPLA	80-680	[c]
1984/87	9	9	NMOR	48-1,240	[c]
1986	2	44	NDiPLA	20-35	46
1987/88	4	127	NDiPLA	40-215	46

[a] Estimated if not known.
[b] NDiPLA: N-nitrosodiiospropanolamine; NMOR: N-nitrosomorpholine; NMDDA: N-nitroso-N-methyldodecylamine; NMODA: N-nitroso-N-methyloctadecylamine; NMTDA: N-nitroso-N-methyltetradecylamine; NDMA: N-nitrosodimethylamine; NDEA: N-nitrosodiethylamine; NDPA: N-nitrosodipropylamine; NDBA: N-nitrosodibutylamine.
[c] FDA unpublished data.

Total Nitrosating Agents. A method was developed at the Food and Drug Administration (FDA) to measure total nitrosating agents in cosmetic products. This method is based on the chemical liberation of nitric oxide from the nitrosating agent followed by its chemiluminescent detection with a Thermal Energy Analyzer (Chou et al., JAOAC Int., in press). Analysis of cosmetics using this method could be used to determine which products have the highest potential for nitrosamine formation. The results of a limited FDA survey of cosmetic products purchased in 1991-92 for nitrosamines and nitrosating agents are shown in Table IV. The results suggest the presence of nitrosating agents in every product analyzed. For the most part, those products containing the highest levels of known nitrosamines were also found to contain the highest levels of nitrosating agents. However no consistent correlation between the total apparent nitrosating agents and nitrosamine level was found. The results suggested that nitrosation potential is a common feature in cosmetic products which may lead to nitrosamine formation if conditions are favorable. Further research is planned in this area.

Nitrosodiethanolamine. NDELA is the nitrosamine found most frequently in cosmetic products and is therefore the nitrosamine consumers are most likely to be exposed to. NDELA is a known carcinogen in laboratory animals (*15,47*); it is absorbed through the skin (*48-53*). The absorption rate is a function of the nature of the cosmetic; absorption is fastest in nonpolar vehicles (*54*).

A summary of the occurrence of NDELA in cosmetic products is shown in Table V. Though NDELA has been found in some cosmetic raw materials (Table I), the data suggests that the primary source of NDELA in cosmetic products is formation in the product from precursors. NDELA is most often found in products containing the raw materials TEA or fatty acid alkanolamides such as cocoamide diethanolamine. The highest levels of NDELA are usually associated with products also containing the preservative BNPD. As discussed earlier, a major amine precursor of NDELA is the secondary amine diethanolamine, found as a contaminant in TEA and fatty acid alkanolamides, although TEA itself can be nitrosated (*22*).

Profile of the Levels of NDELA in Cosmetics. In order to measure progress in reducing the levels of NDELA in cosmetic products, the percentage of cosmetic products in Table V found to contain NDELA above 500 ppb was plotted against time (Figure 1). The 500 ppb reference point was arbitrarily chosen. Based on these results, no trend in the levels of NDELA could be discerned.

Data generated in 1992 by the FDA indicate that NDELA can still be found at parts-per-million levels in some current market cosmetic products. Most of these products contained BNPD and amines such as TEA or a fatty acid alkanolamide such as cocoamide diethanolamine. Despite the known potential for nitrosamine formation when BNPD is used in combination with nitrosatable amines, some cosmetic formulators continue to use this combination of ingredients.

2-Ethylhexyl 4-(N-nitroso-N-methylamino) benzoate. During a 1986 FDA survey of cosmetic products for total nitrosamines, a large unknown peak was observed

Table III. Total Water Soluble Nitrosamines In Cosmetics

| Cosmetic Product | N-Nitrosamine Levels (ppb) | |
	NDELA[a]	Total N-Nitrosamines[b]
Lotion	c	110
Lotion	1,240	1,700
Cream	350	410
Cream	c	c
Make-up	c	370
Shampoo	c	c

Source: Data are from reference 43.
[a] N-Nitrosodiethanolamine.
[b] Calculated as NDELA.
[c] None found.

Table IV. Nitrosamines and Nitrosating Agents in Cosmetic Products

| Product Number | N-Nitrosamine Level (ppb) | | Nitrosating Agents (ppb)[c] |
	NMPABAO[a]	NDELA[b]	
1	3,000	390	600
2	d	270	260
3	d	d	2,150
4	d	d	480
5	4,430	200	480
6	4,240	120	540
7	160	e	660
8	d	e	590
9	350	e	430
10	d	e	860
11	d	e	600
12	7,930	e	3,700
13	d	e	630
14	3,270	e	3,100
15	7,200	e	11,300
16	21,020	e	6,860

[a] 2-Ethylhexyl 4-(N-nitroso-N-methylamino) benzoate.
[b] N-Nitrosodiethanolamine.
[c] Calculated as nitrite.
[d] None found.
[e] Not analyzed because NDELA precursors were not present.

Table V. N-Nitrosodiethanolamine (NDELA) in Cosmetics; 1976-1992

Date Purchased[a]	No. Positive Products	Total No. Products	Levels of NDELA (ppb)	Ref.
1976	20	29	5-47,000	11
1977	10	32	35-130,000	b
1978	58	174	20-42,000	b
1979	28	87	40-25,000	b
1979/80	70	70	5-39	55
1979/83	6	9	140-41,800	56
1980	15	20	100-380	57
1980	30	53	30-4,910	b
1980/81	4	10	27-4,113	58
1981	14	47	110-23,000	b
1981/82	25	145	up to 1,400	45
1982	0	18	c	b
1983	47	47	3-15	42
1983	5	22	350-4,800	b
1984	2	12	140-1,800	b
1985	1	2	700	b
1986	0	2	c	b
1986	8	20	7-2,000	59
1986	16	44	7-2,000	46
1987	16	127	10-235	46
1987	0	6	c	b
1989	0	2	c	b
1991	5	8	120-1,080	b
1992	8	12	210-2,960	b

[a] Estimated if not provided in reference.
[b] FDA unpublished data.
[c] None detected.

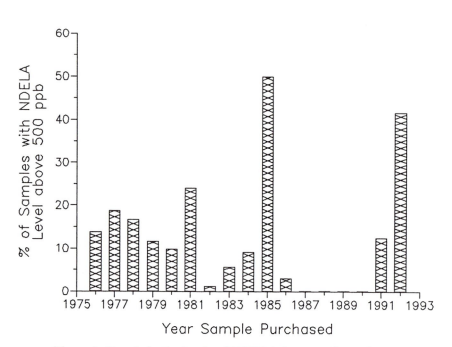

Year Sample Purchased

Figure 1. Trends in the levels of NDELA in cosmetic products.

Table VI. 2-Ethylhexyl 4-(N-nitroso-N-methylamino) benzoate (NMPABAO) in Cosmetic Products

Year	No. Positive Products	Total No. Products	NMPABAO (ppb) Range	Average
1987	19	29	60-2820	674
1988	6	8	170-1300	543
1991	18	38	160-20,520	3,426
1992	14	14	780-5,050	2,132

during gas chromatographic analysis of some sunscreen products containing the sunscreen Padimate O. The peak was isolated and identified as 2-ethylhexyl 4-(N-nitroso-N-methylamino) benzoate (NMPABAO). The carcinogenicity of NMPABAO is uncertain. NMPABAO was reported to be mutagenic in two strains of *Salmonella typhimurium* in the Ames assay (*60*). Long term bioassay is needed to evaluate the toxicity of this compound. Unpublished FDA studies showed that NMPABAO was absorbed through the skin of the hairless guinea pig and through human skin in an *in vitro* test.

The primary source of NMPABAO in sunscreens and cosmetic products appears to be the secondary amine 2-ethylhexyl 4-(N-methylamino) benzoate, which is a contaminant in the sunscreen Padimate O. This secondary amine, which readily nitrosates to give NMPABAO, has been found at levels up to 3% in Padimate O. Padimate O itself can be nitrosated at a rate faster than expected for a tertiary amine and may also be a precursor to NMPABAO (*60*).

The FDA has conducted surveys of sunscreens and cosmetic products for NMPABAO since 1986 (Chou et al., Food and Drug Administration, unpublished data). A summary of these findings is shown in Table VI. In 1989, to address concerns expressed by industry analysts, the analytical method used to determine NMPABAO was evaluated for its potential to form NMPABAO artifactually during analysis. Low levels of artifact formation by the analytical method were observed when nitrite was present at significant levels in the presence of the secondary amine in Padimate O. The analytical method was modified to incorporate both an inhibitor and an amine marker as an indicator of artifact formation (Chou, H.J. et al., JAOAC Int., in press). Market surveys continued in 1991 with the modified analytical method. Surveys showed that many of the cosmetic products that contained Padimate O still contained NMPABAO, some at elevated levels, especially when the product also contained the preservative BNPD. During skin absorption studies with NMPABAO, a question arose about the stability of the nitrosamine when exposed to sunlight, since nitrosamines are known to decompose in the presence of UV light (*61*). Thin films of a sunscreen product containing known amounts of NMPABAO were exposed to simulated and natural sunlight and then analyzed for NMPABAO. The results are shown in Table VII. The data show that NMPABAO is very rapidly decomposed by UV light. The identity and toxicity of the decomposition products have not been determined. It was also noted that, in the less direct sunlight of the month of October, NMPABAO was not completely decomposed, even after 5 minutes. These results suggest that in estimating human exposure to NMPABAO, decomposition due to sunlight may need to be considered. The intended use of the product must also be considered because Padimate O is used in products other than sunscreens, such as hand and face creams.

The Future of Nitrosamines in Cosmetics

The health implication for exposure to nitrosamine contaminants in cosmetics is uncertain. On the one hand, levels of nitrosamines found in cosmetic products are often three orders of magnitude higher than those found in some foods. On the other hand, because many cosmetic products containing nitrosamines are washed off the skin and hair after a very short period of time, absorption through the skin may be minimal. Reducing potential human exposure to nitrosamines

Table VII. Photodecomposition of 2-Ethylhexyl 4-(N-nitroso-N-methylamino) benzoate (NMPABAO) in a Sunscreen Product by Simulated and Natural Sunlight

Light Source	Irradiation Time (min)	NMPABAO (ppb)
	0	7,450
	0.5	1,640
Simulated[a]	1.0	1,340
	3.0	e
	5.0	e
	0	7,520
	1.0	1,170
Simulated[b]	1.5	215
	2.0	e
	0	7,500
	1.0	e
Natural[c]	3.0	e
	5.0	e
	7.0	e
	15.0	e
	0	7,610
	0.3	1,930
	1.0	560
Natural[d]	3.0	500
	5.0	480
	7.0	500
	15.0	530

[a] Simulated sunlight at distance of 85.4 mm from product.
[b] Simulated sunlight at distance of 206 mm from product.
[c] Natural sunlight, September, mid-day.
[d] Natural sunlight, October, mid-day.
[e] None found

to as low a level as is technologically feasible by reducing levels in all cosmetic products is the desirable course of action. The means for reducing nitrosamine formation in cosmetic products are known, and the technology is available.

In 1987, The German Federal Health Office officially recommended against the use of secondary amines in cosmetic products, that fatty acid diethanolamides should contain as low a level of diethanolamine as possible, that TEA with a purity of more than 99%, less than 1% DEA, less than 0.5% monoethanolamine, and less than 50 ppb NDELA should be used (*62*). A recent survey for NDELA in cosmetic products marketed in Germany reported a decline in the levels of NDELA in most products (*46*), suggesting that adoption of such measures may be effective in reducing nitrosamine formation in these cosmetic products.

A number of measures can be taken which can help to reduce nitrosamine formation in cosmetics. These include:
. the elimination of secondary amines as cosmetic ingredients
. the reduction of the levels of secondary amines in cosmetic raw materials
. avoiding the use of formaldehyde-releasing preservatives with raw materials containing secondary amines
. reducing nitrite levels in raw materials
. avoiding contamination of cosmetics and raw materials with oxides of nitrogen
. the use of nitrosation inhibitors

Analysis of finished products by the manufacturer for nitrosamines or nitrosating agents as a part of quality assurance procedures is an important adjunct to any program for nitrosamine reduction in cosmetic products.

Some product reformulation has occurred in the past ten years, and nitrosamines have been reduced or eliminated from some products. Considerable resources have been spent on understanding the mechanisms of nitrosamine formation in cosmetic systems and on means of inhibiting nitrosamine formation. However, there is still room for improvement. New products containing nitrosatable amines and formaldehyde-nitrite-releasing preservatives still appear on the U.S. market. Steps necessary to minimize nitrosamine formation must be communicated to all cosmetic manufacturers and research should continue on the use of nitrosation inhibitors and on the identification of the principal nitrosating agents in cosmetic products. Nitrosamines have been successfully reduced in food products such as bacon, beer, and rubber products. With the information available to cosmetic manufacturers, nitrosamine levels can also be reduced in these consumer products.

Literature Cited
1.   Canas, B.J. Havery, D.C., Joe Jr., F.L., Fazio, T. *J. Assoc. Off. Anal. Chem.* **1986**, *69*, 1020-1021.
2.   Vecchio, A.J., Hotchkiss, J.H., Bissogni, C.A. *J. Food Sci.* **1986**, *51*, 754-756.
3.   Havery, D.C., Hotchkiss, J.H., Fazio, T. *J. Food Sci.* **1981**, *46*, 501-505.
4.   Scanlan, R.A., Barbour, J.F., Chappel, C.I. *J. Agric. Food Chem.* **1990**, *38*, 442-443.
5.   Havery, D.C., Perfetti, G.A., Canas, B.J., Fazio, T. *Food Chem. Toxicol.* **1985**, *23*, 991-993.

6.   Billedeau, S.M., Thompson Jr., H.C., Miller, B.J., Wind, M.L. *J. Assoc. Off. Anal. Chem.* **1986**, *69*, 31-34.

7.   Hindle, R.W., Armstrong, J.F., Peake, A.A. *J. Assoc. Off. Anal. Chem.* **1987**, *70*, 49-51.

8.   Dawson, B.A., Lawrence, R.C. *J. Assoc. Off. Anal. Chem.* **1987**, *70*, 554-556.

9.   Brunnemann, K.D., Genoble, L., Hoffmann, D. *Carcinogenesis* **1987**, *8*, 465-469.

10. Bennett, E.O., Bennett, D.L. *Tribiology Int.* **1984**, *17*, 341-346.

11. Fan, T.Y., Goff, U., Song, L., Fine, D.H., Arsenault, G.P., Biemann, K. *Food Cosmet. Toxicol.* **1977**, *15*, 423-430.

12. Lijinsky, W. In *Chemical Mutagens: Principles and Methods For Their Detection*; Hollaender, A., Ed.,Plenum Press: New York, **1976**, Vol.4; pp 193-217.

13. Magee, P.N., Barnes, J.M. *Adv. Cancer Res.* **1967**, *10*, 163-246.

14. Pour, P., Salmasi, S., Runge, R., Gingell, R., Wallcave, L., Nagel, D., Stepan, K. *J. Natl. Cancer Inst.* **1979**, *63*, 181-190.

15. Ketkar, M.B., Althoff, J., Lijinsky, W. *Cancer Letters* **1981**, *13*, 165-168.

16. Lijinsky, W. Taylor, H.W. *Cancer Res.* **1975**, *35*, 958-961.

17. Mirvish, S.S. *Toxicol. Appl. Pharmacol.* 1975, 31, 325-351.

18. Challis, B.C., Skold, R.O., Svensson, L.C. *8th International Colloquium Tribiology, Tribiology 2000*, 1992 (in press).

19. Challis, B.C., Kyrtopoulos, S.A. *J. Chem. Soc. Chem. Comm.* **1976**, 877-878.

20. Challis, B.C., Kyrtopoulos, S.A. *J. Chem. Soc. Perkin Trans. I* **1979**, 299-304.

21. Keefer, L.K., Roller, P.P. *Science* **1973**, 181, 1245-1247.

22. Lijinsky, W., Keefer, L., Conrad, E., Van de Bogart, R. *J. Natl. Cancer Inst.* **1972**, *49*, 1239-1249.

23. Hecht, S.S. *Drugs Cosmet. Ind.* **1981**, 36-37.

24. Morrison, J.B., Hecht, S.S., Wenninger. J.A. *Food Chem. Toxicol.* **1983**, *21*, 69-73.

25. Schmeltz, I., Wenger, A. *Food Cosmetic. Toxicol.* **1979**, 17, 105-109.

26. Rosenberg, I.E., Gross, J., Spears, T. *J. Soc. Cosmet. Chem.* **1980**, *31*, 237-252.

27. Rosenberg, I. *CTFA Cosmetic J.* **1981**, 30-37.

28. Issenberg, P., Conrad, E.E., Nielsen, J.W., Klein, D.A., Miller, S.E. *IARC Sci. Publ. No.57* **1984**, 43-50.

29. Sommer, H., Loffler, H-P., Eisenbrand, G. *J. Soc. Cosmet. Chem.* **1988**, *39*, 133-137.

30. Motoi, T., Hayashi, S., Kurokawa, M., Yoneya, T., Nishijima, Y. *J. Agric. Food Chem.* **1983**, 31, 1211-1214.

31. Fan, T.Y., Vita, R., Fine, D.H. *Toxicol. Letters* **1978**, *2*, 5-10.

32. Levaggi, D.A., Siu, W., Feldstein, M., Kothny, E.L. *Environ. Sci. Technol.* **1972**, 6, 250-252.

33. Powell, J.B. *J. Soc. Cosmet. Chem.* **1987**, *38*, 29-42.

34. Challis, B.C., Shuker, D.E.G. *Food Cosmet. Toxicol.* **1980**, *18*, 283-288.

35. Kabacoff, B.L., Douglass, M.L., Rosenberg, I.E., Levan, L.W., Punwar, J.K., Vielhuber, S.F., Lechner, R.J. *IARC Sci. Publ. No. 57* **1984**, 347-352.

36. Archer, M.C., Tannenbaum, S.R., Wishnok, J.S. *IARC Sci. Publ. No. 14* **1976**, 141-145.

37. Holland, V.R. *Cosmet. Technol.* **1981**, *3*, 31-2, 34, 36.

38. Cosmetic Ingredient Review Panel *J. Amer. College of Toxicol.* **1984**, *3*, 139-155.
39. Ong, J.T.H., Rutherford, B.S. *J. Soc. Cosmet. Chem.* **1980**, *31*, 153-159.
40. Fellion, Y., De Smedt, J., Brudney, N. *IARC Sci. Publ. No. 31* **1980**, 435-443.
41. Kabacoff, B.L., Lechner, R.J., Vielhuber, S.F., Douglass, M.L. In *N-Nitroso Compounds*; Scanlan R. A. & Tannenbaum, S.R. Eds., ACS Symposium Series No. 174, American Chemical Society: Washington, DC, **1981**, 149-156.
42. Dunnett, P.C., Telling, G.M. *Int. J. Cosmet. Sci.* **1984**, *6*, 241-247.
43. Chou, H.J., Yates, R.L., Wenninger, J.A. *J. Assoc. Off. Anal. Chem.* **1987**, *70*, 960-963.
44. Hecht, S.S., Morrison, J.B., Wenninger, J.A. *Food Chem. Toxicol.* **1982**, *20*, 165-169.
45. Spiegelhalder, B., Preussmann, R. *J. Cancer Res. Clin. Oncol.* **1984**, *108*, 160-163.
46. Eisenbrand, G., Blankart, M., Sommer, H., Weber, B. *IARC Sci. Publ. No. 105* **1991**, 238-241.
47. Pour, P., Wallcave, L. *Cancer Let.* **1981**, *14*, 23-27.
48. Airoldi, L., Macri, A., Bonfanti, M., Bonati, M., Fanelli, R. *Food Chem. Toxicol.* **1984**, *22*, 133-138.
49. Edwards, G.S., Peng, M., Fine, D.H., Spiegelhalder, B., Kann, J. *Toxicol. Letters* **1979**, *4*, 217-222.
50. Lethco, E.J., Wallace, W.C., Brouwer, E. *Food Chem. Toxicol.* **1982**, *20*, 401-406.
51. Lijinsky, W., Losikoff, A.M., Sansone, E.B. *J. Natn. Cancer Inst.* **1981**, *66*, 125-127.
52. Marzulli, F.N., Anjo, D.M., Marbach, H.I. *Food Chem. Toxicol.* **1981**, *19*, 743-747.
53. Preussmann, R., Spiegelhalder, B., Eisenbrand, G., Wurtele, G., Hofmann, I. *Cancer Letters* **1981**, *13*, 227-231.
54. Bronaugh, R.L., Congdon, E.R., Scheuplein, R.J. *J. Investigative Dermatology* **1981**, *76*, 94-96.
55. Telling, G.M., Dunnett, P.C. *Int. J. Cosmet. Sci.* **1981**, *3*, 239-246.
56. Cosmetics Toiletries and Fragrances Association, *Progress Report on the Analysis of Cosmetic Products For Nitrosamines* **1984**.
57. Rollmann, B., Lombart, P., Rondelet, J., Mercier, M. *J. Chromatogr.* **1981**, *206*, 158-163.
58. Klein, D., Girard, A.N., Desmedt, J., Fellion, Y., Debry, G. *J. Cosmet. Toxicol.* **1981**, *19*, 223-235.
59. Sommer, H., Eisenbrand, G. *Z. Lebensm. Unters. Forsch.* **1988**, *186*, 235-238.
60. Loeppky, R.N., Hastings, R., Sandbothe, J., Heller, D., Nagel, D. *IARC Sci. Publ. No.105* **1991**, 244-252.
61. Doerr, R.C., Fiddler, W. *J. Chromatogr.* **1977**, *140*, 284-287.
62. Bundesgesundheitsamt, *Bundesgesundheitsblatt* **1987**, *30*, 114.

RECEIVED March 30, 1993

Chapter 3

N-Nitrosodimethylamine in Nonfat Dry Milk

R. A. Scanlan, J. F. Barbour, F. W. Bodyfelt, and L. M. Libbey

Department of Food Science and Technology, Oregon State University, Corvallis, OR 97331

In 1980, using a detection limit of 0.3 ppb, N-nitrosodimethylamine (NDMA) was found in 51 of 71 commercial nonfat dry milk samples. The range for the positive samples was from 0.3 to 6.5 ppb with a mean for all samples of 0.72 ppb. In 1992, the mean NDMA value for 56 samples of instantized nonfat dry milk samples was 1.05 ppb, the range was 0.1 to 5.3 ppb. A comparison of the NDMA content of regular nonfat dry milk powder and instantized powder showed no additional NDMA formation from instantizing. Several researchers have suggested that NDMA in nonfat dry milk forms during the direct-fire drying process while others have contended that the source was fluid milk. Eight tracking experiments in commercial processing plants detected no NDMA until the drying process began. Furthermore, most of the 62 samples of nonfat dry milk prepared by direct-fire drying contained NDMA while eight of nine samples dried by indirect-fire processes did not contain NDMA. Our results indicate that the main source of NDMA in nonfat dry milk is the direct-fire process used in its manufacture.

Hotchkiss (*1*) has summarized the extensive research regarding nitrosamine formation in foods and beverages that has been conducted over the past several decades. When nitrosamine precursors and nitrosating agents are present, and under certain processing conditions, nitrosamines form in foods. N-Nitrosodimethylamine (NDMA) is most commonly found. The earliest research focused on cured meats to which nitrites had been added but more recent work found that nitrosamine formation is possible when foods are direct-fire dried. Secondary and tertiary amines present in the foods react with oxides of nitrogen produced by the combustion process. For example, as early as 1972, Sen et al. (*2*) reported the presence of NDMA in fish meal that had been direct-fire dried.

Widespread occurrence of NDMA in beer prompted studies to determine its origin (*3*). The direct-fire malting process seemed a likely source. O'Brien et al. (*4*) conclusively demonstrated that NDMA formed in malt during the kilning step where the products of combustion come in direct contact with the malt being dried. Because indirect-fire drying greatly reduced the amount of NDMA formed, the malting industry largely switched to indirect-fire processes for malt manufacture. NDMA levels in beer have been greatly reduced since this change (*5*).

Because of the known link between NDMA formation and direct-fire drying processes, researchers have investigated other dried foods, including nonfat dry milk. Several studies in the early 1980s consistently found NDMA in this product. Libbey et al. (*6*) detected NDMA ranging from 0.4 to 4.5 ppb in six out of seven samples of nonfat dry milk and confirmed its presence in one of the samples by mass spectral analysis. Sen and Seaman (*7*) found NDMA in 11 samples of instant skim milk powders, ranging from 0.3 to 0.7 ppb with an average of 0.4 ppb, while Lakritz and Pensabene (*8*) detected NDMA in nine out of ten samples of nonfat dry milk. Additionally, these researchers (*8*) reported "apparent" NDMA in pasteurized milk (mean, 0.1 ppb), and, therefore, suggested fluid milk as a source of NDMA in nonfat dry milk.

In 1983, Frommberger and Allmann (*9*) reported that of 129 samples of German milk powders, 87% contained less than 0.5 ppb NDMA. The highest sample contained 13.2 ppb. In New Zealand, with a minimum detection limit of 1 ppb, Weston (*10*) found no NDMA in five samples of indirect-fire dried skim milk powder, 20 samples of direct-fire dried skim milk powder, or 10 samples of direct-fire dried buttermilk powder. The next year, Havery et al. (*11*) reported NDMA in 84% of 57 nonfat dry milk samples, with an average of 0.6 ppb. Ten samples were greater than 1 ppb; none was detected in nine of the samples. In the late 1980s, Österdahl (*12*) detected only traces of NDMA in three out of 27 samples of milk powder in Sweden. In a comprehensive report of volatile nitrosamines in foodstuffs and beverages in West Germany in 1991, Tricker et al. (*13*) reported occasional trace levels of NDMA in dried milk powder.

The purpose of this research was to determine the NDMA content of nonfat dry milk currently produced in the United States and compare these values with amounts found 12 years ago. In addition, we felt it important to determine whether the NDMA in nonfat dry milk is simply the result of NDMA contamination in fluid milk, or whether it is formed during the drying process.

Experimental

In 1980, 71 commercially-produced nonfat dry milk samples were obtained through the American Dairy Products Institute (formerly the American Dry Milk Institute). The samples were collected from a number of processing plants throughout the United States. NDMA levels were determined by a vacuum distillation procedure, as described in (*6*). Fifty-gram samples were analyzed in order to produce a minimum detection limit of 0.3 ppb.

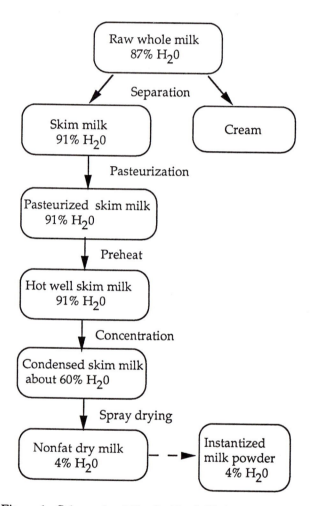

Figure 1. Schematic of Nonfat Dry Milk Manufacturing

In 1992, retail samples of instantized nonfat dry milk were purchased in supermarkets in 13 U.S. cities (Boston, Chester NY, New York City, Washington DC, Atlanta GA, Wichita KS, Sioux City IA, Logan UT, Los Angeles, San Francisco, Corvallis OR, Olympia WA, and Seattle). These samples were analyzed for NDMA according to the procedure of Havery et al. (*11*) with a minimum detection limit of 0.1 ppb.

For the tracking study, samples were obtained in 1981 from two nonfat dry milk manufacturing plants, one in California and one in the Midwest. NDMA was determined in nonfat dry milk samples according to Libbey at al. (*6*) with a minimum detection limit of 0.2 ppb. NDMA levels in fluid milk samples were determined by an atmospheric pressure, steam distillation procedure (described in (*14*)) except that 500 g instead of 300 g samples were used, thereby lowering the minimum detection limit to 0.02 ppb.

An abbreviated representation of the process by which nonfat dry milk is manufactured is depicted in Figure 1. For more information on the process, consult (*5*). In the United States, most, but not all, nonfat dry milk is manufactured by direct-fire, spray-drying processes.

Results

Survey of NDMA in Nonfat Dry Milk. Table I contains a distribution of the NDMA content of 71 commercial nonfat dry milk samples manufactured in 1980. Sixty-two were manufactured by direct-fire, spray-drying processes while nine were prepared by indirect-fire processes. The NDMA values represented in Table I were not corrected for recovery which was approximately 75%. Levels of NDMA ranged from not detectable (<0.3 ppb) to 6.5 ppb with a mean of 0.72 ppb for all samples. The identity of NDMA in three of the samples was confirmed by mass spectrometry, as described in (*6*).

Table I. N-Nitrosodimethylamine in 71 Samples of Nonfat Dry Milk, 1980

Number of samples	*Range (ppb)*
20	nd[a]
37	0.3-1.0
10	1.1-2.0
3	2.1-3.0
1	> 3.0
mean	0.72[b]

[a]Not detected; detection limit 0.3 ppb.
[b]The mean for all 51 positive samples is 1.01 ppb.

Table II reports the distribution of the NDMA content of 56 instantized nonfat dry milk samples purchased at retail outlets in summer 1992. The NDMA values were not corrected for recovery which was approximately 70%. Levels of

NDMA ranged from 0.1 to 5.3 ppb with a mean of 1.05 ppb. The tendency for the 1992 levels to be larger than the 1980 levels was statistically significant according to the Wilcoxon rank sum test (16), in which the one-tailed P-value was 0.003. In order to eliminate a potential bias introduced by different detection limits, all values below 0.3 ppb in the 1992 data set were considered as non-detectable for the Wilcoxon rank sum test.

Table II. N-Nitrosodimethylamine in 56 Samples of Instantized Nonfat Dry Milk, 1992

Number of Samples	Range (ppb)
0	nd[a]
37	0.1-1.0
14	1.1-2.0
3	2.1-3.0
2	>3.0
mean	1.05

[a]Not detected; detection limit 0.1 ppb.

Comparison of NDMA Content of Regular and Instantized Nonfat Dry Milk Powder. This experiment was designed to determine if there was any significant difference between NDMA levels in five samples of regular powders and in their corresponding instantized powders. Results, reported in Table III, show that generally the levels were very similar. Only one sample showed an increase in the level of NDMA when instantized. These results strongly indicate that the amount of NDMA in nonfat dry milk is not further increased during the instantizing process.

Table III. N-Nitrosodimethylamine in Regular and Instantized Nonfat Dry Milk Powder

Sample #	Regular powder (ppb)	Instantized powder (ppb)
1	0.4	0.5
2	0.4	0.3
3	0.7	0.6
4	1.8	3.0
5	nd[a]	nd

[a]Not detected; detection limit 0.3 ppb.

Comparison of NDMA Content of Nonfat Dry Milk Dried by Direct-Fire and by Indirect-Fire Processes.   Sixty-two of the 71 nonfat dry milk samples manufactured in 1980 were dried by the direct-fire process; nine by an indirect-fire method. A comparison of their respective NDMA contents is made in Table IV. Most, but not all, of the samples dried by direct-fire contained NDMA. However, eight of nine samples dried by indirect-fire did not contain NDMA. This result strongly suggests that when NDMA is found in nonfat dry milk it is formed by the direct-fire drying process.

Table IV. N-Nitrosodimethylamine in Direct- and Indirect-Fire Nonfat Dry Milk Samples

Direct-fire		Indirect-fire	
Number of Samples	NDMA	Number of Samples	NDMA
50	positive	1	positive (0.6 ppb)
12	nd[a]	8	nd

[a]Not detected; detection limit 0.3 ppb

The Tracking Study. The results from eight separate tracking experiments conducted in two commercial processing plants are reported in Table V. NDMA was detected in all nonfat dry milk samples. None of the fluid milk samples from which the nonfat dry milk was made contained NDMA at the detection limit of 0.02 ppb.

Table V. N-Nitrosodimethylamine Levels in Tracking Studies

	Midwestern plant			California plant				
				ppb				
Run #	1	2	3	4	5	6	7	8
Raw whole milk	nd[a]	nd	nd	nd	nd	nd	nd	nd
Pasteurized milk	nd	nd	nd	--[b]	--	--	--	--
Hot well skim milk	nd	nd	nd	nd	nd	nd	nd	nd
Condensed milk	nd	nd	nd	nd	nd	nd	nd	nd
NDM[c]beginning	0.6	0.6	0.5	0.7	1.6	0.6	0.7	0.8
NDM, middle	0.5	0.5	0.7	1.0	1.0	0.6	0.6	0.7
MDM, end	0.5	0.6	0.5	0.8	0.8	0.7	0.6	0.6

[a]Not detected; detection limit 0.02 ppb.
[b]Not sampled.
[c]Nonfat dry milk.

When referring to the data in Table V, it is worth noting the significance of the minimum detection limits for NDMA in the fluid milk (0.02 ppb) and in the nonfat dry milk (0.2 ppb samples. Fluid milk is concentrated approximately tenfold when manufactured into nonfat dry milk. Therefore, nonfat dry milk samples that contain 0.2 ppb or more NDMA would have to have been manufactured from fluid milk containing 0.02 ppb NDMA or more, *if* the source of the NDMA in the dry milk was from the fluid milk. Conversely, if there is no detectable NDMA in the fluid milk (< 0.02 ppb), finding NDMA at 0.2 ppb or greater in the nonfat dry milk would strongly suggest that the NDMA was formed during drying rather than by concentration from the fluid milk. Thus, the data in Table III strongly suggest that the NDMA in the nonfat dry milk samples was formed during drying and was not concentrated from NDMA in the fluid milk.

Discussion

Previous researchers have speculated on the sources of NDMA in nonfat dry milk. Several have proposed that NDMA is formed during the direct-fire drying processes of dry milk manufacture (7). Others have suggested that NDMA in nonfat dry milk is concentrated from that already contained in fluid milk (7). However, distinguishing between these possible sources has been difficult--partly due to the insufficiently low detection limits used.

Our results demonstrate that NDMA found in nonfat dry milk is not due to concentration of a contaminant already present in fluid milk. No NDMA was detected in the tracking experiment until the drying stage. Further, NDMA was detected in most of the samples dried by the direct-fire process but no NDMA was detected in eight of nine samples dried by indirect-fire methods. We conclude that the direct-fire process is the main source of NDMA in nonfat dry milk.

A comparison of the results from the 1980 and 1992 surveys indicate that NDMA levels have not decreased in this product. Since NDMA formation occurs as a result of direct-fire drying, the results from this work suggest that NDMA formation in nonfat dry milk could be ameliorated by conversion from direct- to indirect-fire drying processes.

Acknowledgements

The authors express appreciation to the American Dairy Products Institute, Inc., for funding support and help in obtaining samples. This research was also supported in part by Grant No. CA 25002, awarded by the National Cancer Institute, DHHS. The authors are also indebted to Carole Nuckton for help in preparing the manuscript.

Literature Cited

1.   Hotchkiss, J. H. *Cancer Surveys*. 1989, *8*, pp. 295.
2.   Sen, N. P.; Schwinghamer, L. A.; Donaldson, B. A.; Miles, W. F. *J. Agric. Food Chem.* 1972, *20*, pp. 1280.
3.   Spiegelhalder, B.; Eisenbrand, G.; Preussmann, R. In *Proc. VI Inter. Sym. N-Nitroso Compds*. Walker, E.A.; Gricuite, L.; Castegnaro, M.; Borzsonyi,

M; Davis, W. Eds; International Agency for Research on Cancer. Public. No. 31; Lyon, France, 1980, pp. 467-477.
4.   O'Brien, T. J.; Lukes, B. K.; Scanlan, R. A. *Mast. Brew. Assoc. Tech. Quart.* 1980, *17*, pp. 196.
5.   Scanlan, R. A.; Barbour, J. F.; Chappel, C. I. *J. Agric. Food Chem.* 1990, *38*, pp. 442.
6.   Libbey, L. M.; Scanlan, R. A.; Barbour, J. F. *Fd. Cosmet. Toxicol*, 1980, *18*, pp. 459.
7.   Sen, N. P.; Seaman, S. *J. Assoc. Off. Anal. Chem.* 1981, *64*, pp. 1238.
8.   Lakritz, L; Pensabene, J. W. *J. Dairy Sci.* 1981, *64*, pp. 371.
9.   Frommberger, R.; Allmann, H. In *Das Nitrosamin-Problem*, Preussmann, R. Ed.; Verlag Chemie GmbH, Weinheim, 1983, pp. 58-63.
10. Weston, R. J. *J. Sci. Food Agric.* 1983, *34*, pp. 893.
11. Havery, D. C.; Perfetti, G.A.; Fazio, T. *J. Assoc. Off. Anal. Chem.* 1984, *67*, pp. 20.
12. Österdahl, B. -G. *Food Addit. and Contam.* 1988, *5*, pp. 587.
13. Tricker, A. R.; Pfundstein, B.; Theobald, E.; Preussmann, R.; Speigelhalder, B. *Food Chem. Toxic.* 1991, *29*, pp. 729.
14. Scanlan, R. A.; Barbour, J. F.; Hotchkiss, J. H.; Libbey, L. M. *Food Cosmet. Toxicol.* 1980, *18*, pp. 27.
15. Campbell, J. R.; Marshall, R. T. *The Science of Providing Milk for Man*, McGraw Hill, Inc., New York, 1975, pp. 687-720.
16. Snedecor, G. W.; Cochran W. G. *Statistical Methods*, 7th edition; The Iowa State University Press: Ames, IA, 1980; pp. 144.
17. Ellen, G. In *The Significance of N-Nitrosation of Drugs*; Eisenbrand, G.; Bozler, G.; Nicolai, H. v. Eds.; G.F. Verlag, Stuttgart, 1990, pp. 19-46.

RECEIVED June 29, 1993

Chapter 4

Prevention of Nitrosamine Exposure in the Rubber Industry

B. Spiegelhalder and C.-D. Wacker

Department of Environmental Carcinogens, FSP03, German Cancer Research Center, Im Neuenheimer Feld 280, D–6900 Heidelberg, Germany

The occurrence of carcinogenic nitrosamines at workplaces in the rubber and tire industry is still an unsolved problem. Recent air measurements in Germany (1988-1991) showed nitrosamine concentrations up to 41 $\mu g/m^3$. New regulations require levels not greater than 2.5 $\mu g/m^3$. The major source of nitrosamines is the use of certain vulcanization accelerators such as thiurams, dithiocarbamates and sulfenamides. These accelerators are nitrosated during the vulcanization process. The origin of the nitrosating agent are oxides of nitrogen adsorbed on the large surface of inorganic rubber additives, e.g., zinc oxide and carbon black. We investigated two different approaches to prevent the formation of carcinogenic nitrosamines: (1) The nitrosating potential within a rubber mixture can be eliminated or at least drastically reduced by the addition of scavengers, like primary amines, urea or sulfamic acid. (2) Following the concept of "safe amines", the amine moiety of the accelerators does not form carcinogenic nitrosamines, a number of new accelerators with good technical properties was synthesized. Both preventive measures can be used effectively to improve workplace hygiene.

The occurrence of carcinogenic nitrosamines at workplaces in the rubber and tire industry was first described by Fajen et al. in 1979 (*1*). Subsequent studies in other countries (*2-4*) proved that nitrosamines can be found in nearly all factories producing rubber at levels up to >380 $\mu g/m^3$. Recent air measurements in Germany (1988-1991) showed that nitrosamines at workplaces in the rubber industry is still an unsolved problem and concentrations up to 41 $\mu g/m^3$ could be observed. Table I summarizes maximum exposure levels at different work places measured in the early eighties. New regulations on occupational exposure limits in Germany require levels not greater than 2.5 $\mu g/m^3$.

NOTE: Dedicated to Rolf Preussman for his 65th anniversary.

Table I: Maximum exposure levels at different work places in the rubber industry ($\mu g/m^3$)

Location or process	NDMA	NMOR
Raw material handling	0.9	2
Milling, extruding, calendering	2	9
Assembly & building	1	3
Curing	130	380
Inspection & finishing	10	20
Storage & dispatch	19	17

The major source of nitrosamines is the use of certain amine based vulcanization accelerators such as thiurams, dithiocarbamates and sulfenamides. The secondary amine moiety of the accelerators is nitrosated during the vulcanization process. The origin of the nitrosating agent are oxides of nitrogen adsorbed on the large surface of inorganic rubber additives, *e.g.*, zinc oxide and carbon black or nitrosating rubber chemicals. A nitrosation of the released amines can also occur in the air by oxides of nitrogen. Figure 1 is a proposed reaction scheme (*5*). Figure 2 shows the nitrosation reactions of typical accelerators.

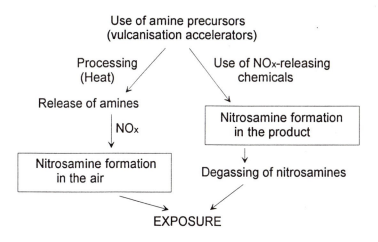

Figure 1: Reaction scheme of nitrosamine formation in the rubber industry

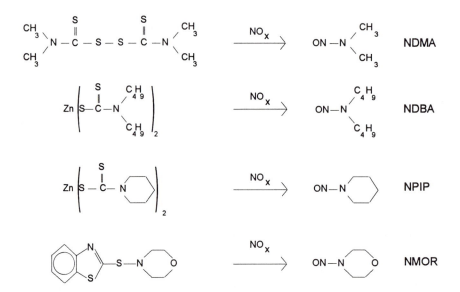

Figure 2: Formation of nitrosamines from their corresponding accelerators

The different approaches to prevent the formation of carcinogenic nitrosamines are listed as follows:

- Avoidance of nitrosating chemicals
  (*e.g.*: "Retarder A", blowing agents)
- Reduction of oxides of nitrogen in workroom air
- Use of amine-free accelerators
  (*e.g.*: peroxide accelerators)
- **Inhibition of nitrosamine formation during vulcanization and/or destruction of nitrosation potential on surface of inorganic rubber additives (carbon black)**
- **Use of accelerators derived from "safe amines"**

In our research program we focused on possibilities to inhibit the nitrosamine formation during the vulcanization process, as well as to synthesize accelerators based on safe amines.

To identify effective inhibitor systems model experiments were carried out in a reaction mixture, which simulates vulcanization conditions in rubber at elevated temperatures (>120° C). Paraffin oil containing nitrosatable rubber chemicals was used (*6*). Using morpholino-2-benzthiazolsulfenamide (MOZ) as a nitrosatable compound inorganic fillers, like zinc oxide, silica or carbon black, could be identified to have a nitrosation potential. Oxides of nitrogen which are adsorbed on the large surface of the inorganic material are probably the nitrosating species. In Figure 3 the nitrosating potential of different inorganic rubber additives is shown. The nitrosamine formed is N-nitrosomorpholine (NMOR). The amount of nitrosamine formed is directly proportional to the amount of NO_x determined as nitrite in ZnO (Figure 4).

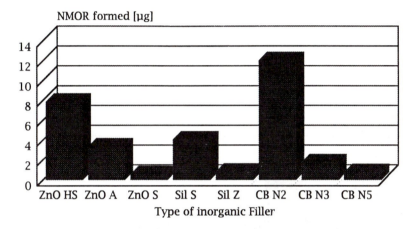

Figure 3: NMOR formation in model nitrosation experiments using different types of fillers (ZnO = zinc oxide, Sil = silica, CB = carbon black)

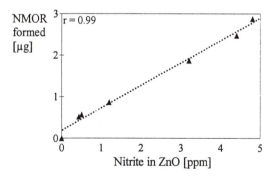

Figure 4: NMOR formation by different ZnO types with different nitrite content.

The nitrosating potential within a rubber mixture can be eliminated, or at least drastically reduced, by the addition of scavengers or inhibitors, like primary amines or diamines, urea or amidosulfonic acid (*6*).

The efficacy of these nitrosation inhibitors can be demonstrated in model experiments simulating the rubber vulcanization. The results for MOZ as nitrosatable accelerator are shown in Figure 5. Mixtures of amidosulfonic acid and urea were most effective in rubber mixtures preventing nitrosamine formation. As an example NMOR concentrations in carbon black reinforced rubber vulcanizates are shown in Figure 6. The inhibitor system amidosulfonic acid / urea was studied in more detail over a range of different relations of the two inhibitors in a rubber containing thiuram and MOZ.

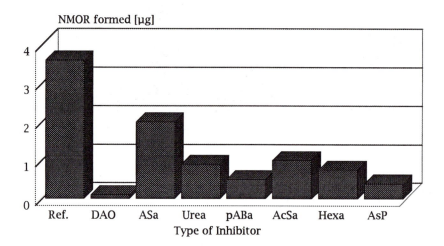

Figure 5: Inhibition of NMOR formation during simulated vulcanization using MOZ as accelerator; Ref = preparation without inhibitor, DAO = diaminooctane, ASa = amidosulfonic acid, pABa = para-amino benzoic acid, AcSa = acetyl salicylic acid, Hexa = hexamethylenetetramine, AsP = ascorbyl palmitate.

The resulting inhibition of the formation of NDMA and NMOR is graphically represented over a range from 0 to 5 phr (parts per hundred rubber) of each inhibitor. It can be seen, that the inhibition of NDMA formation shows a different characteristic than the inhibition of NMOR formation (Figure 7).

The approach to inhibit nitrosamine formation was also investigated by other authors. Chasar (*7*) described the use of alkaline earth oxides and hydroxides (CaO, $Ca(OH)_2$ and $Ba(OH)_2$) and could achieve a 70-95% reduction of nitrosation. The use of Na-hydroxymethane sulfinate by Schmieder et al. (*8*) resulted in nitrosamine levels < 10 ppb in synthetic rubber. Schuster et al. (*9*) showed that α-tocopherol may successfully prevent nitrosamine formation in rubber vulcanizates, especially in black rubber types. Maleic and fumaric acid can also serve as nitrosation inhibitors in rubber (*10*).

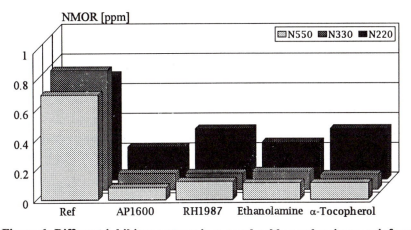

Figure 6: Different inhibitor systems in natural rubber vulcanizates reinforced with different carbon black types and MOZ as accelerator; Ref = without inhibitor, AP1600 and RH1987 = amidosulfonic acid/urea preparations.

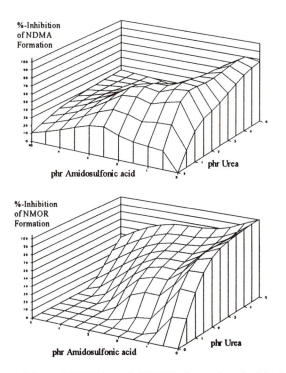

Figure 7: Inhibition of NDMA and NMOR formation in black NBR rubber using thiuram and MOZ as accelerator by different amidosulfonic acid / urea combinations (phr = parts per hundred rubber).

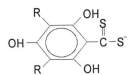

Aromatic hydroxydithiocarbonate

Aromatic hydroxydithiocarbonic acids can serve as nitrosation inhibitors and as accelerators as well (*11*).

The nitrosamine reduction by polyoxymethylene described by Lheureux et al. (*12*) results from scavenging the amine moiety in a vulcanization system containing methylphenylamine derived accelerators.

The chemistry of the miscellaneous inhibition methods of nitrosamine formation can be summarized by the following mechanisms:

- a reduction of oxides of nitrogen to either NO by antioxidants,

- a reduction to nitrogen by scavengers and

- a reaction with the amine moiety of the accelerator or free amine to prevent nitrosation.

The best method for avoiding the formation of nitrosamines may be the use of nitrogen-free accelerators. Compounds like peroxides or thiophosphates, which were suggested as substitutes (*5*) are only of limited to specific rubber mixtures, however In most cases thiurams, dithiocarbamates and sulfenamides are still necessary for rubber to achieve good technological properties. The widely used sulfenamides can in most cases easily be substituted against accelerators containing primary amines (t-butyl-, t-amyl-, cyclohexylamine) instead of secondary amines like morpholine. Some of these substitutes are already commercially available and are used as safe accelerators in the tire industry (*13-16*). Much more difficult to exchange are thiurams and dithiocarbamates. Only dibenzylamine derived accelerators can be obtained commercially for use as substitutes (*17,18*). The wide range of different rubber products require more thiuram and carbamate accelerators to cover the wide range of applications with different needs.

Our aim was to synthesize new accelerators based on "safe amines", which either do not form nitrosamines or which form non carcinogenic nitrosamines. The following structures represent a selection of "safe amines" (Figure 8).

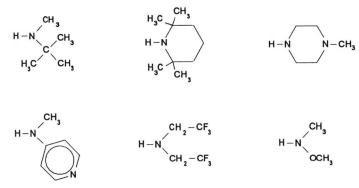

Figure 8: Chemical structures of selected "safe amines".

Following the concept of "safe amines" a number of new accelerators with good technical properties was synthesized (*19,20*). Thiuram and carbamate accelerators based on methylpiperazine are among the most suitable substitutes for recompounding rubber mixtures. As an example for suitability the vulcanization properties of safe accelerators the methylpiperazine derived thiuram and dithiocarbamate are compared against commercial accelerators in Figure 9 and 10.

Vulcasafe ZMP and Vulcasafe MPT are good acting vulcanization accelerator with similar properties as the technical standards. Vulcasafe ZMP and Vulcasafe MPT appear to be good substitutes for tetramethylthiuramdisulfide (TMTD), a widely used thiuram accelerator as well as for some of the traditional dithiocarbamates.

These new accelerators offer the possibility of a "safe" rubber vulcanization without formation of carcinogenic nitrosamines. In the case of the methylpiperazine derivatives the toxicological evaluation is still in progress. The corresponding nitrosamine is probably non carcinogenic or at least a very weak carcinogen. Some of our "safe amine" accelerators are currently being tested by a number of German rubber producers for their applicability in specific rubber formulations.

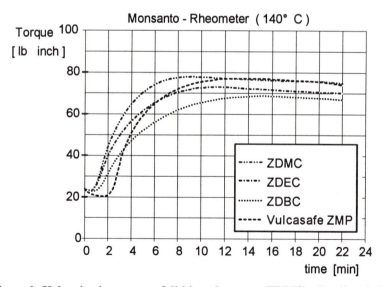

Figure 9: Vulcanization curves of dithiocarbamates; ZDMC = Zn-dimethyl-, ZDEC = Zn-diethyl-, ZDBC = Zn-dibutyl-, Vulcasafe ZMP = Zn-methylpiper-azine-dithiocarbamate.

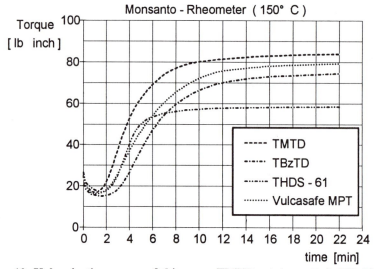

Figure 10: Vulcanization curves of thiurams; TMTD = tetramethyl-, TBzTD = tetrabenzyl-, Vulcasafe MPT = bis-(4-methyl-piperazino)-, THDS-61 = bis-(4-(β-hydroxyethyl)-piperazino)-thiuramdisulphide

Literature Cited

1. Fajen, J. M.; Carson, G. A.; Rounbehler, D. P.; Fan, T. Y; Vita, R.; Goff, U. E.; Wolf, M. H.; Edwards, G. S.; Fine, D. H.; Reinhold,, V.; Biemann, K. *Science* **1979**, *205*, 1262.

2. Daubourg, N.; Coupard, A.; Pepe, A. *Caoutch.Plast.* **1992**, *69*, 103.

3. Rounbehler, D.P.; Reisch, J.; Coombs, J.R.; Fine, D.H.; Fajen, J.M. In *Chemical Hazards in the Workplace*; Choudhary, G., Ed.; American Chemical Society: Washington,D.C., 1981, Vol. 149; pp. 343-356.

4. Spiegelhalder, B.; Preussmann, R. *Carcinog.* **1983**, *4*, 1147.

5. Spiegelhalder, B. *Scand j work environ health* **1983**, *9*, 15.

6. Bauer, A.; Ehrend, H; Menting, K. H.; Preussmann, R.; Spiegelhalder, B. *Eur.Pat.* **1991**, *463465*, 6 pp.

7. Chasar, D. W.; Brecksville, O. H. *KaGuKu* **1992**, *45*, 18.

8. Schmieder, H.; Naundorf, D.; Hühn., G.; Bertram, M. *DDR Pat.* **1991**, *DD 295646 A5*, 2 pp.

9. Schuster, H.; Nabholz, F.; Gmuender, M. *KaGuKu* **1990**, *43*, 95.

10. Thörmer, J.; Scholl, T. *Eur.Pat.* **1992**, *482470*, 11 pp.

11. Hörpel, G.; Haag, H. G.; Nordsiek, K. H. *Ger.Offen.* **1991**, *3941001*, 6 pp.

12. Lheureux, M.; Kuhlmann, T.; Siekermann, V. *KaGuKu* **1990**, *43*, 107.

13. Loadman, M. J. R. *KaGuKu* **1989**, *3*, 201.

14. Lüpfert, S. *KaGuKu* **1989**, *1*, 16.

15. Banerjee, B. *KaGuKu* **1989**, *3*, 217.

16. Gorton, A. D. T.; McSweeney, G. P.; Tidd, B. K. *NR Technology* **1987**, *18*, 1161.

17. Lüpfert, S. *KaGuKu* **1990**, *43*, 780.

18. Hofmann, W. *Kunststoffe* **1990**, *80*, 267.

19. Wacker, C.-D.; Spiegelhalder, B.; Börzsönyi, M.; Brune, G.; Preussmann, R. In *The relevance of N-nitroso compounds to human cancer: Exposure and mechanisms*; Bartsch, H.; O'Neill, I.; Schulte-Hermann, R., Eds.; IARC Scientific Publications: Lyon, 1987, Vol. 84; pp. 370-378.

20. Wacker, C.-D.; Preussmann, R.; Kehl, H.; Spiegelhalder, B. *J. Cancer Res. Clin. Oncol.* **1991**, *117*, S 13.

RECEIVED August 12, 1993

Chapter 5

Blocking Nitrosamine Formation

Understanding the Chemistry of Rapid Nitrosation

Richard N. Loeppky, Yen T. Bao, Jaeyoung Bae, Li Yu,
and Graziella Shevlin

Department of Chemistry, University of Missouri, Columbia, MO 65211

The fundamental chemical principles of blocking inadvertent nitrosamine formation and contamination through the recognition of
structural features which predispose compounds toward rapid nitrosation as well as the development of new blocking agents are reviewed.
Rapid nitrosamine formation from tertiary nitrogen compounds such
as amines, amidines, and gem. diamines is shown to occur through
an electron pair assisted nitrosative solvolysis type mechanism. The
rapid production of nitrosamines from amdinocillin, an amidine containing antibiotic, and hexetidine, a widely used antimicrobial agent,
are presented as examples of new findings. The development and assay of new monomeric and polymeric blocking agents is described.
An advantage of the latter blocking agent is its ability to be removed
from mixtures by physical means. This knowledge and these measures permit significant reductions in nitrosamine contamination to be
realized.

Nitrosamine formation in the human environment continues to be a significant
problem despite the widespread recognition of the possible carcinogenic role of nitrosamines and other N-nitroso compounds. Many effective measures have been
taken to eliminate or significantly reduce the level of nitrosamines in foods and
other commercial products (*1*). In many instances, successful steps to block nitrosamine formation have been based upon an elucidation of the mechanisms by which
they are formed and/or the use of specific blocking agents. These numerous
individual successes, however, have not eliminated the problem of nitrosamine
contamination. Moreover, there is now good evidence that nitrosamines form *in
vivo* from several types of precursors, heightening concern that many tumors could
have their origin in endogenous nitrosation (*2*).

0097–6156/94/0553–0052$08.00/0
© 1994 American Chemical Society

The reasons why nitrosamine contamination and formation have remained a problem are: 1) the ubiquity of the precursors, 2) an incomplete understanding of the chemical mechanisms of formation and the structure types which react most rapidly to generate them, 3) the lack of selective, mild methods for their removal and destruction, and 4) the lack of general and effective blocking agents. The principal routes leading to nitrosamines are summarized diagramatically. The sources of the nitrosating agents are varied and often depend upon the environmental conditions or the chemical composition of specific chemical mixtures and formulations. Nitrate salts are ubiquitous in the macro environment and microorganisms often contain enzymes which effectively reduce nitrate to nitrite. This pathway occurs in the human oral cavity and is a key source of nitrosating activity in the stomach (*3*). Because of the linkage between nitrate and nitrite, nitrosamine formation should always be considered where nitrate levels are high, despite the fact that nitrate and nitric acid are not nitrosating agents. Oxides of nitrogen resulting from pollution are principal sources of nitrosating agents, as well (*4*). Recent evidence suggests that the endogenous generation of NO can lead to nitrosation (*5,6,7*) and it is known that nitrate reducing bacteria can mediate amine nitrosation (*8*). Often, the nitrosating agents are formed inadvertently and their concentrations are small. Nitrous esters (*9,10*), certain nitro compounds (*11*), and specific metal complexes of nitrite or NO (*12*) in addition to common nitrosyl donors (*13*), "NO$^+$", (HNO$_2$, NOX (where X= ONO, ONO$_2$, Cl, Br, I, SCN, SO$_4$H, etc.) can each be effective nitrosating agents. The best procedures for blocking nitrosamine formation are directed at scavenging these agents (*14,15*).

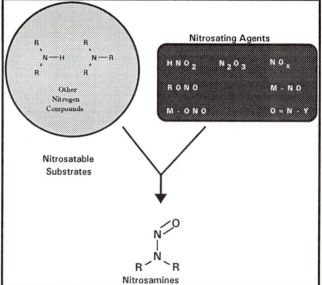

It is well known that secondary amines react rapidly with a variety of nitrosating agents to produce nitrosamines. The rate of this process is very high and often is limited by the molecular encounter of the amine and the nitrosating agent. In aqueous solution, a large percentage of the amine is protonated under the acidic conditions which favor the highest concentrations of the effective nitrosating agent. (*13*) As a result, the nitrosation rate of secondary amines in aqueous solution is a function of their basicity (*16*) (the greater the pK$_b$ the more rapid the nitrosation).

The encounter-controlled nature of the nitrosation step places rather severe limitations on the nature of effective blocking agents. These substances must also react extremely rapidly (encounter controlled) with nitrosating agents. In aqueous solution, their effective concentration cannot be reduced by acid-base reactions which render them unreactive like protonated amines. Because of these limitations, there are situations where either a secondary amine should not be used in a product or extreme measures should be taken to prevent its contamination and interaction with nitrosating agents and their sources. The best situations for blocking nitrosamine formation through the scavenging of nitrosating agents occur in aqueous solution where the pH can reduce the reactivity of the amine (through protonation) without decreasing the reactivity of the blocking agent (*14,15*).

Tertiary nitrogen compounds capable of yielding the secondary amino component of a nitrosamine can also be effective sources of nitrosamines. Nitrosamine formation from, these substances, however, must involve the breakage of a C-N or other bond as is illustrated below for a tertiary amine. This process is often too slow to permit significant nitrosamine formation either during chemical processing or *in vivo*. For example, the generation of significant yields of dibenzylnitrosamine from tribenzylamine in acid requires temperatures as high as 60° C and

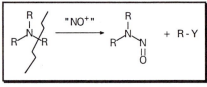

a large excess of nitrous acid (*17*). The mechanism of tertiary amine nitrosation for normal aliphatic tertiary amines is given in Scheme 1. The process involves the reversible generation of a nitrosammonium ion (R_3NNO^+) followed by the rate limiting loss of NOH (*17,18*). More basic amines will form the nitrosammonium ion more slowly and decrease the overall rate of nitrosative dealkylation. Yet it is clear from nitrosamine occurrence data that nitrosamines do form rapidly from some tertiary nitrogen compounds.

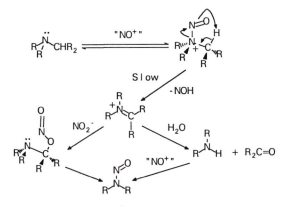

Scheme 1

These reactions must be occurring through pathways different than that taken for the nitrosation of simple aliphatic tertiary amines. One of our goals is the elucidation of the mechanisms and enabling structural requirements which lead to rapid nitrosamine formation from tertiary nitrogen compounds. The knowledge gained in this endeavor has led to the development of new blocking agents and strategies for preventing nitrosamine formation which is the subject of this chapter.

Rapid Nitrosamine Formation for Tertiary Nitrogen Compounds

One of the first demonstrations that nitrosamines could form rapidly from a tertiary nitrogen compound came from Lijinsky's demonstration *(19)* that the common analgesic drug aminopyrine rapidly liberates dimethylnitrosamine when treated with nitrous acid at ambient temperatures (Equation 1). Another observation of this kind was made by Scanlan's group who were investigating the source of dimethylnitrosamine in malted beverages *(20)*. They showed that gramine rapidly generated dimethylnitrosamine when treated with nitrous acid (Equation 2). Our interest in the mechanism of this process led to nitrosation studies on 2-dimethylaminomethylpyrrole *(21)*, a structurally related compound, which also nitrosates very rapidly at 25° C to produce dimethylnitrosamine (Equation 3). Our work has revealed that the latter two compounds nitrosate by two new and distinct mechanistic pathways. While evidence for these reaction paths was provided from investigations of the nitrosation of 2-dimethylaminomethylpyrrole and the Scanlan group' work on gramine, better data has been obtained from a study of the nitrosation of para and meta substituted benzylic N,N-dimethylamines and their heterocyclic counterparts *(22,23)*. The mechanisms given in Schemes 2 and 3 summarize the results of these studies.

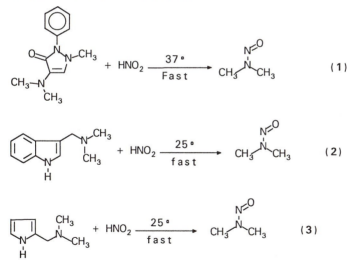

The mechanism depicted in Scheme 2 *(21-23)* has the same first step as the classical nitrosation mechanism (Scheme 1), formation of a nitrosammonium ion. Rapid nitrosamine formation from this intermediate is made possible by the assistance of an electron donating heteroatom, either attached to, or a part of the attached aromatic ring. Electron donation from this moiety permits the SN-1 like fragmentation of the nitrosammonium ion to yield the nitrosamine and a highly stabilized carbocation. Unlike other tertiary amine nitrosations, this pathway requires only a single mole of nitrite per mole of nitrosamine produced and can occur under conditions where the concentration of the nitrosating agent is limited, such as the stomach.

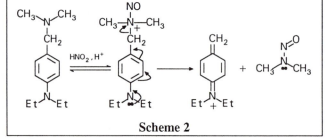

Scheme 2

The pathway shown in Scheme 3 is of chemical interest because it involves the breakage of a C-C bond through ultimate *ipso* substitution on the aromatic ring at the position *para* to the electron donating substituent (*21-23*). The aromatic product of this transformation is a nitro compound which arises through a radical cation intermediate. The nitrosamine forms after the retro-Mannich-like fragmentation of the Wheland intermediate generated from NO_2 and the radical cation. The imminium ion generates the nitrosamine by the same paths shown in Scheme 1.

The reaction paths depicted in Schemes 2 and 3 require powerful electron donating substituents attached to the aromatic ring, or in the case of heterocycles, included in it. We have examined the ability of a number of tertiary amines of structural type **A** or **B** to undergo facile nitrosation at 25° C and only those where **Y**=N are highly reactive. The general order of reactivity observed is **Y**= N>>S>O. Even so, low yields of nitrosamine are obtained from some compounds containing O or S substituents and these substances could engender nitrosamine formation under suitable conditions.

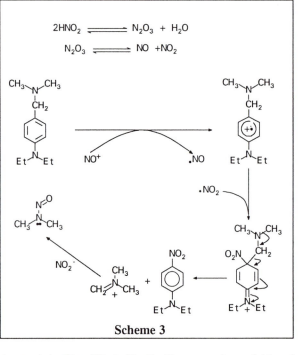

Scheme 3

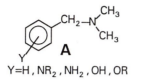

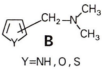

Y=H, NR$_2$, NH$_2$, OH, OR   Y=NH, O, S

One of the most significant accomplishments to arise from this mechanistic investigation was the recognition of general structural characteristics which can give rise to facile nitrosamine formation from tertiary nitrogen compounds (*23*), and how some of these molecular characteristics can be utilized in the development of effective blocking agents. Facile nitrosamine formation from these amines requires the presence of an electron-rich

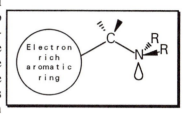

aromatic ring as shown. The nitrosamine is generated through a heteroatom unshared pair assisted nitrosative solvolysis. This mechanistic analysis has led to the prediction that rapid nitrosamine formation could occur from the types of compounds shown in Scheme 4 through the transition states TS1, TS2, and TS3. We have tested several of these hypotheses and find high reactivity toward nitrosation as expected.

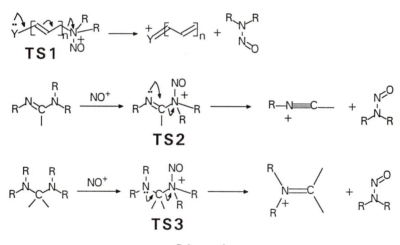

Scheme 4

Tri-N-substituted amidines are very reactive toward nitrosation (*24*), and give not only nitrosamines, but N-nitrosoamides and diazonium ions as products as well (Equation 4). The formation of the latter two types of compounds suggests that *in vivo* nitrosation of amidines could be very deleterious because of the aggressive alkylating characteristics of these products. Although several pathways are operative in the genesis of these products, TS2 is important in generating the nitrosamine. An amidine containing drug, amdinocillin, was subjected to nitrosation conditions at low [NO$_2$], conditions similar to those which prevail in the human stomach (Equation 5). The yields of N-nitrosohomopiperidine, NHPIP, (N-nitrosoperhydroazepine) produced from the incubation of amdinocillin with nitrite at various pH's at 37° are given in the Table I (*25*). It is clear that nitrosamines could be produced from amidines under conditions likely to prevail in the human stomach.

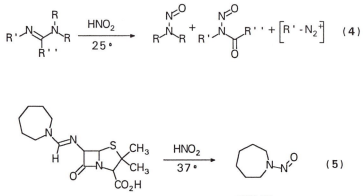

$$(4)$$

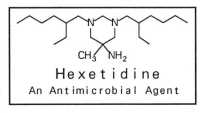

$$(5)$$

Amdinocillin   NHPIP

Table I
Yields of NHPIP Produced from the Nitrosation of Amdinocillin[a]

Condition	HOAc	pH 2.2	pH 3.2	pH 4.2	pH 5.2
% NHPIP	0.3	0.02	0.02	0.007	ND

a. Reaction time 1 hour at 37°C at 0.052 mM Amdinocillin and 0.2 mM NaNO$_2$

Our mechanistic analysis has also suggested that gem. diamines should nitrosate rapidly. Our investigation of this hypothesis has focused on hexetidine, a common antimicrobial agent, and compounds related to it. Our initial studies revealed that hexetidine nitrosates rapidly at 25° C in the presence of limited nitrous acid to produce at least eight nitrosamines. The investigation of the structure of these compounds has been complicated by the nature of the 2-ethylhexyl side chains attached to the nitrogen atoms. These groups are chiral and commercial hexetidine consists of a mixture of diastereoisomers which, when combined with the diastereoisomerism produced by the Z-E isomerism of the N-nitroso group in the products, greatly complicates NMR-based structural characterization. To overcome this problem we examined the nitrosation of a series of model compounds where the N-bound 2-ethylhexyl groups were replaced with methyl groups. Nitrosation of the synthetic intermediate nitro

Hexetidine
An Antimicrobial Agent

gem. diamine is shown in Equation 6. It rapidly reacts with nitrous acid at 25° to give the nitrosamines shown (23,26).

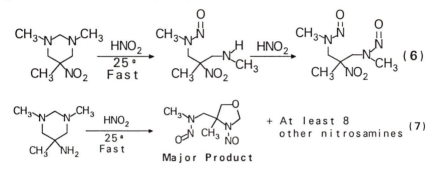

Nitrosation of the methyl analog of hexetidine also rapidly produces a set of N-nitroso bearing products (26). The major product from this mixture is the N-nitrosooxazolidine shown in Equation 7. The structure of this nitrosamine was elucidated by NMR and MS and proved by unambiguous synthesis.

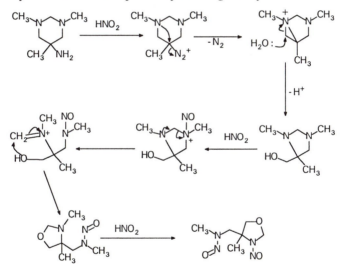

Scheme 5

It has an interesting origin which involves the diazotization of the primary amine as shown in Scheme 5. It is significant to note that the formation of the rearranged nitrosamine involves the generation of an aziridinium ion, an alkylating agent. produced from minimal nitrite and a hexetidine analog.

The nitrosation of hexetidine itself does not give the N-nitrosooxazolidine as a major product. The major nitrosamine formed from the reaction of even limited nitrous acid and hexetidine is shown in Equation 8. The identity of this material is based on spectroscopic data (27). As indicated, numerous products result from the nitrosation of hexetidine. Some of these have been identified by Mende et al. (28).

Our work in this area is continuing. The most striking feature of our research so far is the facility with which nitrosamine formation occurs, demonstrating again the validity of our hypotheses regarding the structural features which permit rapid nitrosamine formation from tertiary nitrogen compounds.

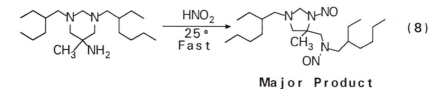

Major Product

Knowledge of the structural features which render a compound particularly vulnerable to nitrosation and the formation of nitrosamines is of value in preventing the formation of nitrosamines in the human environment and reducing the extent of *in vivo* nitrosation. Many times a variety of structurally related compounds are known to possess a desirable property which permits their possible commercialization. Certainly, nitrogen compounds should be examined for their possible ease of nitrosation. Relatively minor structural changes can reduce their proclivity to nitrosate without significantly altering their beneficial commercial properties.

Blocking N-Nitrosation

Our investigation of the structural factors which favor rapid nitrosation have also led us to a new set of effective blocking agents. We have shown in Scheme 3 that electron rich aromatic compounds can undergo nitrosamine formation through attack at the aromatic ring. Our work on the nitrosation of 2-dimethylaminomethylpyrrole (Equation 2) indicated that the pyrrole ring was being extensively degraded by nitrosating agents (*21*). These results suggested that pyrrole and its derivatives might act as effective inhibitors of N-nitrosation. Pyrrole is not basic enough to be protonated at even relatively low pH but reacts rapidly with nitrosating agents to produce an amorphous mixture called nitrosopyrrole black. (Equation 9) A meeting report by Groennen suggested that pyrrole was more effective than many nitrosation inhibitors. (*29*) We have demonstrated that pyrrole is more effective that ascorbic acid in blocking the nitrosation of secondary amines. (*14*) We utilized our assay procedure to determine the relative abilities of various compounds to block the acidic nitrosation of morpholine where the amine was in large excess, a condition which represents practical circumstances and is a stringent test of blocking ability when one considers the nature of the nitrosation process as we have discussed in the introduction to this chapter. The results are summarized in Table II.

The ability of phenolic compounds to block N-nitrosation has been controversial because of reports that they can catalyze N-nitrosation (*28*) While there is no question that mono-hydroxy phenols can catalyze N-nitrosation, this process occurs only when the concentration of the nitrosating agent significantly exceeds the concentration of the phenol, a condition which rarely applies under environmentally or *in vivo* significant situations. It does not occur at all for some poly-hydroxy phenols. We have shown that hydroquinone and catechol are nearly as effective as pyrroles in blocking N-nitrosation (*14*). Pyrroles or phenols attached to electron withdrawing substituents are not effective blocking agents.

Table II
Ability of Various Compounds to Block N-Nitrosomorpholine Formation

Blocking Agents[a]		
97-88% Block	**87-80% Block**	**75-63% Block**
1-Benzyl-2,5-di-methylpyrrole	4-Methylcatechol	Hydrazine
2,5-Dimethylpyrrole	Hydroquinone	3,4-Diaminobenzoic acid
Pyrrogalol	Catechol	Octamethylporphine
Phloroglucinol	1 Benzylpyrrole	1,2,5-Tribenzylpyrrole
Pyrrole	Gallic Acid	Ammonium sulfamate
	Ascorbic Acid	2,5-Diphenylpyrrole
	p-Phenylenediamine	2-Phenyl-3-Benzoylaziridine
	o-Phenylenediamine	
	4-Ethylaniline	

a. The relative ability of each compound to block the nitrosation of morpholine is given. The relative blocking ability decreases as their numerical order within each column increases. Assays were conducted as described in references *14* and *31*.

We have also demonstrated that aziridines having a free N-H are effective at blocking nitrosation (*31*). The products consist of an alkene and N_2O generated from the decomposition of an unstable N-nitrosoaziridine. The attachment of electron withdrawing substituents to the arizidine ring increases their N-nitrosating blocking efficiency by decreasing their basicity. 1,2-Phenylenediamine and structurally related compounds are also effective blocking agents. In the case of the former the product of the scavenging reaction is a benzotriazole, which is inert to further nitrosation (*31*).

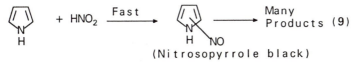

(Nitrosopyrrole black)

A problem associated with the use of pyrroles to block N-nitrosation under some situations is the nature of the products produced from the scavenging of the nitrosating agent by the pyrrole. Pyrrole is thought to undergo C-nitrosation. Maleimide mono- and dioximes are among the few products characterized from the mixture called nitrosopyrrole black produced from the nitrosation. Nitro compounds arising from radical cation intermediates are also likely. There are many commercial applications where the incorporation of any blocking agent into a formulation results in undesirable properties. One must also consider possible adverse toxicological consequences which could arise from the blocking agent itself of products generated from it. In order to overcome these difficulties and provide blocking agents which can be used under different circumstances, we have incorporated pyrroles and hydroquinone into polymers and other solid matrices and examined the ability of these substances to block nitrosation (32,33). The types of materials generated are shown below.polystyrene, polyether, polyethyleneimine, and polyacrylate backbones have been used to attach either pyrrole or hydroquinone scavengers through covalent bonds. Of these modified polymers, the modified polyethylene imines have the greatest hydrophilicity and this property, as well as additional secondary amine sites within the molecule, add significantly to their effectiveness as nitrosation blocking agents in aqueous solution. Aziridine functionality has been incorporated into a polymer with some difficulty by the modification of poly(oxydiethylene maleate). The aziridine groups are incorporated at the C=C sites in the base polymer and the resulting material is an effective blocking agent. (31) We also have explored the use of modified porous glass beads as blocking agents. This composite blocking agent was made by treating controlled-pore glass beads with commercially available trimethoxypropyl substituted polyethyleneimine. Pyrrole groups were then incorporated into the unreacted secondary amine sites on the polymer by means of a Mannich condensation with formaldehyde.

The ability of the various modified polymers and other materials to block N-nitrosation was evaluated in three different kinds of assay procedures (31-33). The first two of these assays employed aqueous solutions of nitrous acid and the third method used an ether mixture produced by the extraction of nitrous acid with ether. This mixture contains N_2O_3 as well as NO and NO_2. The first assay is non-competitive in that the nitrosating mixture is filtered through the solid blocking agent into acid containing a 5-fold excess of morpholine. The quantity of N-nitrosomorpholine produced within five minutes is compared to the blank (no blocking agent or its replacement with Celite). All of the materials assayed were effective blocking agents (90-95%) when this procedure was used. The second procedure permitted the blocking agent to compete directly with a 5-fold excess of morpholine for the available nitrosating agent. Polymers 4, 5 and 7 as well as the modified glass beads effectively competed for the nitrosating agent under these conditions blocking in the range of 60-95% was observed under these stringent conditions. The more hydrophilic polyethyleneimine based polymers were most effective and polyethylene imine without derivatization was also a good blocking   agent. This latter observation has also been made by Eisenbrand and his colleagues (34). They have demonstrated that the nitrosated polyethyleneimine is not mutagenic. Use of the third non aqueous assay procedure revealed that blocking was also effective under these conditions but less so than the aqueous nitrosation. Under non aqueous conditions morpholine is not protonated at all and its higher effective concentration permits more effective competition for the available nitrosating agent.

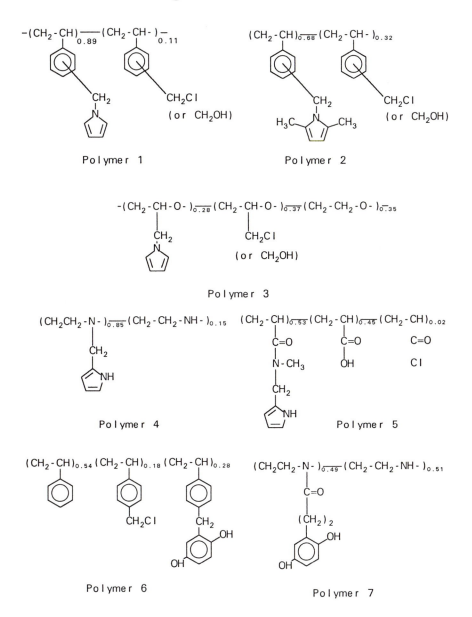

Polymer 1

Polymer 2

Polymer 3

Polymer 4

Polymer 5

Polymer 6

Polymer 7

Conclusions

Despite considerable research into the mechanisms and modes of nitrosamine formation, a great deal remains to be learned regarding the properties of various nitrogen compounds which predispose them to facile nitrosation. In this chapter we have

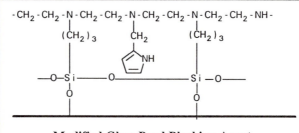

Modified Glass Bead Blocking Agent

concentrated our attention on those processes which rapidly lead to nitrosamines through a nitrosative SN-1 process. We have demonstrated that nitrosamines arise rapidly by this process from electron-rich benzylic and heterobenzylic amines, amidines, and gem. diamines. The possible importance of these pathways has been demonstrated through recent investigations of the nitrosation of commercial compounds, hexetidine and amdinocillin. In another short paper in this volume we show that nitrosamines can also form with ease from dialkyl aromatic amines through radical cation processes. Radical cation pathways are also involved in rapid nitrosamine formation from electron-rich benzylic amines. Recognition of the fundamental chemistry of these nitrosation reactions has led to the development of new agents for blocking N-nitrosation. A particularly promising method involves the incorporation of blocking agents into polymers, which can either remain a part of a formulation or be used to block nitrosamine formation during formulation. Continued research in this area promises even better knowledge and methods for preventing the occurrence and formation of carcinogenic nitrosamines in the human environment.

Acknowledgement: The support of this research by grant R37 CA26914 from the National Cancer Institute is gratefully acknowledged.

References

(1)   Loeppky, R. N. In *Nitrosamin-Probl.*; Preussmann, R., Ed.; Verlag Chem.: Weinheim, Fed. Rep. Ger., 1983; pp. 305-317.
(2)   Ohshima, H.; Bartsch, H. *Cancer Res.* **1981**, *41*, 3658-3662.
(3)   Tannenbaum, S. R.; Sinskey, A. J.; Weisman, M.; Bishop, W. *J. Nat. Cancer Inst.* **1974**, *53*, 79-84.
(4)   Challis, B. C.; Edwards, A.; Hunma, R. R.; Kyrtopoulos, S. A.; Outram, J. R. *IARC Sci. Publ., 19 (Environ. Aspects N-Nitroso Compd.)*, **1978**, 127-42.
(5)   Leaf, C. D.; Wishnok, J. S.; Tannenbaum, S. R. *Carcinogenesis (London)* **1991**, *12*, 537-539.
(6)   Nguyen, T.; Brusnson, D.; Crespi, C. L.; Penman, B. W.; Wishnok, J. S.; Tannenbaum, S. R. *Proceedings of the National Academy of Science* **1992**, *89*, 3030-3034.
(7)   Wink, D. A.; Kasprzak, K. S.; Margos, C. M.; Elespuru, P. K.; Misra, M.; Dunams, T. M.; Cebula, T. A.; Koch, W. H.; Andrews, A. W.; Allen, J. S.; Keefer, L. K. *Science* **1991**, *254*, 1001-1003.
(8)   Leach, S. A.; Challis, B.; Cook, A. R.; Hill, M. J.; Thompson, M. H. *Biochem. Soc. Trans.* **1985**, *13*, 380-381.

(9) Challis, B. C.; Shuker, D. E. G. *J. Chem. Soc., Chem. Commun.* **1979**, 315-316.

(10) Loeppky, R. N.; Tomasik, W.; Millard, T. G. *IARC Sci. Publ., 57 (N-Nitroso Compd: Occurrence, Biol. Eff. Relevance Hum. Cancer)*, **1984**, 353-63.

(11) Schmeltz, I.; Wenger, A. *Food Cosmet. Toxicol.* **1979**, *17*, 105-109.

(12) Croisy, A. F.; Fanning, J. C.; Keefer, L. K.; Slavin, B. W.; Uhm, S. J. *IARC Sci. Publ., 31(N-Nitroso Compd.: Anal., Form. Occurrence)*, **1980**, 83-93.

(13) Williams, D. L. H.. *Nitrosation*; Cambridge University Press: Cambridge, **1988**; p. 96.

(14) Wilcox, A. L.; Bao, Y. T.; Loeppky, R. N. *Chem. Res. Toxicol.* **1991**, *4*, 373-381.

(15) Ellison, G.; Williams, D. L. H. The relative efficiencies of a number of nitrite traps at different acidities and bromide ion concentrations; *J. Chem. Soc., Perkin Trans. II*, **1981**, 699-702.

(16) Mirvish, S. S. *Toxicol. Appl. Pharm.* **1975**, *31*, 325-351.

(17) Smith, P. A. S.; Loeppky, R. N. *J. Am. Chem. Soc.* **1967**, *89*, 1147-1157.

(18) Gowenlock, B.; Hutcheson, R. J.; Little, J.; Pfab, J. *J. Chem. Soc. Perkin Trans. II* **1979**, 1110-1114.

(19) Lijinsky, W.; Greenblatt, M. *Nature (London)* **1972**, *236*, 177-178.

(20) Magino, M. M.; Scanlan, R. A.; O'Brien, T. J. In *N-Nitroso Compounds*; Tannenbaum, S. R.; Scanlan, R. A., Eds.; Am. Chem. Soc.:, 1981; pp. 229-246.

(21) Loeppky, R. N.; Outram, J. R.; Tomasik, W.; Faulconer, J. M. *Tetrahedron Lett.* **1983**, *24*, 4271-4274.

(22) Loeppky, R. N.; Cox, B. C., and Shevlin, G, manuscript in preparation.

(23) Loeppky, R. N.; Shevlin, G.; Yu, L.; *Drug Dev. Eval.*, *16 (Signif. N-Nitrosation Drugs)*, **1990**, 253-66.

(24) Loeppky, R. N.; Yu, L. *Tetrahedron Lett.* **1990**, *31*, 3263-3266.

(25) Yu, L. Ph.D. Dissertation, University of Missouri, Columbia, **1992**.

(26) Loeppky, R. N., Bae, J, and Shevlin, G., *Chem. Res. Toxicol.* **1993**, *6*, in press.

(27) Bae, J., Mende, P., Preussmann, R., Speigelhalder, B., and Loeppky, R. N., *Chem. Res. Toxicol.* **1993**, *6*, in press.

(28) Mende, P.; Wacker, C.-D.; Preussmann, R.; Spiegelhalder, B. *Food Cosmet. Toxicol.* **1993**, *31*.

(29) Groenen, P. J. In *Proceedings of the 2nd international symposium on meat products*(Tinbergen, B. J. and Kroll, B., Eds.) **1976**, p171, Centre for Agricultural Publishing and Documentation, Wegeningen, The Netherlands.

(30) Challis, B. C.; Bartlett, C. D. *Nature (London)* **1975**, *254*, 532-533.

(31) Loeppky, R. N.; Bao, Y. T. In *Nitroso Compounds: Biological Mechanisms, Exposures and Cancer Etiology*; O'Neill, I.K., and Bartsch, H., Ed.; International Agency for Research on Cancer: Lyon, *Technical Report No. 11,* **1992**; pp. 16a-16b.

(32) Bao, Y. T.; Loeppky, R. N. *Chem. Res. Toxicol.* **1991**, *4*, 382-389.

(33) Loeppky, R. N.; Bao, Y. T., U.S. Patent 5,087,671, **1992**; 39 pp.

(34) Eisenbrand, G., Ger. Offen. Patent DE 3939474 A1 6 Jun, **1991**; 3 pp.

RECEIVED August 12, 1993

Chapter 6

Quantitative Aspects of Nitrosamine Denitrosation

D. L. H. Williams

Department of Chemistry, University of Durham, Durham DH1 3LE, England

The kinetics of denitrosation of nitrosamines have been studied under acid conditions in the presence of a sufficient quantity of a nitrous acid trap which ensures that denitrosation to the secondary amine is irreversible. A range of nitrous acid traps, nucleophilic catalysts and solvent systems have been examined for five representative nitrosamine structures in an attempt to establish the most favourable experimental conditions for effecting denitrosation.

Treatment of nitrosamines in acid solution, sometimes in the presence of nucleophiles, leads to the formation of the corresponding secondary amines and free nitrous acid. This reaction (denitrosation) can be made quantitative if a trap for the nitrous acid is added in sufficient concentration to ensure that the reverse reaction (nitrosation) cannot compete. This article describes the effect of (a) different nitrous acid traps, (b) the concentration and nature of the nucleophile, (c) the structure of the nitrosamine, and (d) the solvent, on the rate of denitrosation. The objective in this work is to establish the most favourable reaction conditions available for effecting denitrosation of any given nitrosamine.

Although nitrosamines are generally not naturally-occurring compounds they can readily be formed, particularly from secondary amines and a nitrosating agent. They often crop up as very minor by-products in a whole range of materials including foodstuffs, dyes, detergents, herbicides etc. It is therefore important, given their well-known carcinogenic properties, that procedures are available for their safe destruction. Over the years a number of methods have been examined with this end in view. Two such reactions are given in equations 1 and 2, in which, respectively, nitrosamines are reduced to amines and hydrazines (equation 1), or are subjected to thermal or photochemical degradation (equation 2).

0097–6156/94/0553–0066$08.00/0

$$RR'NNO \xrightarrow{\text{Al,OH}^-} RR'NH + RR'NNH_2 \qquad (1)$$

$$RR'NNO \xrightarrow{\Delta \text{ or } h\nu} RR'N^\bullet + NO^\bullet \qquad (2)$$

Both of these procedures have certain drawbacks and another reaction which has been much used involves denitrosation to the secondary amine using acid catalysts. This is of course the reverse of the reaction by which nitrosamines are generally synthesized (equation 3). In this equation X^- can be Cl^-, Br^-, I^-, SCN^- or H_2O and XNO

$$RR'NNO + H^+ + X^- \rightleftharpoons RR'NH + XNO \qquad (3)$$

$$\downarrow$$

Removed with trap

is merely a carrier of NO^+. Generally, but not always, the equilibrium lies well over towards the nitrosamine. However, if a trap for nitrous acid is added in sufficient quantity to make the trapping reaction much faster than the re-nitrosation, denitrosation can be accomplished quantitatively. This paper describes the quantitative aspects of this reaction (*i.e.*, the effect of different traps and/or different nucleophiles, the effect of changing R and R' and also the results of changing the solvent). The aim throughout is to establish reaction conditions which will allow denitrosation of a range of structures at a reasonable rate.

Nature of Nitrous Acid Traps

In principle any substrate which reacts rapidly and irreversibly with nitrous acid can function as a trap. The following have been used successfully (*1* and references therein) under different experimental conditions:

Cu^I, Fe^{II}, $CO(NH_2)_2$, HN_3, NH_2SO_3H, $NH_2\overset{+}{N}H_3$, RSH, Ascorbic acid, $Fe(CO)_5$, NH_2OH, RNH_2, $C_6H_5SO_2H$.

The relative efficiencies of these traps depend (a) on the acidity of the medium or (b) whether a nucleophilic catalyst (X^-) is present. Nitrosation of the trap will always be acid-catalysed but this may be offset if the trap is protonated and reaction takes place via the un-protonated form. An example of this concerns sulfamic acid and hydrazoic acid. At low acidity ca. 0.1 M, sulfamic acid is the more reactive towards nitrosation, whereas at 1 M acid, hydrazoic acid is the better trap (*2*).

A further complication arises in placing traps in a sequence of
relative reactivity if nucleophilic catalysts X⁻ (of nitrosation)
are present. Scheme 1 shows the general outline mechanism for the
nitrosation of a typical trap S. In many cases, catalysis by X⁻
occurs when the limiting condition $k_2 \gg k_{-1}[X^-]$ applies. However,

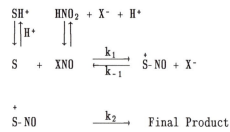

Scheme 1

if the reverse limiting condition applies, i.e. $k_{-1}[X^-] \gg k_2$ then
step k_2 will be rate-limiting and no X⁻ catalysis occurs. The
latter condition is more likely to be the case when there are
powerful electron-withdrawing groups in S which will result in a
large increase in k_{-1}. In principle also the same result might be
achievable at high [X⁻]. Traps which generally show no X⁻ catalysis
include $CO(NH_2)_2$, $NH_2NH_3^3$, NH_2SO_3H and 2,4-dinitroaniline, whereas
traps which do show X⁻ catalysis include NH_2OH, HN_3, anilines
generally (unless they contain powerful electron-withdrawing
groups), thiols and enols.

In general, the following make "good" traps for free nitrous
acid, HN_3, $NH_2NH_3^3$, RSH, $C_6H_5SO_2H$, while $CO(NH_2)_2$, NH_2OH and
sulfanilic acid, make rather "poor" traps. Clearly it is easier to
achieve the limiting condition which ensures irreversibility of
denitrosation with the more reactive traps. As far as I know, enols
have not been used in this way as traps of free nitrous acid,
although it is known that they do react (many at the
diffusion-controlled limit) very rapidly with nitrous acid (*3*).
Thus a ketone with a very high enol content, such as dimedone or
acetylacetone, is predicted to be an excellent trap, whereas simple
ketones, such as acetone with a very low enol content, would be
expected to be a very poor trap.

The Question of Nucleophilic Catalysis in Denitrosation

A more detailed outline mechanism of denitrosation is given in
Scheme 2. It is believed that protonation occurs both at the oxygen
and nitrogen atoms of the nitrosamine but that the pathway to
denitrosation takes place via the N-protonated intermediate. In
general, we would expect protonation to be rapid and attack by the
nucleophile X⁻ to be the rate limiting step. This is borne out in
many cases. The experimentally determined relative reactivities of
a range of nucleophiles is given in Table I.

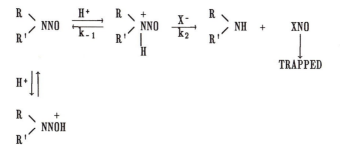

Scheme 2

Table I.   The Relative Reactivities of Nucleophiles in the
Denitrosation of N-methyl-N-nitrosoaniline in Acid
Solution at 31°C

Nucleophile	Relative Reactivity (Cl^- = 1)
Alanine	~ 0
Chloride ion	1
Cysteine	2
Glutathione	3
S-Methylcysteine	35
Bromide ion	55
Methionine	65
Thiocyanate ion	5500
Thiourea	13000
Methylthiourea	14250
Tetramethylthiourea	13500
Iodide ion	15750

These substances show the expected trend and correlate well with
a nucleophilicity parameter where appropriate (*4*). It is evident
that denitrosation will occur more rapidly in the presence of a
reactive nucleophile such as thiourea or iodide ion.
On the other hand, the situation is not quite so straightforward
in that in some denitrosation reactions there is no nucleophilic
catalysis. This has been rationalised as follows: In Scheme 2 when
step k_2 is rate limiting the condition $k_{-1} \gg k_2$ [X^-] applies. If
the reverse situation should obtain however (i.e., k_2 [X^-] $\gg k_{-1}$),
then the rate limiting step is the protonation of the nitrosamine
and no nucleophilic catalysis takes place. This is apparently the
situation with nitrosamides, nitrosoureas, and a nitrososulfonamide.
The rationale is that with these powerful electron-withdrawing
groups k_2 is sufficiently increased to change the limiting
condition. Similarly, the same condition has been achieved for

nitrosamines in the presence of quite high concentrations of a
powerful nucleophile such as thiourea or thiocyanate ion (5).
Further weight is leant to this argument by the observation of a
kinetic isotope effect k_{H2O}/k_{D2O} in the range 1.3 - 1.8 when no X^-
catalysis occurs contrasting with a value of about 0.3 when there is
nucleophilic catalysis. As expected general acid catalysis is also
a feature when there is no X^- catalysis. All of the features
described for denitrosation also appear when the reverse reaction
(i.e., nitrosation) is studied thus satisfying the principle of
microscopic reversibility.

Structural Effects in the Nitrosamine

Most of the work described so far has related to N-nitroso
derivatives of anilines and of amides etc. In order to get a more
general picture we recently quantitatively examined denitrosation
for the following four nitrosamines: dimethylnitrosamine (NDMA),
nitrososarcosine (NSAR), nitrosopyrrolidine (NPYR) and
nitrosoproline (NPRO).

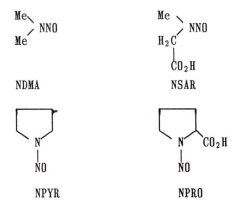

This series extends the range of nitrosamines to include examples of
dialkylnitrosamines and heterocyclic nitrosamines and also examines
the effect of an electron-withdrawing substituent (CO_2H in this
case).
    In water at high acid concentration and at high concentrations
of SCN^- or $SC(NH_2)_2$, denitrosation of both NDMA and of NPYR was so
slow at 25° that it virtually was not measureable. This represents
a major change from the alkyl-arylnitrosamines. However, when the
electron-withdrawing CO_2H group is introduced as in NSAR and NPRO,
denitrosation did occur (at comparable rates) in aqueous acidic
solution in the presence of thiourea as a catalyst, but again at
rates much less than for the aromatic systems studied. The effect
of electron-withdrawing groups is thus quite clear, producing an
increase in the rate of denitrosation, no doubt by increasing the
value of k_2 (Scheme 2) to such an extent as to outweigh any
reduction of the protonation equilibrium constant brought about by
the CO_2H substituent.

Solvent effects

Ethanol. In ethanol solution using HCl or H_2SO_4 as the acid catalyst and ascorbic acid as the "nitrous acid" trap, both NDMA and NPYR failed to undergo denitrosation to any measurable extent at 25°. However, both NSAR and NPRO reacted, albeit rather slowly and without nucleophilic catalysis (see Table II) and with a solvent kinetic isotope effect k(EtOH) / k(EtOD) of 1.6-1.8.

Table II. Rate Constants for the Reaction of NPRO in Ethanol
          Containing H_2SO_4 (1M) and Ascorbic Acid

Thiourea /M	$k_0/10^{-5}$ s^{-1}
0.010	7.9
0.025	8.8
0.040	9.1
0.050	8.3

Acetonitrile. Reactions were not particularly rapid in this solvent, but did take place, for the more reactive nitrosamines, (e.g. NPRO). Typically the half-life was ca. 1 h at 25°C in H_2SO_4 (0.6 M) using thioglycolic acid (0.1 M) as the nucleophile. This solvent appears to have no advantage over ethanol and was not studied further.

Acetic acid. Denitrosation occurred quite readily in various acetic acid-water mixtures, particularly in the presence of a nucleophile such as Br^- and a trap such as sodium azide. The solvent effect was examined for the denitrosation of N-methyl- N-nitrosoaniline in 10%, 33%, 67% and 80% (by volume) acetic acid-water mixtures. The results showing the bromide ion dependence are given in Figure 1. Clearly there is bromide ion catalysis throughout and a marked increase in the rate of denitrosation as the water content of the solvent is reduced. Another point to note is that the relative reactivities of the nucleophiles change as the solvent composition changes, going from the sequence $Cl^- < Br^- < SCN^- < SC(NH_2)$ in water (and 10% acetic acid) to $Cl^- < SCN^- < SC(NH_2)_2 < Br^-$ in 80% acetic acid. This is not surprising since one of the factors governing nucleophilicity in polar solvents is the solvation of the nucleophile. This will change significantly as the solvent changes from water to 80% acetic acid. The absolute reactivity of thiourea changes very little over this range, as does that of SCN^- where protonation is another factor which comes into play. For Br^- and Cl^-, however, there is quite a marked increase in reactivity, no doubt brought about by the reduction in solvation, but the limiting situation where the nucleophilicity of the two halide ions is reversed is not achieved in 80% acetic acid. As expected,

denitrosation is found to be acid catalysed in this solvent, as
occurs in each solvent system studied.
These results suggest that the most effective reaction medium
for bringing about denitrosation is 80% acetic acid-water containing
added sulfuric acid, sodium azide and sodium bromide. We have
examined the rates of denitrosation of a number of the less reactive
nitrosamines under these conditions. Reaction occurs for NDMA,
NSAR, NPYR and NPRO with the reactivity order NPRO > NSAR >> NPYR ≃
NDMA, although none are as reactive as the aromatic derivatives.

Conclusions

We have examined a range of nitrosamine structures under different
experimental conditions of solvent, acidity and nucleophile
concentration. We find that the most efficient system is 80% acetic
acid-water solvent containing added sulfuric acid, sodium bromide
and sodium azide (or any one of a range of nitrous acid traps). As
far as the structure of the nitrosamines is concerned, we find that
the aromatic (alkyl or aryl-aniline derivatives) are the more
reactive. Dialkyl- and heterocyclic nitrosamines are very
unreactive unless there is an electron-withdrawing group close to
the amino nitrogen atom. It is clear that phenyl groups are
activating, presumably by an electron-withdrawing effect which
assists in the rate-limiting nucleophilic attack. Some quantitative
idea of relative reactivity can be obtained from Table III which
gives the absolute reactivity, measured by the product of the
equilibrium constant for protonation (K) and the rate constant (k_2)
for nucleophilic attack by thiourea for reaction in aqueous acid.

Table III. Quantitative Estimates of Denitrosation Reactivity
towards Thiourea in Water Containing Sulfuric Acid, and
Sodium Azide (as the Nitrous Acid Trap)

Nitrosamine	$k_2 K/dm^6 mol^{-2} s^{-1}$
Ph_2NNO (ND)	3.0
$PhN(Me)NO$ (NMNA)	0.52
(NPRO) [pyrrolidine ring with N, CO_2H, NO]	1.0×10^{-3}
$HO_2CCH_2N(Me)NO$ (NSAR)	6.4×10^{-4}
(NPYR) [pyrrolidine ring with N, NO]	Too slow to measure
Me_2NNO (NDMA)	Too slow to measure

Reproduced with permission from reference 1.

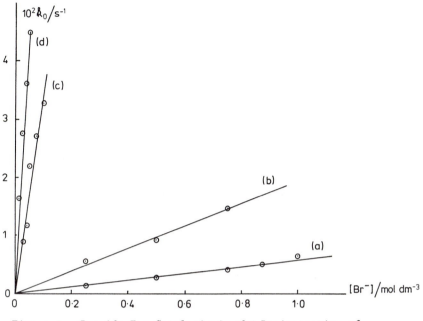

Figure 1. Bromide Ion Catalysis in the Denitrosation of
N-methyl-N-nitrosoaniline (NMNA) in (a) 10% acetic acid-water,
(b) 33% acetic acid-water, (c) 67% acetic acid-water and
(d) 80% acetic acid-water.
Reproduced with permission from reference *1*.

A similar trend of reactivity occurs in 80% acetic acid-water
solvent.
    The results described here provide a quantitative basis for
denitrosation procedures and support what is known qualitatively
regarding the most appropriate system for effecting denitrosation
i.e. the procedure of using HBr/HOAc in boiling EtOAc - under these
conditions denitrosation of all nitrosamines occurs readily.

Literature cited

1. Dix, L.R.; Oh, S.M.N.Y.F.; Williams, D.L.H. *J. Chem. Soc.*,
   Perkin Trans. 2. **1991**, pp 1099-1104
2. Fitzpatrick, J.; Meyer, T.A.; O'Neill, M.E.; Williams, D.L.H. *J.*
   Chem. Soc., Perkin Trans. 2. **1984**, pp 927-932
3. Roy, P.; Williams, D.L.H. *J. Chem. Res (S).* **1988**, pp 122-123
4. Thompson, J.T.; Williams, D.L.H. *J. Chem. Soc., Perkin Trans. 2.*
   1977, pp 1932-1937
5. Al-Kaabi, S.S.; Hallett, G.; Meyer, T.A.; Williams, D.L.H. *J.*
   Chem. Soc., Perkin Trans 2. **1984**, pp 1803-1807

RECEIVED November 3, 1993

Chapter 7

Peptide Nitrosations

**Brian C. Challis[1,2], Neil Carman[1], Maria H. R. Fernandes[2],
Benjamin R. Glover[2], Farida Latif[2], Pravin Patel[1],
Jaswinder S. Sandhu[1,2], and Shabaz Shuja[1]**

[1]Chemistry Department, Open University, Milton Keynes, MK7 6AA,
United Kingdom
[2]Chemistry Department, Imperial College, London SW7 2AZ, United
Kingdom

All peptides undergo nitrosation at the terminal primary amino
group to generate a diazopeptide and at the peptide N-atoms to
produce an N-nitrosopeptide. In general, using both mildly
acidic nitrite and gaseous NO_2, diazopeptides form more readily
than N-nitrosopeptides, but are less stable. Diazopeptide
decomposition is strongly acid-catalysed to give substitution,
elimination and rearrangement products characteristic of highly
reactive carbocation intermediates. For peptides bearing
selected amino acid constituents, the carbocation intermediate
interacts with a neighbouring group to produce cyclic products
with potential to act as alkylating agents. These
transformations are exemplified by the artificial sweetener
Aspartame which produces substantial amounts of a substituted
β-lactone on nitrosation under simulated gastric conditions.

In common with other amino compounds, peptides react with nitrosating agents
under a range of conditions. Various products may arise depending on the
amino acid constituents of the peptide. However, two general reactions
(Scheme 1) always proceed, irrespective of these constituents. The first (path
1) involves the terminal amino group to give a primary nitrosamine intermediate
(I) which rapidly transforms to a relatively unstable diazopeptide (II). The
second (path 2) proceeds at the peptide N-atoms to give a more stable
N-nitrosopeptide (III). Under a wide range of conditions, however, the
N-nitrosopeptide (III) will also transform via cleavage of the peptide linkage to
a diazopeptide (IV), analogous to (II) but shorter. Thus, both general
nitrosation reactions of peptides eventually lead to a diazo product.

Peptides containing proline, cysteine, tyrosine, tryptophan, arginine,
lysine, glutamine and asparginine amino acid constituents undergo additional,
specific nitrosation reactions. Thus, proline and tryptophan form specific
N-nitroso derivatives, cysteine is oxidised via an S-nitroso intermediate to
generate disulphide (cystine) linkages, substitution occurs in the phenolic ring
of tyrosine, and arginine, lysine, glutamine and asparginine may undergo

deamination of their side-chain amino groups. All of these additional reactions have been observed, sometimes with protected amino acids if not with actual peptides. Peptides *N*-terminal in proline, for example, produce the corresponding *N*-nitrosoprolylpeptide (an α-substituted nitrosamine) in substantial yields(*1,2*); and tyrosylproteins give 3-nitrotyrosine and 3,4-dihydroxyphenylalanine (dopa) products, both of which can be rationalised by transformations of a 3-nitrosotyrosine intermediate(*3*).

Scheme 1

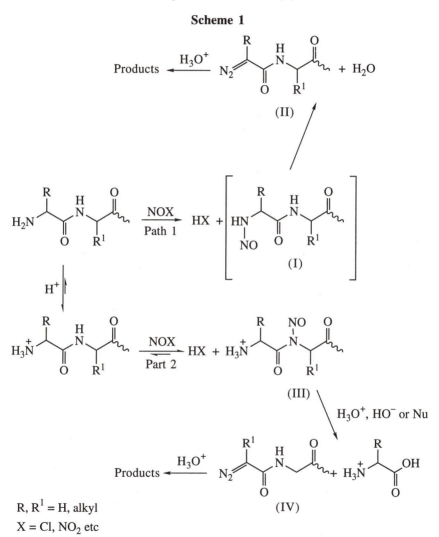

R, R^1 = H, alkyl

X = Cl, NO$_2$ etc

Our interest in peptide nitrosation relates to its potential involvement in dietary-related cancers. Nitrosation proceeds in the stomach(*4*) and is likely to involve peptides and proteins because of their common dietary occurrence.

Interestingly, the incidence of colon cancer correlates with protein intake(*5*), and both diazopeptides and *N*-nitrosopeptides exhibit a spectrum of cytotoxic activity(*6*). For example, *N*-(2-diazoacetyl)glycine compounds usually give a mutagenic response(*7,8*) and *N*-(2-diazoacetyl)glycinamide and the analogous hydrazide induce dose-dependent pulmonary adenomas and leukaemias in mice(*9,10*). Also, the two *N*-nitrosopeptides tested thus far are mutagenic with dose-dependent responses for Ames, sister chromatid exchange, dominant lethal and micronucleus tests(*11-13*).

This paper focuses on diazopeptides, which form much more readily than *N*-nitrosopeptides under physiological conditions(*14*). Furthermore, both the *N*-nitrosation pathways of Scheme 1 ultimately lead to a diazopeptide product, so much of the chemistry relevant to the biological consequences of the two pathways should be identical. Recent progress in the synthesis, formation and decomposition of diazopeptides is reported along with the identification of decomposition products. Attention is drawn to the formation of novel cyclic products with potential cytotoxicity and exemplified for the nitrosation of the artificial sweetener Aspartame under simulated gastric conditions.

Synthesis and Formation of Diazopeptides

The synthesis of diazopeptides was first reported by Curtius in 1904 for the ethyl ester of tetraglycine using nitrite in aqueous acetic acid: the *N*-(2-diazoacetyl)trisglycine ethyl ester was obtained as a crystalline compound. Subsequently, other *N*-(2-diazoacetyl)glycine peptides were prepared and isolated except those from the native peptides with an underivatised terminal carboxylic acid(*16-18*). This early work also established that diazopeptides are labile in mild acid with ready replacement of the diazo group by nucleophiles such as NH_3, Cl^- and I^-(*16-18*). Improved methods for the preparation of Curtius-type compounds have been developed. The best of these methods uses an immiscible solvent (eg. methylene chloride) to continuously extract the diazopeptide from the aqueous acid phase as it forms(*19*).

A general procedure for synthesising diazopeptides from a wider range of substrates including native peptides is a recent development(*20*). This utilizes aprotic nitrosation by liquid N_2O_4 (NO_2) in an organic solvent at -40°C with added base (Et_3N or NaOAc) and dessicant (Na_2SO_4) to maintain non-acidic, anhydrous conditions (Scheme 2) similar to that used earlier for the synthesis of *N*-nitrosamides(*21*).

After suitable work-up including silica column chromatography, the diazopeptides were obtained as pale yellow oils in yields of 40-50%(*20*). Native peptides with underivatised terminal carboxylic acids were solubilised in the organic solvent by prior conversion to a tetra-n-butylammonium salt, and their diazoderivatives after similar work-up were obtained as calcium salts in 20-30% yield by precipitation from ethanol using $CaCl_2$ (Scheme 2)(*20*). These aprotic nitrosation procedures seem suitable for all but alanyl- and phenylalanyl- peptides.

Scheme 2

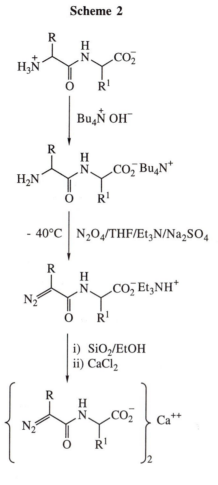

All of the diazopetides are characterised by strong UV absorptions at λ_{max} *ca.* 250-260 nm (log ε 4-4.35), a much weaker visible absorption at λ_{max} *ca.* 380 nm which accounts for their yellow colouration, and strong ir. absorption at ν_{max} *ca.* 2100 cm^{-1}(20). In the ^{1}H-nmr spectra, the most obvious consequence of peptide diazotisation is pronounced deshielding of protons α and β to the diazo group: from 0.83-1.50 ppm for the α-H of N-(2-diazoacetyl)peptides and 0.39-0.60 ppm for the β-H of other diazopeptides(20). In electron-impact mass spectra, the diazopeptide molecular ions are either unobservable or extremely weak because of facile loss of N$_2$. Good results are obtained, however, with fast atom bombardment techniques in the positive mode for diazopeptide esters and the negative mode for the calcium salts, where MH$^+$ and M-H$^+$ ions, respectively, are intense (*ca.* 80%)(20).

Formation by Aqueous Nitrite. There have been few mechanistic studies of diazopeptide formation in aqueous media but pathways elaborated for the nitrosation of other amino compounds (especially amino acids) should apply. Thus, in acidified nitrite, one or another of several nitrosating agents (e.g. N_2O_3, NOCl, NOSCN) may react with the neutral peptide in a rate-limiting step (Scheme 3)(22). The resultant *N*-nitroso ammonium ion intermediate (V) rapidly loses H_3O^+ to generate the diazopeptide (VI), which in turn has only a transient existence under the acidic reaction conditions. With concentrated nitrite (> 10 mM), N_2O_3 is the principal nitrosating agent, as observed by Kurosky and Hofmann(23) for isoleucylvaline, valylvaline and both terminal and side chain (lysine) primary amino groups of several chymotrypsins. In HCl with dilute nitrite (< 1mM), however, NOCl is the principal nitrosating agent for both glycylglycine and its ethyl ester(14). Native peptides should also react with nitrosating agents at the terminal carboxylate moiety to form a reactive acyl nitrite ester (VII) able to generate diazopeptides via both intramolecular and intermolecular transnitrosations (Scheme 4). The full scope of these reactions has not been ascertained but an intramolecular transnitrosation would explain the multiple order peptide dependence observed for the diazotisation of glycylglycine with dilute nitrite (< 1mM) in $HClO_4$ (Challis, Glover and Pollock, to be published). Similar discrepancies are not apparent in HCl, which suggests that gastric nitrosation is most likely to involve NOCl (and possibly NOSCN in smokers) with first order dependences on nitrite and peptide and negligible acid catalysis below pH 3.

Scheme 3

$$HNO_2 + HX \xrightleftharpoons{\text{fast}} NOX + H_2O$$

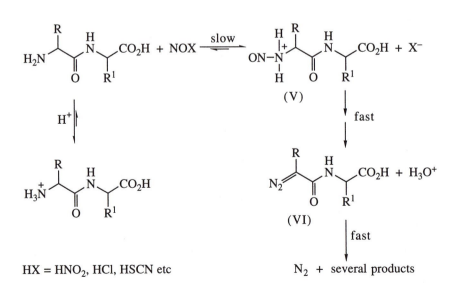

HX = HNO_2, HCl, HSCN etc N_2 + several products

Scheme 4

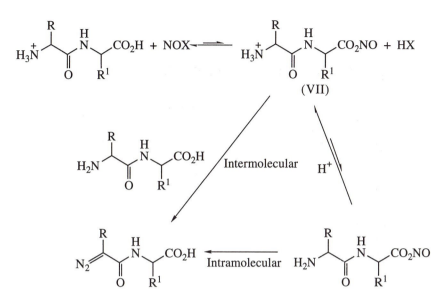

Formation by Gaseous Nitrogen Oxides. The high nitrosating potency of gaseous nitrogen oxides is exploited in the recent, successful synthesis of diazopeptides by aprotic nitrosation (vide supra). The facility of similar reactions in aqueous media is less obvious where peptide diazotisation must compete with the rapid hydrolysis of the nitrogen oxides to nitrite and nitrate salts. Nonetheless, in aqueous buffers, saliva or blood above pH 6, peptides are rapidly diazotised on treatment with gaseous NO_2(23). In aqueous buffers, yields depend on peptide concentration, NO_2 level, pH and temperature. For 0.1M glycylglycine and 200 ppm NO_2 at 25°C, about 3.8% of the NO_2 forms *N*-(2-diazoacetyl)glycine at pH 6.8, rising to 11.5% at pH 9.6 following the titration curve for the terminal primary amino group of the peptide(24). These reactions probably involve (Scheme 5) nitrosation of the peptide anion (VIII) by an NO_2 dimer, followed by rapid loss of H_3O^+. There is no evidence of appreciable, concurrent nitrosation of the peptide *N*-atom at pH > 6, but the gaseous NO_2 is competitively hydrolysed to a mixture of nitrite and nitrate salts(23).

Aside from small differences related to the basicity (pK_A) of the terminal primary amino group, structurally different peptides give similar yields of diazo products to glycylglycine (Table I). Some of the other diazopeptides, however, are much less stable (see below). As nitrosation by NO_2 must involve a dimer, diazopeptide formation ought to diminish with NO_2 dilution. For 0.1M glycylglycine in aqueous buffer at pH 8.35 and 25°C, this effect is only evident below 20 ppm NO_2, and even at 2 ppm *ca.* 2.3% of the sparged NO_2 forms *N*-(2-diazoacetyl)glycine (Figure 1)(24).

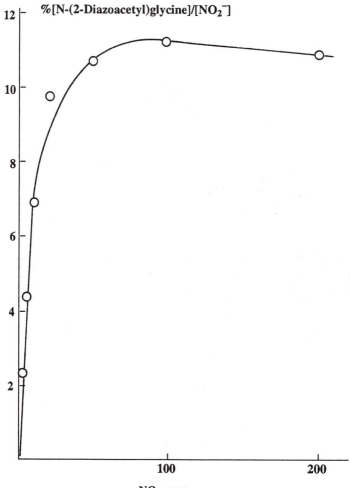

Figure 1.   Effect of gaseous NO_2 concentration on the formation of
N-(2-diazoacetyl) glycine at pH 8.35 and 25 °C. Reproduced with permission
from reference (*24*).

Scheme 5

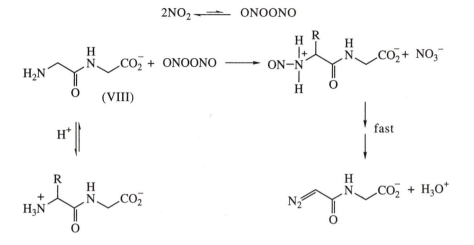

Table I Diazopeptide Yields for Reaction of 0.1M Peptide at pH 9.1 and 25°C with 200 pm Gaseous NO_2

Peptide	% Diazopeptide[a]
Glycylglycine	8.7
Triglycine	8.2
Glycyl-L-alanine	8.4
Glycyl-L-leucine	7.9
Glycyl-L-serine	8.7
L-Serylglycine	3.4
L-Leucylglycine	1.0
L-Alanylglycine	0.95
L-Alanyl-L-alanine	0.81

[a]Relative to total NO_2 sparged.

Gaseous NO_2 interacts rapidly with the medium at 25°C to give mainly nitrite in serum and plasma, and nitrate in whole blood. No low MW diazo products are detectable in these media probably because many different diazopeptides form, each at extremely low concentration. Diazopeptides are detected, however, from small peptides added to serum and whole blood, in yields not substantially different from reactions in aqueous buffers(24). Thus, blood does not seem to inhibit the diazotisation of peptides by gaseous NO_2. Since NO_2 is largely retained on inhalation(25) and its interaction with erythrocytes is well-documented($26,27$), a potential causal role exists for diazopeptides in urban- and tobacco- related cancers.

Stability of Diazopeptides

In common with other diazo compounds, diazopeptides are relatively unstable particularly in acidic media. In aqueous acidic buffers (HA $\rightleftharpoons$ H_3O^+ + A^-), their decomposition follows equation 1 and is therefore general acid catalysed. For most diazopeptides, equation 1 refers to decomposition via an $A-S_E2$ pathway in which H^+-transfer (step k_1 of Scheme 6) is rate-limiting. For N-(2-diazoacetyl) peptides where R=H, however, equation 1 relates to an $A-2$ pathway in which bimolecular decomposition of the diazonium ion intermediate (IX), step k_2 of Scheme 6, is rate-limiting. Consequently, the latter reactions are also catalysed by nucleophiles such as Cl^-, Br^- and SCN^-(24).

$$\text{Rate} = [\text{Diazopeptide}]\{k_{H_3O+}[H_3O^+] + k_{HA}[HA]\} \tag{1}$$

Scheme 6

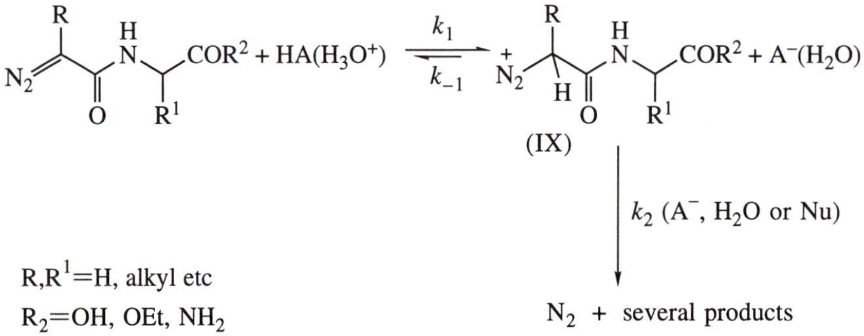

R,R^1=H, alkyl etc
R$_2$=OH, OEt, NH$_2$

k_2 (A$^-$, H$_2$O or Nu)

N$_2$ + several products

At 25°C - 37°C, diazopeptides decompose very rapidly (t$_{\frac{1}{2}}$ < 10 sec) at pH < 3, but are moderately stable (t$_{\frac{1}{2}}$ 0.2 - 5h) at neutral pH (Table II). The fastest decompositions apply to compounds bearing electron donating substituents (R) adjacent to the diazo group, which stabilise the carbocation generated by expulsion of N_2 from the diazonium ion (IX). It follows that the diazo derivatives of of glycyl, seryl and threonyl peptides should be among the most stable.

In general, diazopeptides are more stable in lipophilic media, serum and blood than in aqueous buffers (Table II): for example, in serum at 37°C N-(2-diazoacetyl)glycine and N-(2-diazopropanoyl)glycine decompose with t$_{\frac{1}{2}}$ ca. 20h and 0.3h, respectively.

Table II   Decomposition of Diazopeptides at 25 °C in Phosphate Buffer (pH 7.5), Serum (pH 7.2-7.6) and Plasma (pH 7.2). Initial [Diazopeptide] about 10^{-4} M

Diazopeptide	$10^5 \, k \, s^{-1a,b}$		
	Buffer	Serum	Plasma
$(N_2digly^-)_2Ca^{++}$	2.12	2.78	
N_2digly ethyl ester	0.45		
$(N_2trigly^-)_2 \, Ca^{++}$	0.43		
$(N_2pentagly^-)_2 \, Ca^{++}$	0.41	2.14	1.95
N_2gly leu	1.72	0.68	
N_2gly ser	1.47		
N_2gly ala	2.58		
N_2ala gly	815	23.8	37.9
N_2ala ala	891	43.3	
N_2leu gly	184	5.56	

[a]Rate = k[Diazopeptide]
[b]Data from reference (*24*)

Diazopeptides, themselves, seem too unstable under normal gastric conditions (pH < 4, 37°C) to elicit appreciable genetic damage. A different conclusion may apply, however, to blood, saliva and the achlorohydric stomach. Further, some cyclic decomposition products of diazopeptides with alkylating properties (e.g. (XI), (XIII) and (XXI), vide infra) are sufficiently stable under normal gastric conditions to be candidate carcinogens.

Decomposition Products

Nitrosation is a standard procedure for deaminating primary amino compounds including peptides. As is evident above, the initial diazopeptide product is isolable but it readily expels N_2 especially under acidic conditions to generate a reactive carbocation intermediate. It follows that diazopeptides are alkylating agents and that peptide nitrosation will lead to a spectrum of substitution, elimination and rearrangement products characteristic of carbocation intermediates.

For glycylpeptides, where elimination and rearrangement reactions are impossible, substitution products resulting from the alkylation of water and other nucleophilic substances predominate, as shown in Table III for both the diazotisation of glycylglycine ethyl ester and the decomposition of authentic *N*-(2-diazoacetyl)glycine ethyl ester in 1M HCl and 1M HOAC buffer at 37°C(*14*). For *N*-(2-diazo-4-methylvaleroyl)glycine ethyl ester, however, decomposition in 0.1M $HClO_4$ at 37°C produces a mixture of substitution (by H_2O), elimination and rearrangement (by 1,3 hydride shift) products in similar yields.

Table III   Products and % Yield for the Nitrosation of Glycylglycine Ethyl Ester and the Decomposition of *N*-(2-Diazoacetyl)glycine Ethyl Ester and *N*-(2-Diazo-4-methylvaleroyl)glycine Ethyl Ester

Reaction	Product and % Yield[a]		
	X = HO	X = Cl	X = AcO
in 1M HCl[b]	42	46	–
in 1M AcOH[b]	68	–	31
in 1M HCl[b]	45	47	–
in 1M AcOH[b]	68	–	31
in 0.1M $HClO_4$[c]	42	29	17
Y =			

[a] Relative to HNO_2 or Diazopeptide
[b] From reference (*14*)
[c] Challis and Shuja, to be published.

The formation of cyclic products by intramolecular interactions between carbocations and neighbouring nucleophiles is well-known and reported for the diazotisation of some α-amino acids [(eg. L-glutamine(*28, 29*), L-cysteine(*30*)]. Similar reactions are likely for the diazotisation of seryl, threonyl, cysteinyl, asparginyl, glutamyl, aspartyl, glutaminyl, lysyl and arginyl peptides. These are of interest in the present context insofar as some of the cyclic products are likely to be more robust alkylating agents than their diazopeptide precursors.

Some of these expectations have been realised for the diazotisation of cysteinyl, threonyl and aspartyl dipeptides. For example, the diazotisation of L-cysteinylglycine methyl ester (X) by NaNO$_2$ in dilute HCl at 0° yields the thioepoxide (XI) in ca. 44% yield (equation 2); decomposition of *N*-(2-diazo-3-hydroxybutanoyl)glycine ethyl ester (XII) in 0.01M HClO$_4$ at -40°C to 25°C gives the ketone (XIV) in nearly quantitative yield probably via rearrangement [1,2 hydride shift] of the epoxide (XIII) (equation 3); finally, diazotisation of L-aspartyl-L-phenylalanine methyl ester produces a β-lactone as described below.

Three other neighbouring group interactions may be common to all peptide diazotisations irrespective of their amino acid constituents. The first involves the terminal carboxylic acid of native dipeptides to produce the diketomorpholine (XV), which is the major product for the nitrosation of glycylycine in dilute acid (equation 4). The second presumably involves the peptide *N*-atom to form an α-lactam intermediate (XVI) which is hydrolysed to an iminodialkanoic acid (XVII) and then further nitrosated (Scheme 7). *N*-Nitroso iminodialkanoic acids (XVIII) are minor products (0.1-10%) from the aqueous nitrosation of many di- and larger- peptides. First reported by Pollock(*31*) using high reactant concentrations, *N*-nitroso iminodialkanoic acids have subsequently been detected at low level with 10 μM NaNO$_2$(*32*) and by incubating native gastric aspirates with nitrite(*33*), confirming the ease with which peptides undergo diazotisation in aqueous acid and gastric juice. The third neighbouring group interaction relates to cyclization to a triazole (XIX) under alkaline conditions (pH > 10) (Scheme 8). This interesting reaction was discovered by Curtius and Thompson(*34*) for *N*-(2-diazoacetyl)glycine ethyl ester but is probably common to all diazopeptides. The cyclization is reversible in acid and the reverse reaction (i.e. conversion of triazoles to diazoamides) was studied extensively by Dimroth and his colleagues(*35*) at the turn of the century. For triazoles derived from diazopeptides, however, conversion to diazopeptides proceeds only in very concentrated acids (Challis, Glover and Pollock, unpublished), and there is no evidence that triazoles react with genetic material under physiological conditions.

Nitrosation of L-Aspartyl-L-phenylalanine Methyl Ester (Aspartame)

Currently L-aspartyl-L-phenylalanine methyl ester (XX) which has been extensively used as a synthetic, low-calorie sweetening agent (Aspartame) is a common dietary dipeptide. Our studies suggest that Aspartame should readily diazotise under gastric conditions and then generate the β-lactone, *N*-(1'-methoxycarbonyl-2'-phenyl)ethyloxetan-2-one-4-carboxamide (XXI) via a neighbouring group interaction of the aspartyl carboxylic acid with the incipient carbocation centre (Scheme 9). As some β-lactones are

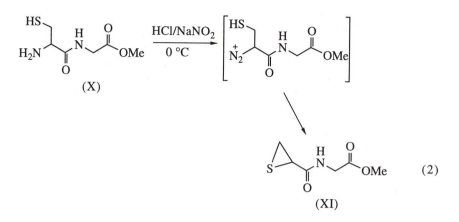

(2)

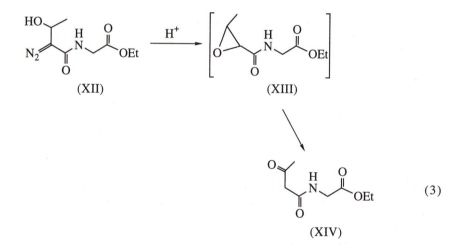

(3)

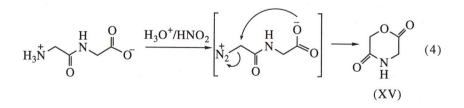

(4)

Scheme 7

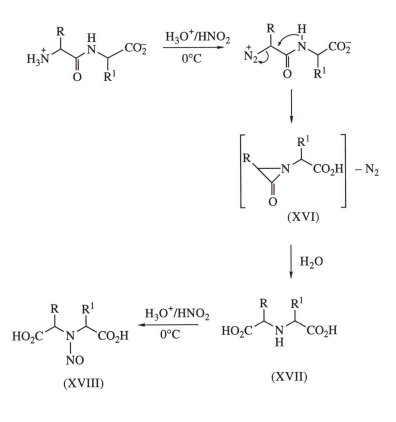

(XVI)

(XVIII)

(XVII)

Scheme 8

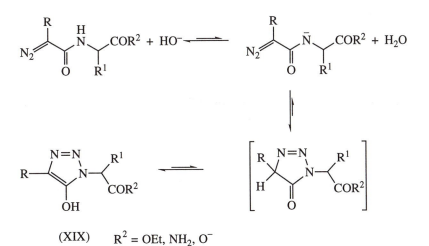

(XIX) $R^2 = OEt, NH_2, O^-$

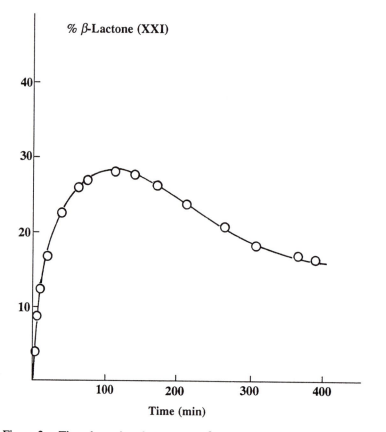

Figure 2. Time dependent formation of β-lactone (XXI) from Aspartame and sodium nitrite in dilute HCl at pH2 and 37 °C .

carcinogenic(*36*), and the nitrosation of Aspartame is known to produce an unidentified, alkylating mutagen(*37*), confirmation of Scheme 9 is of potential relevance to dietary-related cancers.

Scheme 9

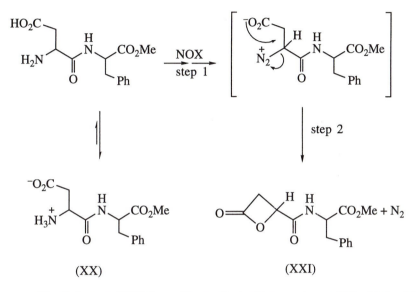

(XX)   (XXI)

The β–lactone (XXI) is readily synthesised by reaction at 0°C of 0.5 mM Aspartame with 1.5 mM $NaNO_2$ in a biphasic reaction solution of aqueous 1M HCl and dichloromethane.   After separation of the organic layer, appropriate work-up, silica column chromatographic separation and recrystallisation from ether, the β–lactone (XXI) is obtained as a crystalline solid in 38% yield.   It shows the expected spectroscopic properties including a strong infra-red absorption at v_{max} = 1883 cm^{-1}, which is very characteristic of the β–lactone structure.

The formation and decomposition of β–lactone (XXI) is conveniently followed by capillary gc as shown for 50 mM Aspartame reacting with 2 mM $NaNO_2$ in 0.01M HCl (pH 2) at 37°C in Figure 2.   This has the characteristic shape of sequential formation and decomposition reactions and the concentration of β–lactone (XXI) passes through a maximum of 27% (relative to $NaNO_2$) at ca. 120 min.   The formation rate of (XXI) in dilute HCl with mM $NaNO_2$ follows equation 5, and k_f increases rapidly with acidity as shown in Figure 3.   These

$$\text{Rate} = k_f[\text{Aspartame}][\text{NaNO}_2] \qquad (5)$$

kinetic dependences are consistent with Scheme 9 where NOCl is the reagent and step 1 is rate-limiting.   Independent measurements show that the decomposition (step 2) of β–lactone (XXI) follows pseudo first order kinetics (equation 6) with catalysis by acids, bases and nucleophiles.

$$\text{Rate} = k_d[(\text{XXI})] \qquad (6)$$

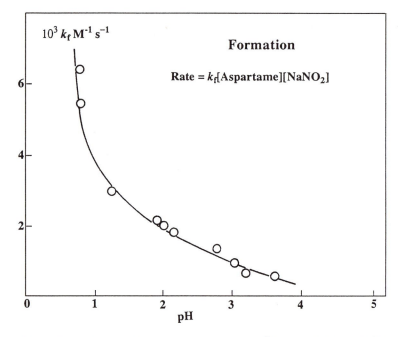

Figure 3.   pH Dependency for the formation of β-lactone (XXI) at 37 °C.

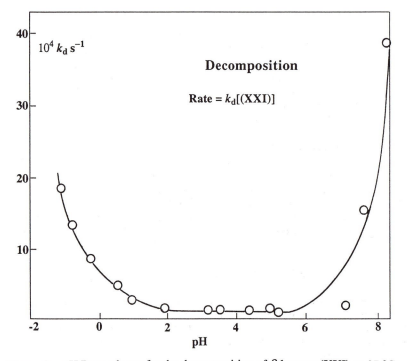

Figure 4.   pH Dependency for the decomposition of β-lactone (XXI) at 37 °C.

The variation of k_d with pH for aqueous buffers at 37°C has an inverted bell shape (Figure 4) similar to that observed earlier for β-propiolactone(*38*). It seems likely that β-lactone (XXI) will also form by endogenous nitrosation in the stomach following the ingestion of food and drink containing Aspartame as a synthetic sweetener. It is probably the unidentified mutagen from the nitrosation of Aspartame reported by Meier et al(*37*), and its chemical properties are very similar to β-propiolactone, itself, a known carcinogen(*36*).

Acknowledgments

We thank Pollock and Pool International Ltd, the Cancer Research Campaign, the Ministry of Agriculture, Fisheries and Food, the Science and Engineering Research Council and the Open University for their support of this work. We also thank Shirley Foster for preparing the camera ready copy.

Literature Cited

1.   Kubacka, W.; Libbey, L. M.; Scanlan, R. A., J. Agric. Food Chem., **1984**, *32*, 401-404.
2.   Tricker A. R.; Perkins, M. J.; Massey, R. C.; McWeeny, D. J., *Food Add. Contam.*, **1984**, *1*, 307-312.
3.   Knowles, M. E.; McWeeny, D. J.; Couchman, L.; Thorogood, M., *Nature*, **1974**, *247*, 288-289.
4.   Oshima, H.; Bartsch, H., *Cancer Res.*, **1981**, *41*, 3658-3662.
5.   Armstrong, B. K.; Doll, R., *Int. J. Cancer.*, **1975**, *15*, 617-619.
6.   Challis, B. C., In *Cancer Surveys*; Forman, D.; Shuker, D. E. G. Eds; OUP, Oxford, **1989**, *Vol 8*, 363-384.
7.   Banfi, E.; Tamaro, M.; Pani, B.; Monti-Bragadin, C., *Boll. Inst. Siero. Milan*, **1974**, *531*, 632-635.
8.   Pani, B.; Babudri, N.; Bartoli-Klugmann, F.; Venturini, S.; de Fant, I., *Mutat. Res.*, **1980**, *78*, 375-379.
9.   Brambilla, G.; Cavanna, M.; Parodi, S.; Caraceni, C. E.; *Boll. Soc. Ital. Biol. Sper.*, **1970**, *46*, 227-230.
10. Brambilla, G.; Cavanna, M.; Parodi, S.; Baldini, L., *Eur. J. Cancer*, **1972**, *8*, 127-129.
11. Challis, B. C.; Hopkins, A. R.; Milligan, J. R.; Massey, R. C.; Anderson, D.; Blowers, S. D., *Toxicol Letters*, **1985**, *26*, 89-93.
12. Anderson, D.; Phillips, B. J.; Challis, B. C.; Hopkins, A. R.; Milligan, J. R.; Massey, R. C., *Food Chem. Toxicol.*, **1986**, *24*, 289-292.
13. Blowers, S. D.; Anderson, D. *Food Chem. Toxicol.*, **1988**, *26*, 785-790.
14. Challis, B. C.; Glover, B. R.; Pollock, J. R. A. In *The Relevance of N-Nitroso Compounds to Human Cancer*; Bartsch, H.; O'Neill, I.; Schulte-Hermann, R. Eds.; IARC, Lyon, **1987**, *IARC Scient. Publ.[n] No. 84*, 345-350.
15. Curtius, T., *Chem. Ber.*, **1904**, *37*, 1285-1300.
16. Curtius, T.; Callan, T., *Chem. Ber.*, **1910**, *43*, 2447-2457.
17. Curtius, T.; Darapsky. A., *Chem. Ber.*, **1906**, *39*, 1373-1378.
18. Curtius, T.; Thompson, J., *Chem. Ber.*, **1906**, *39*, 1379-1388.
19. Looker, J. M.; Carpenter, J. W., *Canad. J. Chem.*, **1967**, *45*, 1727-1734.

20.   Challis, B. C.; Latif, F., *J. Chem. Soc. Perkin Trans 1*, **1990**, 1005-1009.
21.   White, E.H., *J. Amer. Chem. Soc.*, **1955**, *77*, 6008-6010.
22.   Williams, D. L. H., *Nitrosation*, C.U.P., Cambridge, **1988**.
23.   Kurosky, A.; Hofmann, T., *Canad. J. Biochem.*, **1972**, *50*, 1282-1296.
24.   Challis, B. C.;Fernandes, M. H. R.; Glover, B. R.; Latif, F., In *The Relevance of N-Nitroso Compounds to Human Cancer*; Bartsch, H.; O'Neill, I.; Schulte-Hermann, R. Eds.; IARC, Lyon, **1987**, *IARC Scient. Publ.*[n] *No. 84*, 308-314.
25.   Goldstein, E.; Peck, N. F.; Parkes, N. J.; Hines, H. H.; Steffey, E. O.; Tarkington, B., *Amer. Rev. Respir. Diseases*, **1977**, *115*, 403-412.
26.   Keise, M., In *Methemoglobinemia- A Comprehensive Review*, CRC Press, Cleveland, Ohio, **1974**, 72-73.
27.   Doyle, M. R.; Hoekstra, J. W., *J. Inorg. Biochem.*, **1981**, *14*, 351-358.
28.   Howard, A. J.; Austin, J., *J. Chem. Soc.*, **1961**, 3278-3282; *ibid.*, *J. Chem. Soc.*, **1961**, 3284-3289.
29.   Markgraf, J. H.; Davis, H. A., *J. Chem. Ed.*, **1990**, *67*, 173-176.
30.   Maycock, C. D.; Stoodley, R. J., *J. Chem. Soc. Perkin Trans. 1*, **1979**, 1852-1855.
31.   Pollock, J. R. A., *Food Chem. Toxicol.*, **1985**, *23*, 701-704.
32.   Tricker, A. R.; Preussman, R., *Carcinogenesis*, **1986**, *7*, 1523-1526.
33.   Outram, J. R.; Pollock, J. R. A., In *N-Nitroso Compounds: Occurrence, Biological Effects and Relevance to Human Cancer*, O;Neill, I. K.; Von Borstel, R. C.; Miller, C. T.; Long, J.; Bartsch, H., Eds.; IARC, Lyon, **1984**, *IARC. Scient. Publ.*[n] *No. 57*, 71-76.
34.   Curtius, T.; Thompson, J., *Chem. Ber.*, **1906**, *39*, 4140-4144.
35.   Dimroth, O., *Ann.*, **1910**, *373*, 336-346 and references cited therein.
36.   Lawley, P. D, In *Chemical Carcinogens 2nd Edn.*; Searle, C. E., Ed; *ACS monograph 182*, ACS, Washington, DC, **1984**, 428.
37.   Meier, I.; Shephard, S. E.; Lutz, W. K., *Mutat. Res.*, **1990**, *237*, 193-197.
38.   Long, F. A.; Purchase, M., *J. Amer. Chem. Soc.*, **1950**, *72*, 3267-3271.

RECEIVED July 28, 1993

Chapter 8

Nitrosatable Secondary Amines

Exogenous and Endogenous Exposure and Nitrosation In Vivo

A. R. Tricker[1], B. Pfundstein, and R. Preussmann

Institute of Toxicology and Chemotherapy, German Cancer Research Center, Im Neuenheimer Feld 280, D-6900 Heidelberg, Germany

Mean dietary exposure to N-nitrosodimethylamine (NDMA), N-nitroso-pyrrolidine (NPYR) and N-nitrosopiperidine (NPIP) is 0.20-0.31 µg/day; exposure to the secondary amines dimethylamine (DMA), diethylamine (DEA), pyrrolidine (PYR) and piperidine (PIP) is 7.4 mg/day. These amines are also found in human saliva, gastric juice, blood, urine and faeces. Patient groups with chronic bacterial infections of the urinary tract (paraplegics, Bilharzia and bladder augmentation patients) and healthy controls excreted a mean of 40.5-49.7 mg DMA/day, 19.4-23.8 mg PYR/day and 26.1-31.7 mg PIP/day. In patient groups a mean of 0.43-1.88 µg NDMA/day, 0.13-0.39 µg NPYR/day and 0.21-0.30 µg NPIP/day was excreted in urine but not in a control group. The endogenous formation of amines, and volatile N-nitros-amines under some clinical conditions is considerably higher than the exogenous exposure.

The potential for the endogenous gastric formation of N-nitroso compounds has been evaluated in kinetic models (1) using data obtained from either animal feeding studies (2) or calculations based on the predicted intake of precursor amines from the diet (3). These studies suggest that endogenous formation and exposure to dialkyl and heterocyclic volatile N-nitrosamines is not of sufficient magnitude to present a potential health risk. However, biological monitoring of specific drug nitrosation products formed in man and excreted in urine (4,5) provides strong evidence that endogenous nitrosation can provide a substantial additional exposure to N-nitroso compounds. As in the case of N-nitrosodimethyl-amine (NDMA) formation from endogenous nitrosation of aminopyrine (4), animal studies show that sufficient endogenous nitrosation occurs for biological monitor-

---

[1]Current address: Wissenschaftliche Abteilung, Verband der Cigarettenindustrie, Königswintererstrasse 550, 53227 Bonn, Germany

ing of 7-methylguanine DNA adducts excreted in urine (6) and to induce liver tumors in rats (7).

Theoretical studies suffer from two major limitations: (i) the lack of available analytical data on the exposure to both nitrosatable amines in the diet and their occurrence in biological fluids such as gastric juice, and (ii) the fact that endogenous nitrosation is not limited to chemical reactions involving dietary nitrate, nitrite and amines occurring under acidic conditions in the stomach. Extragastric nitrosation may also result from reactions involving bacteria (8), yeasts (9), alveolar (10) and peritoneal macrophages (11) in the oral cavity (9), urinary bladder (12,13), sigmoid colon (14) and possibly the vaginal vault (15). The following studies evaluate the daily dietary exposure to nitrosatable secondary amines and volatile N-nitrosamines, and their endogenous occurrence in human saliva, gastric juice, blood, urine and faeces.

Methods and Materials

Exogenous exposure to secondary amines and volatile N-nitrosamines. In a dietary survey, 38 alcoholic drinks and 215 foods prepared for human consumption using standard culinary practices were analysed for secondary amines (16,17) and volatile N-nitrosamines (18). Food samples were purchased from supermarkets in Heidelberg, Germany. The choice of samples was based on the 1988 German Nutritional Report (19) which was used to calculate the mean daily intake of individual foodstuffs. A 10 day duplicate diet survey was also performed, diet samples were collected in plastic containers, homogenized and frozen at -20°C pending analysis.

Biomonitoring studies. Both smoking and nonsmoking volunteers participating in the studies consumed their normal diets without additional vitamin supplementation. Alcohol consumption was forbidden in investigations in which urine samples were collected. Urine samples were collected from spine-injured paraplegic patients (13), Bilharzia (12) and bladder augmentations patients (20). A control group was selected from candidates not taking any form of pharmaceutical drug treatment. Urine (24 hour and individual samples) was collected by direct urination into sterile plastic containers containing 10 g sodium hydroxide. Saliva samples were collected (3-4 hours after breakfast) by expectoration into sterile plastic vials containing 10 M sodium hydroxide (100 µl). Gastric juice samples (5-10 ml) were collected by endoscopy. The pH was immediately measured using a glass electrode prior to stabilization by addition of 10 M sodium hydroxide (200 µl). Blood (peripheral) samples were collected using heparinized sterile tubes. Plasma was obtained by centrifugation at 3000 g for 15 min at 4°C and stabilized by addition of 10 M sodium hydroxide (200 µl). Faeces samples (50-100 g) were weighed into plastic bottles containing 0.5 M sodium hydroxide (50 ml) and vigorously shaken to produce a homogenous slurry. Sodium hydroxide stabilization was used for all samples to prevent artefact N-nitrosamine formation and to prevent bacterial growth and/or bacterial reduction of nitrate to nitrite. All samples were stored at -20°C prior to analysis using previously reported methods for the analysis of nitrate, nitrite, secondary amines and volatile N-nitrosamines (21).

Results and Discussion

The exogenous exposure to secondary amines and volatile N-nitrosamines is presented in Table I. Good agreement on the mean daily intake was obtained from the dietary survey after correction for the daily intake according to the German Nutrition Report and the 10 day duplicate diet survey. However, caution must be excercised in the estimation of exposure to food constituents due to uncertainties in the average food consumption within a population. According to the German Nutrition Report, adult women have a total average consumption of 1.8 kg/day while our duplicate diet survey showed an average daily consumption of 2.4 kg/day for a 60 kg female adult.

The available data from food surveys on volatile N-nitrosamines published over the last decade (reviewed in *18*) indicate that: (i) the dietary intake of NDMA is 0.5 μg/day or less in most countries, and (ii) the current levels are about one-third of those found a decade ago. The latter is true for The Netherlands, Japan and Germany where sufficient data is available to evaluate the dietary situation. This is the first dietary survey which calculates a mean daily intake for secondary amines. Previous estimates by Shephard *et al.* (*3*) underestimate the mean exposure to the most commonly encountered amines (1.7 mg DMA/day, 0.3 mg PYR/day and 0.06 mg PIP/day) and overestimate the exposure to other secondary amines (0.6 mg MBzA/day and 0.3 mg MEA/day).

The endogenous exposure to nitrate, nitrite and amines expressed in μg/ml (ppm) in saliva, faeces, and blood are presented in Table II. Only DMA, PYR and PIP are regularly present in these biological samples. DEA was occasionally found in saliva and faeces but not in blood. Individual samples were occasionally found to contain trace levels of EMA and DBA (data not shown).

In the oral cavity, reduction of nitrate to nitrite in the salivary glands (*22*) produces the highest effective nitrite concentrations in the human body. The presence of oral bacteria, yeasts (*9*) and increased concentrations of the nitrosation catalyst thiocyanate in the saliva of smokers (*23*) could theoretically catalyse the nitrosation of secondary amines naturally present in saliva. However, the presence of N-nitrosamines in 'fasting' saliva has not been reported. Ingestion of food increases not only the salivary flow but also increases the levels of nitrite-reactive species in the oral cavity. The concentrations of secondary amines in the diet are even less than those found in saliva, thus considerably reducing the chance of endogenous nitrosation occurring in the oral cavity during the consumption of a normal diet.

In the lower gastrointestinal tract, the biological half-life of nitrate (ca. 20 mins) and nitrite (ca. 5 mins) is relatively short due to both bacterial metabolism and absorption (*24,25*). Bacterial decarboxylation of amino acids produces a continuous source of amines within the gastrointestinal tract (*26*). These amines are almost completely absorbed (*27*). Our results show that only low concentrations of secondary amines are excreted in faeces while nitrate and nitrite were not detected. Contrary to an earlier investigation by Wang *et al.* (*28*) reporting both NDMA and NDEA to be present in normal human faeces, we did not find the presence of volatile N-nitrosamines. The only reported situation in which nitrosamine formation occurs in the lower intestinal tract is in the sigmoid colon of ureterosigmoidostomy

Table I. Exogenous Exposure to Secondary Amines and Volatile N-Nitrosamines from the Diet

| | Mean daily dietary exposure[1] | | |
| | Food survey | | 10 Duplicate |
	Men	Women	diets (women)
Secondary amines (mg/day)[2]			
DMA	4.4	3.5	3.6 (0.4-15.1)
PYR	1.5	1.3	1.2 (0.88-4.2)
PIP	1.3	1.2	0.5 (0.22-0.7)
DEA	0.22	0.16	0.03 (ND[3]-0.2)
DBA	0.09	0.06	ND
MOR	0.03	0.02	ND
EMA	0.01	0.01	0.06 (ND-0.2)
DPA	0.003	0.003	ND
MBzA	0.0004	0.0004	ND
N-Nitrosamines (µg/day)			
NDMA	0.28	0.18	0.15 (ND-0.6)
NPIP	0.015	0.015	ND
NPYR	0.011	0.011	ND

[1]Calculated from the 1988 German Nutritional Report (*19*).
[2]Secondary amines abbreviated as: Dimethylamine (DMA), pyrrolidine (PYR), piperidine (PIP), diethylamine (DEA), dibutylamine (DBA), morpholine (MOR), methylethylamine (EMA), dipropylamine (DPA)and methylbenzylamine (MBzA). Volatile N-nitrosamines abbreviated as: N-nitrosodimethylamine (NDMA), N-nitrosopiperidine (NPIP) and N-nitrosopyrrolidine (NPYR).
[3]ND, not detected. Limit of detection 10 µg/kg for amine and 0.05 µg/kg for volatile N-nitrosamine analysis.

patients (*14*). In these patients, the bladder is surgically removed and the ureters implanted into the sigmoid colon producing an enriched mixture containing urinary nitrate and secondary amines together with nitrate-reducing faecal bacteria. In blood, nitrosatable secondary amines and nitrate were detected. While we did not analyse for nitrosamine due to the small sample volumes collected, a previous study has shown the presence of NDMA in blood of patients with chronic renal failure (*29*). The presence of NDMA in blood most probably results from exogenous dietary nitrosamine exposure and/or formation and adsorption from the gastrointestinal tract prior to first-pass hepatic metabolism. However, the possibility that NDMA is formed in blood or tissue as a result of NO synthase activity cannot be neglected.

The presence of nitrate, nitrite and secondary amines in gastric juice as a function of pH is shown in Figure 1. The pH of 50 gastric juice samples ranged between pH 1.1 and 8.1 of which 56 % were acidic (pH<4.0) and only 2 samples had a pH>7.0 (data not shown). Nitrate was present in all of 50 samples at a mean concentration of 16.0 ± 22.5 μg/ml (range 0.9-72.0 μg/ml), a general decreasing trend in nitrate concentration was observed with increasing gastric pH. Nitrite was detected in 13 of 50 gastric juice samples at a mean concentration of 0.40 ± 1.26 μg/ml (range ND-8.0 μg/ml). Only 2 gastric juice samples with pH<4.0 contained free nitrite (0.1 and 0.6 μg/ml). At pH>4.0, 11 of 32 gastric juice samples contained 0.60 ± 1.55 μg/ml (range ND-8.0 μg/ml) nitrite. DMA was found in all 50 gastric juice samples at a mean concentration of 0.82 ± 0.82 μg/ml (range 0.20-4.20 μg/ml). Lower concentrations of 0.06 ± 0.11 μg/ml (range ND-0.45 μg/ml) DEA were detected in 19 samples, 0.17 ± 0.14 μg/ml (range ND-0.70 μg/ml) PYR in 43 samples and 1.31 ± 2.44 μg/ml (range ND-15.8 μg/ml) PIP in 48 samples. A marked decrease in amine concentration was found with increasing gastric pH. Isolated gastric juice samples contained EMA and DBA at concentrations of less than 0.06 μg/ml.

In the fasting stomach, the only secondary amines of any consequence which occur in gastric juice are DMA, PYR and PIP. Chemical nitrosation is controlled by two basic physical properties: (i) the dissociation of nitrite in aqueous media for which the acid-nitrite equilibrium has a pKa of 3.4, and (ii) the basicity of the amine. Protonated amines are unreactive to the nitrite-derived nitrosating agent N_2O_3. As a result, amine nitrosation rates typically show an optimum at pH 3.0-3.5 due to complex interactions of decreasing concentrations of N_2O_3 and increasing concentrations of free amines with increasing pH (*30*). For strongly basic secondary amines such as PYR (pKa 11.3), PIP (pKa 11.2) and DMA (pKa 10.7), the effective concentration of the unprotonated amine which can participate in chemical nitrosation under acidic conditions is small. At present, the extent of nitrosation in the gastric compartment cannot be assessed. However, endogenous nitrosation of aminopyrine (*8*) and piperazine (*9*) in man and the monitoring of their nitrosation products in urine show that a range of amino precursors do undergo nitrosation in vivo resulting in considerable transient levels of N-nitrosamines after exposure to high concentrations of a single nitrosatable amine.

The mean 24 hour levels of nitrate, nitrite, secondary amines and their corresponding N-nitroso derivatives in urine of three different patient groups prone to chronic bacterial infections of the urinary tract, and in a comparative German

Table II. Nitrate, Nitrite and Secondary Amines in Saliva, Blood and Faeces
(μg/ml)[1]

	Saliva (n=20)	Blood (n=10)	Faeces (n=8)
NO_3^-	35.5 ± 17.4	3.21 ± 2.74	ND[2]
	(5.8-92.0)	(1.4-8.1)	
NO_2^-	8.9 ± 6.3	ND	ND
	(1.6-23.0)		
DMA	0.18 ± 0.05	0.91 ± 0.84	0.41 ± 0.46
	(ND-0.30)	(0.15-2.40)	(0.10-1.35)
DEA		ND	0.03 ± 0.04
	(0.03)		(ND-0.10)
PYR	0.21 ± 0.23	0.36 ± 0.58	0.55 ± 0.67
	(ND-0.65)	(ND-2.20)	(0.25-1.30)
PIP	0.49 ± 0.37	1.86 ± 1.02	0.39 ± 0.24
	(ND-1.30)	(0.45-2.70)	(0.20-0.70)

[1]Mean ± S.D. (range of individual results). Secondary amines abbreviated as shown
in Table I.
[2]ND, not detected. Limit of detection 1.0 μg/ml for nitrate and nitrite, 0.01 μg/ml
for secondary amine analysis.

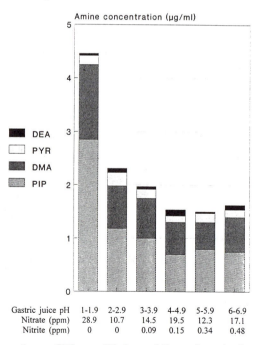

Figure 1. Concentrations of Nitrate, Nitrite and Secondary Amines in Gastric Juice
as a Function of pH.

control group of healthy subjects, are given in Table III. Urine contains the highest concentrations of both secondary amines and nitrate within the human body. Under normal conditions, in the absence of bacterial infections, nitrosation within the urinary tract does not occur. In the three patient groups with bacterial colonisation of the urinary tract, excretion of NDMA, NPIP and NPYR most probably occur as a result of bacterial mediated nitrosation or as an indirect result of the infection which stimulates NO production. Aerobic bacterial cultures showed a complex

Table III. Daily Excretion of Nitrate, Nitrite, Secondary Amines and Volatile N-Nitrosamines in Urine[1]

	Control group (n=40)	Patient groups with bacterial infections of the urinary tract		
		Paraplegic patients (n=30)	Bladder augmentations (n=12)	Bilharzia patients (n=50)
NO_3 (mg)	62.9±34.8 (29.8-106)	47.0±33.4 (1.6-147)	65.1±16.0 (31.9-96.4)	158.1±166.7 (16.0-855)
NO_2 (mg)	(1.30)	10.4±13.2 (ND[2]-45.9)	0.8±1.0 (ND-2.6)	3.36±4.02 (ND-45.0)
DMA (mg)	40.5±22.3 (25.8-119)	47.8±24.4 (12.3-125)	49.7±19.6 (21.1-118)	37.8±17.2 (9.8-80.7)
DEA (mg)	(1.2, 1.9)	(1.2, 1.9)	(0.5)	0.69±1.75 (ND-9.04)
PYR (mg)	20.4±13.1 (8.6-37.3)	19.4±13.8 (1.5-73.6)	23.8±15.3 (7.4-45.0)	21.9±23.2 (1.4-78.8)
PIP (mg)	26.1±27.3 (3.8-96.7)	26.4±19.1 (3.3-103)	31.7±17.4 (6.6-98.1)	24.2±29.7 (0.6-117.1)
NDMA (μg)	(0.15)	0.65±0.69 (ND-2.70)	0.43±0.38 (ND-1.40)	1.88±4.63 (ND-32.1)
NDEA (μg)	ND	(2.15)	ND	0.10±0.18 (ND-1.90)
NPYR (μg)	ND	0.39±0.50 (ND-1.95)	0.13±0.32 (ND-1.05)	0.16±0.25 (ND-0.90)
NPIP (μg)	ND	0.25±0.44 (ND-2.15)	0.30±0.48 (ND-1.40)	0.21±0.18 (ND-0.60)

[1]Mean ± S.D. (range of individual results). Secondary amines and volatile N-nitrosamines abbreviated as shown in Table I.
[2]ND, not detected. Limit of detection 1.0 μg/ml for nitrate and nitrite, 0.01 μg/ml for secondary amine, and 0.05 μg/l for volatile N-nitrosamine analysis.

bacterial flora including nitrate-reducing bacteria in urine samples from both paraplegic patients and patients with bladder augment-ations. Bacterial cultures were not obtained from Bilharzial patients. However, other studies confirm that chronic bacterial infections are usually present in such patients *(31)*.

In summary, our results show the concentrations of nitrosatable secondary amine precursors to volatile N-nitrosamines in the diet and in different compartments of the human body in which endogenous nitrosation may occur. The endogenous formation of these amines is far greater than the exogenous exposure from the diet. The principal site of amine formation is in the lower gastrointestinal tract from which amines are almost totally absorbed and circulated in blood. From blood, amines may diffuse into acidic gastric juice where on protonation there is little net transport from the gastric lumen back into blood. As a result, most amines enter an enterogastric cycle prior to renal excretion.

Acknowledgements

The authors are grateful to Dr. D. Stickler, School of Pure and Applied Biology, UWIST, Cardiff, Wales, Prof. M. H. Mostaffa, Institute of Graduate Studies and Research, University of Alexandria, Alexandria, Egypt, and Dr. T. Kälble, Urological Department, University of Heidelberg, Germany for collection of biological fluids used in these studies.

Literature cited

1. Licht, W.R.; Deen, W.M. *Carcinogenesis (Lond.)*, **1988,** *9,* 2227-2237.
2. Ohshima, H.; Mahon, G.A.T.; Wahrendorf, J.; Bartsch, H. Cancer Res., **1983,** *43,* 5072-5076.
3. Shephard. S.E.; Schlatter. Ch.; Lutz, W.K. *Food Chem. Toxicol.,* **1987,** *25,* 91-108.
4. Spiegelhalder, B.; Preussmann, R. *Carcinogenesis (Lond.)*, **1975,** *6,* 545-548.
5. Tricker, A.R.; Kumar, R.; Siddiqi, M.; Khuroo, M.S.; Preussmann, R. *Carcinogenesis (Lond.)*, **1991,** *12,* 1595-1599.
6. Gombar, C.T.; Zubroff, J.; Strahan, G.D.; Magee, P.N. *Cancer Res.,* **1983,** *43,* 5072-5075.
7. Lijinsky. W.; Taylor, H.W.; Synder, C.; Nettesheim, P. *Nature (Lond.)*, **1973,** *244,* 176-178.
8. Leach, S.A.; Thompson, M.H.; Hill, M.J. *Carcinogenesis (Lond.)*, **1987,** *8,* 1907-1912.
9. Krogh, P.; Hald, B.; Holmstrup, P. *Carcinogenesis (Lond.)*, **1987,** *8,* 1543-1548.
10. Huot, A.E.; Hacker, M.P. *Cancer Res.,* **1990,** *50,* 7863-7866.
11. Miwa, M.; Stuehr, D.J.; Marletta, M.A.; Wishnok, J.S.; Tannenbaum, S.R. *Carcinogenesis (Lond.)*, **1987,** *8,* 955-958.
12. Tricker, A.R.; Mostafa, M.H.; Spiegelhalder, B.; Preussmann, R. *Carcinogenesis (Lond.)*, **1989,** *10,* 547-552.
13. Tricker, A.R.; Stickler, D.J.; Chawla, J.C.; Preussmann, R. *Carcinogenesis (Lond.)*, **1991,** *12,* 943-946.

14. Tricker, A.R.; Kälble, T.; Preussmann, R. *Carcinogenesis (Lond.),* **1989,** *10,* 2379-2382.
15. Harrington, J.S.; Nunn, J.R.; Irwig, L. *Nature (Lond.),* **1973,** *241,* 49-50.
16. Pfundstein, B.; Tricker, A.R.; Preussmann, R. *J. Chromatogr.,* **1991,** *539,* 141-148.
17. Pfundstein, B.; Tricker, A.R.; Theobald, E.; Preussmann, R.; Spiegelhalder, B. *Food Chem. Toxicol.,* **1991,** *7, 733-739.*
18. Tricker, A.R.; Pfundstein, B.; Theobald, E.; Preussmann, R.; Spiegelhalder, B. *Food Chem. Toxicol.,* **1991,** *11,* 729-732.
19. Ernährungsbericht 1988; Deutsche Gesellschaft für Ernährung; Frankfurt a.M., **1988.**
20. Gröschel, J.; Reidash, G.; Kälble, T.; Tricker, A.R. *J. Urol.,* **1992,** *147,* 1013-1016.
21. Tricker, A.R.; Pfundstein, B.; Kälble, T.; Preussmann, R. *Carcinogenesis (Lond.),* **1992,** *13,* 563-568.
22. Spiegelhalder, B.; Eisenbrand, G.; Preussmann, R. (1976) *Food Cosmet Toxicol.,* **1976,** *14,* 545-548.
23. Fan, T.Y.; Tannenbaum, S.R. *J. Agric. Food Chem.,* **1973,** *21,* 237-240.
24. Saul, R.L.; Kabir, S.H.; Cohen, Z.; Bruce, W.R.; Archer, M.C. *Cancer Res.,* **1981,** *41,* 2280-2283.
25. Schultz, D.S.; Deen, W.M.; Karel, S.F.; Wagner, D.A.; Tannenbaum, S.R. *Carcinogenesis (Lond.),* **1985,** *6,* 847-852.
26. Barker, H.A. *Ann. Rev. Biochem.,* **1981,** *50,* 23-40.
27. Ishiwata, H.; Iwata, R.; Tanimura, A. *Food Chem. Toxic.,* **1984,** *8,* 649-653.
28. Wang, T.; Kakizoe, T.; Dion, P.; Furrer, R.; Varghese, A.J.; Bruce, W.R. *Nature (Lond.),* **1978,** *276,* 280-281.
29. Dunn, S.R.; Simenhoff, M.L.; Lele, P.S.; Goyal, S.; Pensabene, J.W.; Fiddler, W. (1990) *J. Natl. Cancer Inst.,* **1990,** *82,* 783-787.
30. Mirvish, S.S. *Toxicol. Appl. Pharmacol.,* **1975,** *31,* 325-352.
31. Hicks, R.M.; Walters, C.L.; El-Sabai, I.; El-Aaser, A.; El-Merzabani, M.M.; Gough, T.A. *Proc. R. Soc. Med.,* **1976,** *70,* 413-416.

RECEIVED July 19, 1993

Chapter 9

Improved Methods for Analysis of *N*-Nitroso Compounds and Applications in Human Biomonitoring

B. Pignatelli, C. Malaveille, P. Thuillier, A. Hautefeuille, and H. Bartsch

International Agency for Research on Cancer, 150 Cours Albert-Thomas, 69008 Lyon, France

A sensitive group-selective method for analysis of total N-nitroso compounds (NOC) has been developed and applied to investigate their role in gastric cancer etiology. Total NOC in gastric juice (GJ) were not elevated in subjects with gastric precancerous conditions or omeprazole-induced achlorhydria. However, the presence of precursors of nitrosation-dependent direct mutagens and a lowered anti-oxidant defense and reduced nitrite scavenging-ability were found in GJ of subjects with elevated cancer risk. Unknown non-volatile NOC were prevalent among total NOC occurring in body fluids and some foods. An improved HPLC-photohydrolysis-colorimetry method was set up and validated with 28 reference NOC in order to separate and detect hitherto unknown NOC in biological fluids and food extracts.

Humans are exposed to a wide range of nitrosatable compounds and nitrosating agents (1,2) which can react in vivo to form carcinogenic N-nitroso compounds (NOC). Nitrite, nitrate and nitrosating agents can also be synthesized endogenously in reactions mediated by bacteria, activated macrophages or other cell types through the enzyme nitric oxide synthase (3-7). Carcinogenic NOC formed in the stomach have been suggested to be involved in gastric cancer causation. Gastric achlorhydria permits the colonization of the stomach by bacteria that convert nitrate to nitrite, which may lead to intragastric formation of NOC (8). Positive correlations between the presence of precancerous conditions of the stomach, high intragastric pH, elevated bacterial count and nitrite concentrations have been found consistently (reviewed in 9). However, group determination of total NOC in human gastric juice (GJ) by two methods (10-12) gave contradictory results and the reported concentrations showed wide inter-laboratory variations. As both methods may overestimate or underestimate the true NOC levels (12-14), we have developed an improved analytical

0097–6156/94/0553–0102$08.00/0
© 1994 American Chemical Society

procedure and applied it to patients with and without precancerous
conditions of the stomach, who live in areas with contrasting gastric cancer risk and to volunteers before and after treatment for
inhibition of gastric acid secretion. As ascorbic acid, a known
anti-oxidant and inhibitor of nitrosation, protects against gastric cancer (15-17), the relationships between precancerous conditions of stomach and gastric levels of nitrite, total NOC and
ascorbic acid were also investigated. Combined application of the
methods for the analysis of total NOC, volatile nitrosamines (VNA)
and N-nitrosamino acids (NAA), revealed a high prevalence of unknown NOC in human body fluids and in some food extracts. HPLC-
photohydrolysis-colorimetry method for trace analysis of NOC was
validated and can be used in combination with bacterial mutagenicity assays for further identification of hitherto unknown
genotoxic NOC.

Improved Method for Group-Selective Analysis of Total NOC in Human Body Fluids

As humans are exposed to a variety of NOC precursors, hitherto unknown non-volatile NOC (and their metabolites) are expected to
occur in body fluids. We have modified previous methods (10-12)
to develop an improved group-selective method for the determination of total NOC in GJ (18). It is based on their chemical denitrosation with HBr and chemiluminescence detection of the released
nitric oxide using a thermal energy analyser (TEA). Sequential
addition of CH_3COOH and HBr permits to distinguish NOC from acid-
labile TEA responsive compounds (TAC). The GJ sample, treated
with sulphamic acid to destroy nitrite, is injected directly into
refluxing ethyl acetate containing either CH_3COOH - 0.1% (v/v) HCl
for determining TAC or HBr for the simultaneous determination of
TAC and NOC. The amount of nitric oxide released in each case is
measured by a TEA detector (Figure 1). The difference between the
two determinations represents the concentration of total NOC in
the sample. The amount of TAC and (TAC + NOC) can be calculated
using nitrite and nitrosodiphenylamine as respective standards.
The TAC concentration can greatly influence the response of total
TEA-responsive compounds. The presence of up to 6% water in ethyl
acetate or of nitrate up to 1000 µmol/l did not affect the determinations. The method was adapted for analysing aqueous samples
containing high concentrations of nitrate (e.g., human urine)
(19). Nitrate was removed by an anion-exchange procedure without
a significant loss of various added reference NOC and unidentified
urinary NOC.
     This method is highly selective for NOC, reproducible (coef-
ficient of variation from triplicate measurements 5-10%), sensitive (detection limit 0.01 µmol/l), requires only a few ml of
sample and no prior extraction. Suitable techniques for stabili-
zing human GJ and urine samples, and the influence of time and
storage conditions on NOC concentrations have been established
(18,19). The absence of artefactual nitrosation due to sample
handling has been verified. The method distinguishes NOC from
most other TEA-responsive species such as pseudonitrosites, ni-
trosothiols, alkyl nitrites and is also applicable to the analysis
of faeces (20) and food extracts (21).

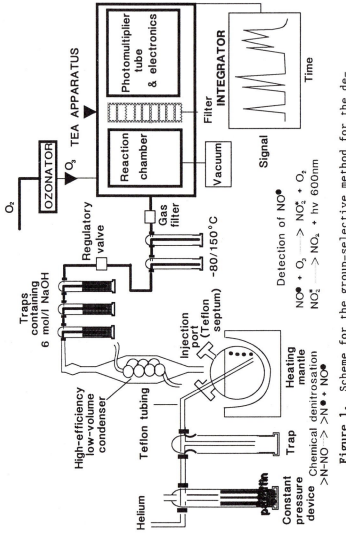

Figure 1. Scheme for the group-selective method for the determination of total NOC.

Levels of NOC, Direct Genotoxins, and their Precursors in GJ of
Patients with and without Precancerous Lesions of the Stomach and
Living in Areas with Contrasting Gastric Cancer Risk.

A study was performed to examine whether elevated risk of gastric
cancer is associated with high levels of total NOC, their
precursors and nitrosation-dependent direct genotoxins in GJ (22).
Patients (N = 207) were classified according to histologically
confirmed diagnosis of the stomach, country of origin and pH of
their GJ (Table I). They were living in three areas with up to
8-fold variations in gastric cancer risk: one in Colombia
(Nariño), where the gastric cancer incidence is among the highest
in the world, and two others in the UK and France, where risks are
lower. Concentrations of total NOC (18) and direct genotoxic
activity (23) were determined in GJ of fasting subjects before and
after in vitro nitrosation. Nitrite was analyzed colorimetrically
(24).
      In agreement with earlier findings, an acidic gastric pH
($\leq$ 4) was strongly and significantly associated with normal
gastric mucosa or moderate gastritis. Among patients without
gastric precancerous lesions, GJ with elevated pH was seen much
more frequently in Colombia, as compared to the other countries
(Table I). Colombian patients with precancerous lesions had the
highest prevalence of elevated gastric pH (Table I). The nitrite
concentrations in GJ (range 1-472 µmol/l) were highest for
Colombian patients with precancerous lesions and for subjects with
GJ of pH>4 (Figure 2A). In contrast, levels of total NOC (range
$\leq$ 0.01 - 8.0 µmol/l) did not differ between countries nor between
GJ samples grouped according to histopathological diagnosis or
according to a pH below or above 4 (Figure 2B). Total NOC levels
increased with original nitrite concentrations at a greater rate
in acidic GJ than in those with pH > 4. The data together suggest
that acid-catalysed nitrosation contributes at least as much as
other nitrosation pathways to intragastric NOC formation.
      In vitro nitrosation of GJ with excess $NaNO_2$ (at pH 1.5 for
60 min at 37°C) increased the concentration of total NOC up to
several thousand-fold, with a maximum of 1330 µmol/l (Figure 2C).
High NOC levels after in vitro nitrosation were not associated
with higher gastric cancer risk and the NOC concentration in
nitrosated GJ increased with the original pH of GJ only for French
samples (Figure 2C). These findings suggest that both the level
and probably the nature of some of the substances in GJ are
different between French and Colombian subjects. After acid-
catalysed nitrosation, all GJ from France and Colombia exhibited
genotoxic activity that was more elevated in GJ with an original
pH above 4 than in acidic GJ; the highest genotoxicity was seen in
Colombian GJ (Figure 2D). Thus, genotoxicity of nitrosated GJ was
dependent on its original pH and patients from the highest gastric
cancer risk area had the highest levels of precursors of nitrosation dependent genotoxins in their GJ.
      The results of this study do not support the idea that intra-
gastric total NOC levels are elevated in subjects with precancerous stomach conditions or in those living in a high-risk area for
stomach cancer but are consistent with intragastrically formed

Table I.   Distribution of Patients According to their Gastric pH,
           Country of Origin & Histological Diagnosis

Diagnosis (Group)	Country of origin	No	% of gastric juice samples with pH ≤ 4	> 4
Normal gastric mucosa	UK	18	89	11
or	France	27	85	15
superficial gastritis (I)	Colombia	32	56	44
Reflux gastritis (II)	UK	9	78	22
Diffuse interstitial (III)	UK	12	92	8
gastritis	France	17	71	29
Chronic atrophic gastritis	UK	17	53	47
with or without intestinal	France	37	52	48
metaplasia or dysplasia	Colombia	38	29	71
(IV)				
Total	UK	56	77	23
	France	81	66	34
	Colombia	70	41	59

Table II.   Influence of Treatment with Omeprazole on Gastric
            Acidity and Bacterial Flora in Healthy Volunteers[a]

Treatment[b]	pH[c]	Total bacterial counts[c] $(x\ 10^4/ml)$	Nitrate-reducing bacterial counts[c] $(x\ 10^4/ml)$
None/placebo	1.9 (1.3-5.8)	<1 (<1-500)	<1 (<1-300)
1 week omeprazole	3.9* (1.8-6.4)	16* (<1-5200)	10 (<1-2200)
2 weeks omeprazole	4.3* (1.3-6.8)	40* (<1-3300)	35* (<1-3100)

[a]N = 14;   [b]20 mg omeprazole daily;   [c]Medians with ranges in
brackets;   *p<0.05 vs placebo

nitrite-derived direct mutagens (nitrosamides or diazonium compounds) having a role in gastric cancer etiology.

Nitrite, Bacterial Flora and NOC in GJ of Volunteers before and after Treatment with Anti-acid Secretory Drugs

Decreased gastric acidity resulting from treatment with anti-acid secretory drugs has been hypothesized to allow proliferation of nitrite producing bacteria in the stomach, possibly leading to in situ formation of carcinogenic NOC. The effect of inhibition of acid secretion following treatment by omeprazole on gastric colonization by nitrate-reducing bacteria and intragastric NOC formation was examined. Fourteen healthy volunteers received a placebo for one week, followed by omeprazole (20 mg/person daily) for two weeks. Bacteriological analysis and nitrite (24) and total NOC (18) measurements were performed in GJ at the end of each week after a 24 h recording of gastric pH. Treatment with omeprazole, led to a higher (median) gastric pH, predisposing to increased gastric bacterial proliferation including nitrate-reducing bacteria (Table II). In these healthy subjects, no consistent trend was observed for elevation of nitrite or total NOC concentrations following omeprazole treatment (Figure 3). A similar study is being performed in patients with duodenal or gastric ulcer or reflux oesophagitis before and after treatment with omeprazole or other anti-acid secretory drugs.

Pro-oxidant Status and Nitrite and NOC Levels in GJ of Patients with or without Premalignant Conditions of the Stomach

Simultaneous measurements were performed of ascorbic acid, vitamin C (sum of ascorbic acid and dehydroascorbic acid), nitrite and total NOC levels in GJ and of vitamin C in plasma from 56 subjects with or without gastric precancerous conditions that varied in severity (25). The pH was highest in GJ of chronic atrophic gastritis patients (Table III). In chronic gastritis (with or without atrophy) high nitrite concentrations were associated with high pH but gastric nitrite or total NOC concentrations were not related to gastric histology (Table III). In GJ of patients with chronic gastritis, in particular those with atrophic gastritis and intestinal metaplasia, significantly lowered ascorbic acid and vitamin C levels were observed, although vitamin C plasma levels were not significantly different from those of subjects with normal mucosa (Figure 4). No relationships between intragastric levels of ascorbic acid, nitrite and total NOC concentrations in GJ was detected. The presence of gastritis, in particular chronic atrophic gastritis with intestinal metaplasia, was highly significantly associated with H. pylori infection (Figure 4). H. pylori which is capable of activating human monocytes leading to an increased production of reactive oxygen intermediates (26) has emerged as one of the risk factors in gastric cancer etiology (27,28).

In conclusion, premalignant conditions were associated with H. pylori infection, a lowered antioxidant defense state and

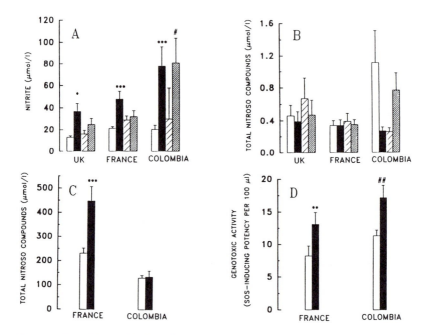

Figure 2. GJ analysis of fasted patients according to gastric pH, histopathological diagnosis and country of collection before (A,B) and after (C,D) acid-catalysed in vitro nitrosation. Values are plotted as mean ± standard error. For characteristic of subjects, see Table I. ▨ Group I without, ▨ group IV with precancerous lesions of the stomach; ☐ GJ pH<4; ■ GJ pH>4; p values of within country comparisons of group IV vs group I: #p<0.02 ##p<0.006 and of GJ pH>4 vs GJ pH<4: *p<0.002 **p<0.0003 ***p<0.0001

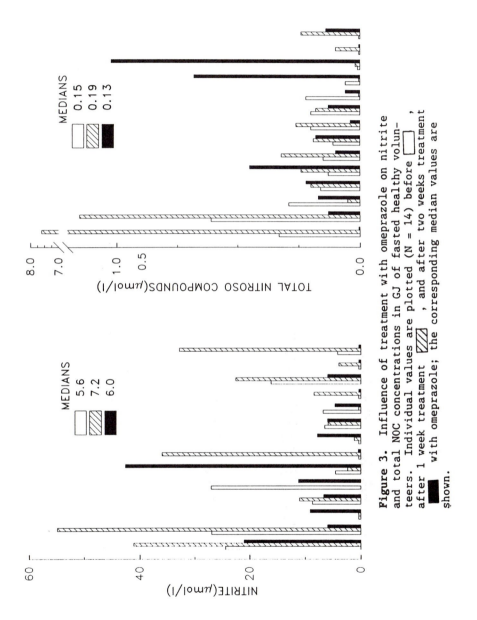

Figure 3. Influence of treatment with omeprazole on nitrite and total NOC concentrations in GJ of fasted healthy volunteers. Individual values are plotted (N = 14) before ☐, after 1 week treatment ▨, and after two weeks treatment with omeprazole ▮; the corresponding median values are shown.

Table III. Gastric Juice Analyses According to Gastric Pathology
            of Fasted Patients

	Normal histology	Chronic superficial gastritis	Chronic atrophic gastritis
Number of patients	12	18	17
pH[a]	2.0 (1.7-4.2)	2.1 (1.5-6.6)	3.1* (2.0-7.3)
Nitrite[a] (μmol/l)	10.5 (4.7-19.5)	10.9 (4.05-48.4)	17.2 (6.3-21.7)
total NOC[a] (μmol/l)	0.40 (0.02-4.46)	0.16 (0.02-0.68)	0.22 (0.02-3.04)

[a]Medians with ranges in brackets; *p<0.01; data from ref. 25

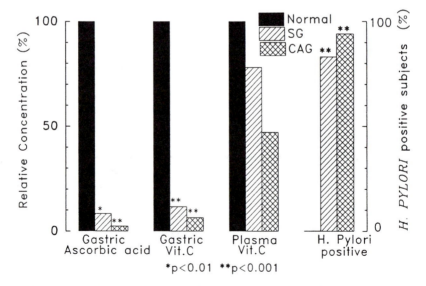

*p<0.01 **p<0.001

Figure 4. GJ and plasma analysis of fasted patients according
to gastric histopathology. Data from ref. 25.

reduced nitrite-scavenging ability but not with elevated gastric nitrite or total NOC levels in fasting GJ.

Prevalence of Unknown Non-Volatile NOC among Total NOC in Human GJ, Urine and Some Food Extracts

Except for some VNA, the nature of NOC present or formed after nitrosation in GJ is unknown. Total NOC, VNA and NAA were analysed in pooled GJ after acid-catalyzed nitrosation in vitro (29). About 40-50% of total NOC was not extractable in organic solvents. NOC extractable with CH_2Cl_2 accounted for only 1-2% of total NOC, in which the VNA identified accounted for less than 10%. NOC extractable with $MeOH:CH_2Cl_2$ represented 10-17% of the total NOC, of which 1-10% were NAA, the most abundant being N-nitrosoproline, accounted for 2-6%. Of the total NOC, 15-16% was extractable in ethyl acetate. These results revealed that a preponderance of nonvolatile, unknown NOC of varying polarity is formed in nitrosated GJ. Urine samples (N = 15) from proline-dosed healthy volunteers (500 mg proline/person) were analysed for total NOC and NAA (19). Concentrations of total NOC ranged from 0.2 to 2.6 μmol/l. Nitrosoproline accounted for <10% to 16% and the sum of the NAA for 5-27% of the total NOC. Thus, the structures of the majority of urinary NOC are unknown.

The high gastric cancer mortality in the Fujian province of China has been associated with the consumption of certain salted fermented fish products. The levels of NOC and genotoxins present, before and after nitrosation, in fish sauce samples collected in this high-risk area were investigated (21). Total NOC concentrations before nitrosation ranged from 0.2 to 16 μmol/l (8.8 - 704 μg (N-NO)/l) and, after nitrosation at pH 2 and pH 7, they rose by up to 4800- and 100-fold, respectively. In genotoxic nitrosated samples, 40-50% of total NOC was not extractable into organic solvents; VNA accounted for 1-2% and NAA for 8-16% of total NOC. The identity of compounds measured as total NOC remains to be investigated.

Concentrations of total NOC in smoked and unsmoked fried bacon varied from 430 to 6800 μg (N-NO)/kg (30). The sum of VNA, N-nitrosothiazolidines and NAA accounted for, on average, 16% of the total NOC. Again it appears that the identity of the majority of the NOC in bacon is unknown. Total NOC were detected in 42% of 170 beer samples (31) in concentrations up to 569 μg (N-NO)/kg, nitrosodimethylamine ranging from 0.1 to 1.2 μg/kg. Most of these were highly polar NOC of unknown identity (31).

In conclusion, known NOC constitute only a minor fraction of total NOC that are formed or occur in body fluids and in human diets. Therefore, methods and studies are warranted to identify hitherto unknown genotoxic NOC (and their metabolites).

HPLC-Photohydrolysis-Colorimetric Method for Analysis of NOC in Human Body Fluids and Foods

A new selective and sensitive method for separating and detecting polar nitrosamines, nitrosamides and non-volatile genotoxic NOC of unknown structures based on an earlier procedure (32) has been

established, improved and validated with mixtures of 28 reference
NOC (Figure 5). NOC are first separated by HPLC and then photoly-
tically cleaved by UV irradiation. The resulting nitric oxide is
transformed through oxidation and hydrolysis into nitrite ion
which is detected spectrometrically (λ = 546 nm) by post-column
formation of an azodye with Griess reagent. The photohydrolyser
(32) was kindly provided by D. Shuker (IARC) and technical impro-
vements were achieved (e.g. post-column mixing chamber, heat-
controlled oven for azodye formation). The photolytic cleavage of
NOC can also be performed in a commercially available photoreactor
(Knauer, Berlin, Germany) in which a knitted teflon tubing coil is
placed around a low-pressure UV lamp. As only compounds libera-
ting nitric oxide after UV irradiation can be detected the method
has a high selectivity. Reversed phase HPLC (column ODS C_{18}, 10
μm) was shown to separate various NOC (Figure 5) that include
nitrosamides, bulky alkylnitrosamines and NOC carrying hydroxyl
and carboxylic groups. Various organic solvents (CH_3CN, MeOH,
CH_3COCH_3) and acids (H_3PO_4, CF_3COOH, CH_3COOH) were tested as
modifiers of the aqueous mobile phase. Acetonitrile proved to give
the highest sensitivity which increased in proportion to the
percentage of CH_3CN in the mobile phase. Phosphoric acid was most
efficient in increasing the sensitivity and allowed the detection
of NAA as well as all other NOC at low levels. A reduced flow
rate (0,5 ml/min. vs up to 1.2 ml/min.) increased the yield of
cleaved NOC and led to a higher sensitivity. Under these HPLC
conditions, a satisfactory separation of complex NOC mixtures
could be achieved. Chromatograms of the HPLC-photohydrolysis-
colorimetry analysis of a mixture of 17 various NOC and of 6 NAA
are shown in Figures 6A and B respectively. Sensitivity was 1 to
10 ng (0.006-0.05 nmol) depending on the individual NOC
(Table IV).
    Application of this method to the analysis of NOC in a CH_2Cl_2
extract of nitrosated GJ revealed the presence of three unknown
NOC which were different from the 8 reference NOC separated under
the same conditions (Figure 7A). Similarly, in addition to nitrosopyrrolidine, 6 NOC of unknown structures were separated by HPLC
analysis of the CH_2Cl_2 extract of a genotoxic nitrosated fish
sauce sample (Figure 7B). These two examples illustrate the use-
fulness of the method to separate, detect and for further characterization of unknown NOC. Independently, a photolytic interface
has been developed (33) allowing the sensitive detection of NOC
through the measurement by chemiluminescence/TEA of nitric oxide
liberated by UV irradiation of the effluent containing HPLC-
separated NOC.
    In conclusion, the HPLC-photohydrolysis-colorimetry method
offers a powerful analytical tool for trace analysis of NOC that
are found in body fluids and food extracts. When combined with
short-term mutagenicity tests, this method should permit the
isolation and identification of a number of hitherto unknown NOC
and investigations of their biological relevance.

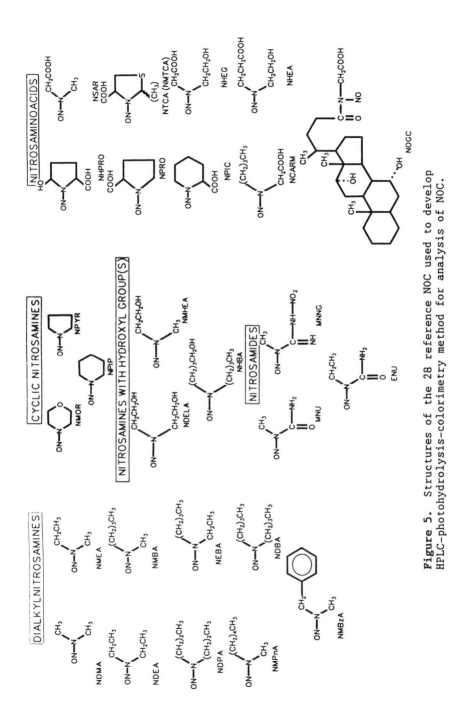

Figure 5. Structures of the 28 reference NOC used to develop HPLC–photohydrolysis–colorimetry method for analysis of NOC.

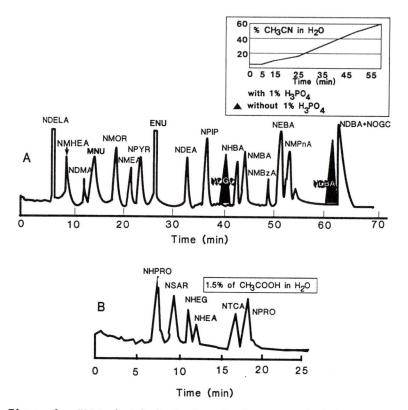

Figure 6. HPLC-photohydrolysis-colorimetry method for analysis of NOC. Chromatograms of a mixture of 17 NOC, 0.5 – 1.7 nmol/inj. (A) and of 6 nitrosaminoacids, 0.6 – 2 nmol/inj. (B). For abbreviations: see Figure 5. Flow rates (HPLC and post-column Griess reagent delivery): 0.8 ml/min.

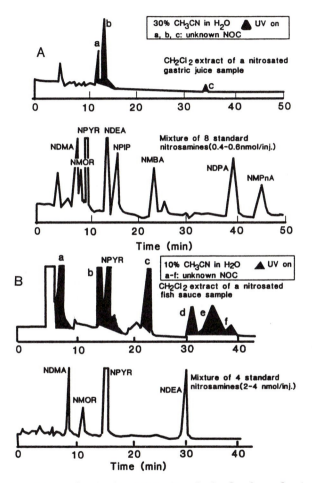

Figure 7. NOC analysis by HPLC-photohydrolysis-colorimetry in nitrosated gastric juice (A) and in nitrosated fish sauce (B). For abbreviations: see Figure 5. Flow rates (HPLC and post-column Griess reagent delivery): 0.8 ml/min.

Table IV. Sensitivity of the HPLC Post Column Photohydrolysis-
Colorimetry Method for Analysis of NOC

Type of NOC	No	Limit of detection	
		nmol	ng
Dialkylnitrosamines	9	0.01 - 0.05	1.5 - 4
Cyclic nitrosamines	3	0.01 - 0.02	1.5 - 2.5
Dialkylnitrosamines with hydroxyl groups	3	0.02 - 0.04	3.0 - 5.0
Nitrosamides	3	0.006-0.008	<1
Nitrosamino acids	6	0.02 - 0.05	3.0 - 10
Nitrosohydroxyamino acids	4	0.02 - 0.04	4.0 - 6.0
All	28	0.006- 0.05	<1 - 10

Acknowledgments

The authors express their gratitude to the US National Cancer
Institute for partial support of this work under grant No. 45 591.
They thank C.S. Chen, University of Nebraska Medical Center,
Omaha, USA; A. Rogakto & N. Muñoz, IARC and B. Moulinier &
R. Lambert, E. Herriot Hospital, Lyon France; P. Correa,
Louisiana State University Medical Center, New Orleans, USA; B.
Ruiz, Fundacion para la Educacion Superior, Cali, Colombia; G.M.
Sobala, C.J. Schorah & A.T.R. Axon, University of Leeds, U.K.; D.
Armstrong, F. Viani, A.L. Blum & M. Fried, Polyclinique Médicale
Universitaire, Lausanne, Switzerland; H. Siegrist, J.P., Idström &
C. Cederberg, Astra Hässle, Mölndal, Sweden.

Literature Cited

1. National Research Council National Academy Press, The Health
    Effects of Nitrate, Nitrite and N-Nitroso Compounds,
    Washington D.C., 1981; (Part 1).
2. Shephard, S.E.; Schlatter CH.; Lutz W.K. Food Chem. Toxicol.
    1987, 25, 91.
3. Calmels, S.; Ohshima, H.; Rosenkranz, H.; McCoy E.; Bartsch,
    H. Carcinogenesis 1987, 8, 1085.
4. Miwa, M.; Stuehr, D.J.; Marletta, M.A.; Wishnok, J.S.;
    Tannenbaum, S.R. Carcinogenesis 1987, 8, 955.
5. Marletta, M.A. Chem. Res. Toxicol. 1988, 1, 249.
6. Leaf, C.D.; Wishnok, J.S.; Tannenbaum, S.R. Cancer Surveys
    1989, 8, 323.
7. Billiar, T.R.; Curran, R.D.; Stuehr, D.J.; Stadler, J.;
    Simmons, R.C.; Murray, S.A. Biochem. Biophys. Res. Commun.
    1990, 168, 1034.
8. Correa, P. Cancer Res. 1988, 48, 3554.

9. Bartsch, H.; Ohshima, H.; Pignatelli, B.; Calmels, S. Cancer Surveys 1989, 8, 335.
10. Walters, C.L.; Hart, R.J.; Smith, P.L.R. In "Environmental Carcinogens Selected Methods of Analysis"; Preussmann, R.; O'Neill, I.K.; Eisenbrand, G.; Spiegelhalder, B.; Bartsch H. Eds.; IARC Sci. Public. No 45; IARC: Lyon, France, 1983; p. 295.
11. Walters, C.L.; Downes, M.J.; Edwards, M.W.; Smith, P.L.R. Analyst, 1978, 103, 1127.
12. Bavin, P.M.G.; Darkin, D.W.; Viney, N.J. In "N-Nitroso Compounds: Occurrence and Biological Effects"; Bartsch, H.; O'Neill, I.K.; Castegnaro, M.; Okada, M., Eds.; IARC Sci. Public. No 41, IARC: Lyon, France, 1982, p. 337.
13. Walters, C.L.; Smith P.L.R.; Reed, P.I. In "N-Nitroso Compounds: Occurrence, Biological Effects and Relevance to Human Cancer"; O'Neill, I.K.; Von Borstel, R.C.; Miller, C.T.; Long, J.; Bartsch, H. Eds.; IARC Sci. Public. No 57, IARC: Lyon, France, 1984, p. 113.
14. Smith, P.L.R.; Walters, C.L.; Reed, P.I. Analyst 1983, 108, 896.
15. Boeing, H. J. Cancer Res. Clin. Oncol. 1991, 117, 133.
16. Mirvish, S.S. In "Achievements, Challenges and Prospects for the 1980's"; Burchenal J.H.; Oettgen, H.F. Eds; Grune and Stratton: Orlando FL, 1981, Vol. 1, p. 557.
17. Bartsch, H.; Ohshima, H.; Pignatelli, B. Mut. Res. 1988, 202, 307.
18. Pignatelli, B.; Richard, I.; Bourgade, M.-C.; Bartsch, H. Analyst 1987, 112, 945.
19. Pignatelli, B.; Chen, C.S.; Thuillier, P.; Bartsch, H. Analyst 1989, 114, 1103.
20. Pignatelli, B.; Chen, C.S.; Thuillier, P.; Bartsch, H. In "The Significance of N-nitrosation of Drugs", Eisenbrand, G.; Bozler, G.; Nicolai, H.V. Eds; Gustav Fischer-Verlag: Stuttgart, 1990, (Drug Development Vol. 16), p 123.
21. Chen, C.S.; Pignatelli, B.; Malaveille, C.; Bouvier, G.; Shuker, D.; Hautefeuille, A.; Zhang, R.F.; Bartsch, H. Mut. Res. 1992, 265, 211.
22. Pignatelli, B.; Malaveille, C.; Rogatko, A; Hautefeuille, A.; Thuillier, P.; Muñoz, N.; Moulinier, B.; Berger, F.; De Montclos, H.; Lambert, R.; Correa, P.; Ruiz, B.; Sobala, G.M.; Schorah, C.J.; Axon, A.T.R.; Bartsch, H. The European J. of Cancer 1993 (in press).
23. Quillardet, P.; Hofnung, M. Mutat. Res. 1985, 147, 65.
24. Green, L.C.; Wagner, D.A.; Glogowski, J.; Skipper, P.L.; Wishnok J.S.; Tannenbaum, S.R. Anal. Biochem. 1982, 126, 131.
25. Sobala, G.M.; Pignatelli, B.; Schorah, C.J.; Bartsch, H.; Sanderson, M.; Dixon, M.F.; Shires, S.; King, R.F.G.; Axon, A.T.R. Carcinogenesis 1991, 12, 193.
26. Mai, U.E.; Perez-Perez, G.I.; Wahl, L.M.; Blaser, M.J.; Smith, P.D. J. Clin. Invest. 1991, 87, 894.
27. Correa, P. N. Engl. J. Med. 1991, 35, 1170.
28. Forman, D. J.N.C.I. 1991, 83, 1702.

29. Pignatelli, B.; Malaveille, C.; Chen, C.-S.; Ohshima, H.;
    Hautefeuille, A.; Thuillier, P.; Muñoz, N.; Bartsch, H.;
    Moulinier, B.; Berger, F.; Lambert, R.; De Montclos, H. In
    "Relevance to Human Cancer of N-Nitroso Compounds, Tobacco
    Smoke and Mycotoxins"; O'Neill, I.K.; Chen, J.; Bartsch, H.
    Eds., IARC Sci. Public. No 105; IARC: Lyon, France, 1991, p.
    172.
30. Massey, R.C.; Key, P.E.; Jones, R.A.; Logan, G.L. Food
    Additives and Contam. 1991, 8, 585.
31. Massey, R.C.; Dennis, M.J.; Pointer, M.; Key, P.E. Food
    Additives and Contam. 1990, 5, 605.
32. Shuker, D.E.G.; Tannenbaum, S.R. Anal. Chem. 1983, 55, 2152.
33. Conboy, J.J. ; Hotchkiss, J.H. Analyst 1989, 114, 155.

RECEIVED June 29, 1993

NITRIC OXIDE CHEMISTRY AND BIOCHEMISTRY

Chapter 10

DNA Damage and Cytotoxicity Caused by Nitric Oxide

Steven R. Tannenbaum[1,2], Snait Tamir[1], Teresa de Rojas-Walker[1], and John S. Wishnok[2]

[1]Department of Chemistry and [2]Division of Toxicology, Massachusetts Institute of Technology, Cambridge, MA 02139

Nitric oxide is formed by many types of cells in the body for the purpose of intercellular communication (brain, cardiovascular system) or as part of the immune or inflammatory response system (macrophages, endothelial cells). The chemistry of NO in oxygenated biological systems is very complex both in number of chemical species and in number of parallel and consecutive reactions. Damage to DNA in mammalian cells is caused by at least two major pathways; one arising from reaction of NO with molecular oxygen and one arising from reaction of NO with superoxide. Reaction with O_2 yields N_2O_3 which can either a.) nitrosate secondary amines to form carcinogenic or mutagenic N-nitrosamines or b.) nitrosate primary amines on DNA bases. The latter reaction yields deaminated bases from adenine, cytosine, 5-methylcytosine, and guanine. The resulting mutations are predominately GC --> AT and occur as hotspots in CpG dinucleotides. Reaction with superoxide leads rapidly to peroxynitrite anion (O_2NO^-) that, when protonated, decomposes via intermediates such as reactive conformers of peroxynitrous acid. The spectrum of base changes and mutations arising from these pathways are similar to those normally associated in biological systems with the iron-catalyzed Haber-Weiss reaction.

In the past few years there has been an explosion of interest in a small highly reactive gaseous molecule that unpredictably occupies a central role in mammalian physiology. That molecule is nitric oxide ($NO\cdot$), a free radical better known for its formation in combustion processes or for its momentary occurrence in the microbial pathways of denitrification in soils and in the ocean than for its biological activity.

The endogenous synthesis of nitrate by mammals was demonstrated to occur via oxidation of trivalent nitrogen and to be greatly stimulated by infection or inflammation (1-3). Studies by Marletta and co-workers (4,5) with murine macrophages disclosed a novel enzymatic pathway that involved the oxidation of

0097–6156/94/0553–0120$08.00/0

a guanido nitrogen of L-arginine to nitric oxide. The NO· thus generated undergoes subsequent oxidations and is ultimately excreted as urinary nitrate (6). Numerous other investigations concerning problems of cardiovascular physiology, neuronal signaling, endotoxic shock, sexual function, etc. have resulted in the discovery of a large family of NO· synthases (NOS; EC 1.14.23). These NOS include both constitutive and inducible forms and may be membrane bound or cytosolic depending on cell type; the biochemistry and physiology of these enzymes have been the subject of numerous reviews (7-11). It is now known that these enzymes are a member of the cytochrome P_{450} superfamily, all of which contain an NADPH diaphorase and FMN and FAD binding sites (12). The cell-signalling capability of NO· is related to its ability to diffuse from one cell to a neighboring cell. At the same time, the propensity of NO· to rapidly react with O_2 and superoxide ($O_2 \cdot^-$) leads to potentially toxic nitrogen oxides (N_2O_3, peroxynitrite) which may either be useful (e.g., destroy invading organisms) or cause unwanted collateral damage (cytotoxicity, genotoxicity) to normal neighboring cells of the organism. Cells with cytotoxic potential release NO· on the order of 5×10^3 to 5×10^4 molecules/cell-sec (13,14). In situations where infection and/or inflammation may continue over months or longer target cells will be exposed to impressive quantities of highly reactive chemical species. The discussion that follows will describe some of these reactants and the nature of some of their more important chemical reactions.

Cytotoxic Effects of Nitric Oxide

The toxic effects of nitric oxide have been investigated in tumor cells (15-17), microbes (14,15,18-21); and lymphocytes (13,19-23). Activation of macrophages to their tumoricidal levels, by exposure to lymphokines or bacterial products, occurs in a sequence of well defined steps which lead to the release of different factors (such as hydrogen peroxide, tumor necrosis factor, interleukin-1 and NO·) known to induce several biochemical changes in their target cells leading to cell death (22-25). Although many of these agents are cytotoxic to target cells, NO· is often found to be the primary mediator of macrophage-induced cytostasis (15,25,26). Nitric oxide may also be cytotoxic for the cells that produce it or for neighboring cells (27,28). The exact mechanism or mechanisms of cell-killing by NO· remains an open question.

Analysis of the possible molecular targets of nitric oxide has shed considerable light on some mechanisms of the toxic action of NO· (Table I). Known nitric oxide targets are diverse and include Fe-S proteins, tyrosyl radical proteins, thiol proteins and other macromolecules which can be activated or inhibited by reacting with NO·. The consequences of reacting with NO· may include oxidative injury associated with iron loss and progressive and long-lasting inhibition of enzymes with iron-dependent catalytic activity that plays a major role in cytostatic and respiration-inhibiting effects.

Chemistry of DNA Damage by NO·

Although a variety of mechanisms may ultimately prove to be involved in nitric oxide-related damage to DNA, most attention to date has been focused on three types of reaction, i.e., 1.) indirect interactions with electrophiles arising from metabolism of endogenously-formed N-nitrosamines, 2.) direct nitrosation of primary amines on nucleic acid bases or 3.) attack by active species arising from reactions of nitric oxide with endogenous oxygen radicals. Each of these is addressed in one of the following sections.

Endogenous Formation of N-Nitroso Compounds. This topic has been studied extensively during the past 20 years and will be only summarized here (*29-36*). N-nitrosamines are well-known chemical carcinogens that are metabolized to strongly alkylating electrophiles that react with DNA at several nucleophilic sites; this has been demonstrated both in cell cultures and in animals and with preformed and endogenously-formed N-nitroso compounds. The metabolism of the simplest representative dialkylnitrosamine, N-nitrosodimethylamine, is outlined in Scheme 1. Most dialkyl and alkylaryl nitrosamines appear to metabolize by similar pathways although the details depend on the structure. Unsymmetrical dialkylnitrosamines, for example, may be oxidized on both alkyl groups resulting in two types of DNA alkylation. There is good evidence for the initial enzymatic oxidation and for methylation of DNA at several sites. The alkylation spectra vary from agent to agent, but the N-7 and O-6 positions of guanine and the N-3 position of adenine typically predominate; for a review see reference (*37*). The exact nature of the alkylation step itself is not known in detail, and probably varies with the structure of the nitrosamine, but it presumably involves one or more of the intermediates shown in the scheme.

Amine nitrosation in the laboratory has been done traditionally by reaction with acidic nitrite as shown in Scheme 2. The actual nitrosating species for amines is N_2O_3, which arises via the equilibria shown in Scheme 3. The overall rates for amine nitrosation reactions go through maxima near pH 3.5, reflecting increasing concentrations of nitrosating species and decreasing concentration of nitrosatable free amine as the medium becomes more acidic (*38*). It was thus initially intriguing to easily detect nitrosamine formation under the neutral or slightly basic conditions found in nitrate/nitrite-producing cell cultures.

Among the first experiments, following the observation of nitrite/nitrate production by activated macrophages, was to add an amine - morpholine - to the medium (Scheme 4). N-nitrosomorpholine was detected, and shown to be formed independently of added nitrite. This ruled out the possibility that the macrophages or something produced by them were catalyzing nitrosation by the nitrite itself, and suggested that a common intermediate was involved in nitrosation and in nitrite/nitrate production (*5,9,39*).

As detailed in the Introduction, this intermediate is nitric oxide and the resulting nitrosations are straightforward examples of chemical pathways that were elucidated in the 1970's and 1980's largely by Challis and co-workers (*40-45*). Reaction of nitric oxide with oxygen initiates a series of equilibria (Scheme 5) that produces N_2O_3, the same amine-nitrosating agent formed from acidic nitrite as shown in Scheme 3. At high concentrations of HNO_2, a ternary process can also lead to KNO_3 and NO (46), but this pathway is probably of

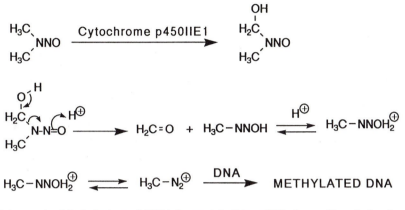

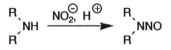

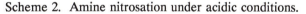

Scheme 1. Methylation of DNA by metabolites of N-nitrosodimethylamine.

Scheme 2. Amine nitrosation under acidic conditions.

Scheme 3. Equilibria arising from aqueous acidic nitrite.

Macrophages + LPS/Interferon

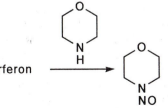

Scheme 4. Nitrosation of morpholine by activated macrophages.

minor importance under physiological conditions where the levels of the components of these equilibria are low. This series of reactions leads in principle to both N_2O_3 and N_2O_4, each of which can react with water. The reaction of water with N_2O_4 produces equimolar nitrate and nitrite while that with N_2O_3 produces only nitrite. In most cellular systems for nitric oxide production *in vitro*, nitrite levels are generally higher than nitrate levels (*4,9,47*); this is also observed when nitric oxide at comparably low concentrations (1 -10 μM) is added directly to buffer or to cell-culture media (unpublished). N_2O_3 formation under these conditions therefore appears to predominate over N_2O_4 production (also see below) and the following discussions will consequently focus primarily on N_2O_3.

N_2O_3 is an effective nitrosating agent for primary and secondary amines under non-acidic conditions. Reaction with secondary amines to form N-nitrosamines has been demonstrated in aqueous and organic solutions *in vitro* and for a number of nitric-oxide generating cell systems in culture, including bacteria (*48,49*) and macrophages (*5,39*). There is good evidence that this process occurs in animals (*50,51*) and suggestive evidence that it occurs in humans (*1,3,6,52,53*).

Nitrosation of secondary amines in the presence of appropriate enzyme systems, i.e., in cell cultures or *in vivo*, can thus result in indirect DNA damage via alkylation as summarized in Scheme 1.

Nitrosation of primary amines produces alkylating intermediates without the requirement for metabolism. If the primary amine is on a DNA base, this may lead to direct DNA damage as described in the next section.

Deamination of DNA Bases. Nitrosative deamination is a well-known consequence of the reaction of primary amines with acidic nitrite (*54*). The nitrosating species is N_2O_3 generated from nitrous acid, so this reaction will also occur under non-acidic conditions when N_2O_3 is generated in the presence of primary amines, e.g., by reaction of nitric oxide with oxygen as summarized above. The overall reaction rates are in fact higher based upon N_2O_3 at neutral or basic pH than at acidic pH because of the increased concentration of free amine under these conditions.

Nitrosation of primary amines (Scheme 6) results in rapid deamination via diazonium ions or diazohydroxides (the analogous intermediates to those arising from metabolism of dialkylamines as shown above for NDMA in Scheme 1). Deamination by this mechanism is possible in principle with most purines and pyrimidines; an exocylic amino group is the principal structural requirement. The details of this pathway are shown for adenine in Scheme 7, and the overall transformations for adenine, guanine, cytosine, and 5-methylcytosine are summarized in Figure 1.

The potential consequences of these reactions vary from nucleoside to nucleoside (summarized in Table II). Deamination of guanine to xanthine, for example, may result in abasic sites in DNA and consequent single strand breaks or misrepair. Methylation of cytosine to form 5-methylcytosine and subsequent deamination to thymine could result in a G-C --> A-T transition and then to mutagenesis. Reaction with the aryl diazonium ion of the base undergoing

Table I. Cytostatic/Cytotoxic Activity of NO ·

Molecular target	Mechanism of action	Consequences
Fe-S proteins	Enzyme inactivation aconitase, NADPH-, Succinate-ubiquinone oxidoreductase	inhibition of the Mitochondrial respiratory chain and citric cycle
Tyrosyl radical protein	Ribonucleotide reductase	Inhibition of DNA synthesis
Sulfhydryls Protein thiols	S-nitrosylation	Inactivation of SH-dependent dehydrogenases
Non-heme Fe	Ferritin	Fe release
RNA	Deamination	Inhibition of protein synthesis
DNA	Deamination	DNA damage and mutations

Table II. Mutations that Potentially arise from Deamination of DNA bases

Conversion	Type of Mutation	Ref.
5-Methylcytosine --> thymine	G·C --> A·T	(77)
Cytosine --> uracil	G·C --> A·T	(78)
Adenine --> Hypoxanthine	A·T --> G·C	(79)
Guanine --> xanthine --> apurinic site	G·C --> T·A	(80)
Adenine --> hypoxanthine --> apurinic site	G·C --> T·A	(80)

$$NO + O_2 \rightleftharpoons O_2NO\cdot$$

$$O_2NO\cdot + NO \rightleftharpoons 2\,NO_2$$

$$2\,NO_2 \rightleftharpoons N_2O_4$$

$$NO + NO_2 \rightleftharpoons N_2O_3$$

$$N_2O_3 + H_2O \rightleftharpoons 2\,H^{\oplus} + 2\,NO_2^{\ominus}$$

$$N_2O_4 + H_2O \rightleftharpoons 2\,H^{\oplus} + NO_2^{\ominus} + NO_3^{\ominus}$$

Scheme 5.　Reactions arising from nitric oxide and oxygen.

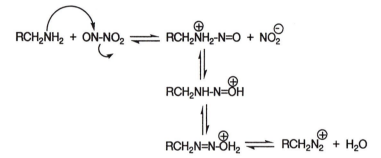

Scheme 6.　Nitrosation of primary amines.

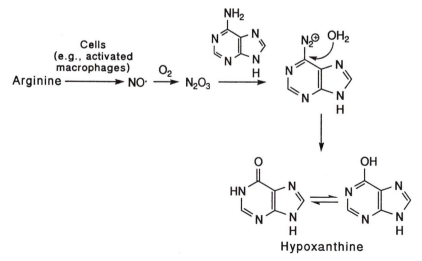

Hypoxanthine

Scheme 7.　Nitric oxide-related deamination of adenine to form hypoxanthine.

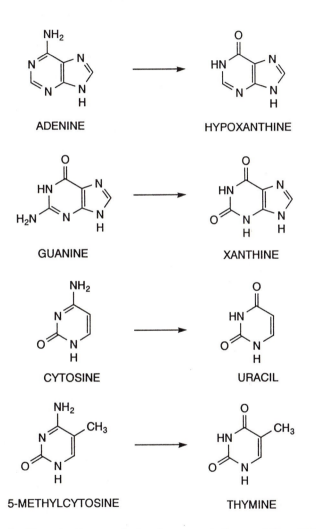

Figure 1. Deamination products of representative nucleic acid bases.

deamination with a nucleophilic site on an adjacent macromolecule could lead to crosslinking (Scheme 8) with other nucleic acids or with proteins (55-58). The strongest evidence in support of the deamination mechanism of DNA damage is probably the direct observation of deamination products following treatment of cells or bare DNA with nitric oxide in the presence of oxygen. Mutations caused by nitric oxide treatment of *S. typhimurium*, for example, appear to be associated largely with conversion of 5-methylcytosine to thymine (59). Nitric oxide-induced mutations in a human lymphoblastoid cell line (TK6) were accompanied by the formation of xanthine and hypoxanthine that presumably arose via deamination of guanine and adenine, respectively (60). Treatment of calf thymus DNA, yeast RNA, and bovine liver transfer RNA with nitric oxide *in vitro* also resulted in substantial yields of xanthine and hypoxanthine; the yields observed from the nucleic acids were higher than those observed directly from guanine and adenine (60). Thus, deamination of nucleic acid bases clearly results from treatment of nucleic acid bases, nucleic acids themselves, or cells in culture, with nitric oxide and this deamination appears to be directly related to mutagenesis and cytotoxicity.

Oxygen Radicals Related to Nitric Oxide. The deamination reactions summarized above are based firmly on well-known nitrosation chemistry and are directly although perhaps not definitively supported by the chemistry and related biochemistry observed in several model systems. DNA damage by oxygen radicals that may be formed from nitric oxide has been more difficult to establish in detail. This is due partly to the difficulty of distinguishing possible oxidative damage from the more certain deaminative process and partly to the complexity of the radical-generating cascade of reactions that can potentially participate.

Although the specific mechanisms of nitric oxide-related radical reactions with DNA may be incompletely understood, there is good evidence that such reactions occur. For example, tissue damage that accompanies several nitric oxide-related processes, e.g., reperfusion of ischemic or hypoxic brain (61), can be attenuated by superoxide dismutase, indicating that the superoxide anion radical is involved. Superoxide, however, appears to be too unreactive to account directly for this tissue damage, and secondary products, e.g., hydroxyl radical arising from superoxide via Fenton- or Haber-Weiss-like chemistry (i.e., reactions of iron with H_2O_2; 62-64), have been suggested as the active species (65-67; Scheme 9). While this pathway cannot be completely ruled out, Beckman and coworkers argue cogently that it is probably of minor significance *in vivo*. They note that the Fenton pathway requires generation of hydrogen peroxide from two molecules of superoxide, reduction of ferric iron to ferrous iron by a third superoxide, and, finally, reaction of ferrous ion with hydrogen peroxide to generate hydroxide ion, ferric iron, and hydroxyl radical. In addition to the complexity of this mechanism, the reactant concentrations are low *in vivo*, the ferric-ferrous reduction by superoxide is slow, and ferric ion can be depleted by rapid reaction with other reductants, e.g., ascorbic acid (68). In a similar context, Wink and coworkers present evidence for oxidants other than hydroxyl radical during the reaction of nitrosamines with the Fenton reagent (69). The reactions themselves are nonetheless consistent with those expected from hydroxyl radicals, and there is increasing evidence that the crucial process is generation of hydroxyl radical-like compounds (*70*) by decomposition of peroxynitrite formed by the direct reaction of superoxide and nitric oxide (*68,70-72*; Scheme 10).

Nitric oxide and superoxide are generated simultaneously by many types of cells including Kupffer cells (*73-75*), neutrophils (*76*), endothelial cells (*77*) and macrophages (*71*); peroxynitrite has been measured directly in the macrophage system and appears to account for a substantial percentage of the nitric oxide (*71*), although the accuracy of the analytical method has been questioned (*78*).

Peroxynitrite, once formed, can undergo a number of possible reactions at different pH (see reference (*72*) for a recent review of peroxynitrite chemistry). Peroxynitrite, with a pK_a of 6.8, is protonated in acidic solution to form peroxynitrous acid that then decays rapidly to predominately nitrate. The mechanism has not been completely elucidated but appears to involve a combination of isomerization and radical formation-recombination (*72*; Scheme 10). In base, nitrite is the major product, along with molecular oxygen, with reported yields between >75% (*79*) and essentially 100% (*72*). The reaction in base is slow unless metal ions, e.g., cupric ion, are present (*72,80*).

Peroxynitrite is bacteriocidal to *E. coli* in a dose-dependent manner with an LD_{50} of 250 μM at pH 7.4 (*74*). This effect increases with increasing pH, presumably because of slower decomposition of peroxynitrite (*81*). Chelators such as diethylenetriamine pentaacetic acid, ethylene diamine tetracetate, and desferrioxamine slightly enhance the activity as does the hydroxyl scavenger dimethylsulfoxide; other hydroxyl scavengers, i.e., mannitol, benzoate, and ethanol have no effect. NO_2 production increases along with the bacteriocidal activity and may therefore contribute to this effect (*82*).

There is thus currently some controversy concerning whether peroxynitrite releases hydroxyl radical *per se* or gives rise to an unstable reactive intermediate, e.g., a conformational or geometric isomer of the peroxynitrite anion (*68,78,82,83*), that yields products consistent with those expected from direct reactions with the hydroxyl radical and/or with substances, e.g., NO_2, that are co-generated with hydroxyl radical by homolysis of peroxynitrous acid. The potential subtleties of the mechanism, however, should not obscure the biological reality, i.e., that genotoxic damage is associated with the concurrent generation of nitric oxide and superoxide.

Significance

The collection of changes that might be induced in DNA by the mechanisms described above could lead to several types of mutations which are known to be important in human cancer genes. A great deal of attention has been paid recently to the p53 tumor suppressor gene which is of predominant importance for a number of human cancers including colon, liver, breast and lung (*84*). In many cases GC --> AT transitions are found as hot-spots in the mutation spectrum, and these could arise via methylation at C followed by deamination to T induced by NO•.

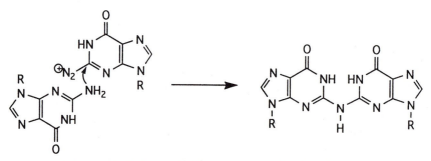

Scheme 8. G-G crosslinking via nitrosative deamination.

$$2O_2^- + 2H^+ \longrightarrow H_2O_2 + O_2$$

$$Fe^{3+} + O_2^- \longrightarrow Fe^{2+} + O_2$$

$$Fe^{2+} + H_2O_2 \longrightarrow Fe^{3+} + HO^- + HO\cdot$$

Scheme 9. Generation of hydroxyl radicals via the Fenton reaction.

$$O_2^- + NO\cdot \xrightleftharpoons{\text{Fast}} ONO_2^- \xrightleftharpoons{H^+} ONO_2H \longrightarrow \left[\begin{array}{l}\text{Hydroxylating}\\ \text{Species}\end{array}\right]$$

$$H^+ + NO_3^-$$

Scheme 10. Reaction of superoxide with nitric oxide.

In tissue undergoing an inflammatory reaction, both the infiltrating and resident cell populations produce a time-dependent mixture of nitrogen oxide radicals and oxygen radicals. These different radical species interact forming new reactive intermediates which may contribute to DNA damage. Although the flux of radicals per unit time is low, an inflammatory condition that continues for years becomes a significant risk factor for carcinogenic cell transformation. The relative contributions of various radical species to the carcinogenic process must be assessed through both chemical analysis of DNA and through genetic and molecular biological analysis of mutations in surviving exposed cells.

In summary, the chemistry of DNA damage by nitric oxide, although potentially complex with respect to the overall mechanisms and the actual alterations of the nucleic acid, might arise from only two fundamental processes, i.e., reaction with oxygen to form nitrosating species and reaction with superoxide to form peroxynitrite. This overview is depicted below (Scheme 11) and formulates a testable hypothesis for inflammation-induced DNA damage, i.e., the proposed pathways can be intercepted at several points by substances that can inhibit formation of or compete for the several reactive species. For example, an important test of this hypothesis would be to see whether or not the undesirable effects of nitric oxide could be reduced by compounds such as antioxidants.

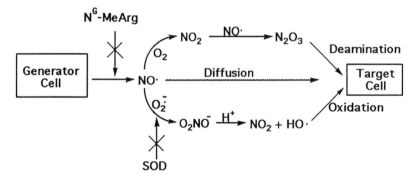

Scheme 11. Summary of pathways for DNA damage via nitric oxide.

Acknowledgments. This work was supported by NIEHS Grants CA26731 and ES02109. The contents of this manuscript are based solely on the opinions of the authors and do not necessarily represent the official views of the National Cancer Institute.

Literature Cited

1.   Green, L.C., Wagner, D.A., Ruiz de Luzuriaga, K., Istfan, N., Young, V.R. and Tannenbaum, S.R. *Proc. Nat'l. Acad. Sci. USA* **1981**, *78*, 7764-7768.
2.   Wagner, D.A., Schultz, D.S., Deen, W.M., Young, V.R. and Tannenbaum, S.R. *Cancer Res.* **1983**, *43*, 1921-1925.
3.   Wagner, D.A., Young, V.R. and Tannenbaum, S.R. *Proc. Nat'l. Acad. Sci. USA* **1983**, *80*, 4518-4521.
4.   Stuehr, D.J. and Marletta, M.A. *Proc. Nat'l. Acad. Sci. USA* **1985**, *82*, 7738-7742.
5.   Iyengar, R., Stuehr, D.J. and Marletta, M.A. *Proc. Nat'l. Acad. Sci. USA* **1987**, *84*, 6369-6373.
6.   Leaf, C.D., Wishnok, J.S. and Tannenbaum, S.R. *Biochem. Biophys. Res. Comm.* **1989**, *163*, 1032-1037.
7.   Moncada, S., Palmer, R.M.J. and Higgs, E.A. *Pharm. Rev.* **1991**, *43*, 109-142.
8.   Lancaster, J.R.,Jr. *Amer. Scient.* **1992**, *80*, 248-259.
9.   Marletta, M.A. *Chem. Res. Toxicol.* **1989**, *1*, 249-257.
10. Snyder, S.H. and Bredt, D.S. *Scient. Amer.* **1992**, *266*, 68-77.
11. Ignarro, L.J. *Pharmacol. Toxicol.* **1990**, *67*, 1-7.
12. White, K.A. and Marletta, M.A. *Biochem.* **1992**, *31*, 6627-6631.
13. Marletta, M.A., Yoon, P.S., Iyengar, R., Leaf, C.D. and Wishnok, J.S. *Biochem.* **1988**, *27*, 8706-8711.
14. Hibbs, J.B., Taintor, R.R., Vavrin, Z. and Rachlin, E.M. *Biochem. Biophys. Res. Comm.* **1988**, *157*, 87-94.
15. Stuehr, D.J. and Nathan, C.F. *J. Exp. Med.* **1989**, *169*, 1543-1556.
16. Keller, R., Geiges, M. and Keist, R. *Cancer Res.* **1990**, *50*, 1421-1425.
17. Nathan, C.F. and Hibbs, J.B.,Jr. *Current Opin. Microbiol.* **1991**, *3*, 65-70.
18. Adams, L.B., Hibbs, J.B.,Jr., Taintor, R.B. and Krahenbuhl, J.L. *J. Immunol.* **1990**, *144*, 2725-2729.
19. Nathan, C.F. *Ann. Inst. Pasteur Immunol.* **1986**, *137C*, 345-351.
20. Hibbs, J.B.,Jr., Taintor, R.B. and Vavrin, Z. *Science* **1987**, *235*, 473-476.
21. Nathan, C. *FASEB Journal* **1992**, *6*, 3051-3064.
22. Drysdale, B.E., Zacharchuk, C.M. and Shin, H.S. *Prog. Allergy* **1988**, *40*, 111-161.
23. Nathan, C.F. *Fed. Proc.* **1982**, *41*, 2206-2211.
24. Urban, J.L., Shepard, H.M., Rothstein, J.L., Sugarman, B.J. and Schreiber, H. *Proc. Nat'l. Acad. Sci. USA* **1986**, *83*, 5233.
25. Onozaki, K., Matsushima, K., Kleinerman, E.S., Saito, T. and Oppenheim, J.J. *J. Immunol.* **1985**, *135*, 314.
26. Hibbs, J.B.,Jr., Taintor, R.B., Vavrin, Z. and Rachlin, E.M. *Biochem. Biophys. Res. Comm.* **1989**, *157*, 87-94.
27. O'Connor, K.J. and Moncada, S. *Biochim. Biophys. Acta* **1991**, *1097*, 227-231.

28. Palmer, R.M.J., Bridge, L., Foxwell, N.A. and Moncada, S. *Brit. J. Pharmacol.* 1992, *105*, 11-12.
29. Lijinsky, W., Keefer, L., Conrad, E. and Van de Bogart, R. *J. Nat'l. Cancer Inst.* 1972, *49*, 1239-1249.
30. Pignatelli, B., Scriban, R., Descotes, G. and Bartsch, H. *Carcinogenesis* 1983, *4*, 4914-494.
31. Spiegelhalder, B. and Preussmann, R. *Carcinogenesis* 1985, *6*, 545-548.
32. Ohshima, H. and Bartsch, H. *Cancer Res.* 1981, *41*, 3658-3666.
33. Tahira, T., Tsuda, M., Wakabayashi, K., Nagao, M. and Sugimura, T. *Gann* 1984, *75*, 889-894.
34. Bartsch, H., Ohshima, H. and Pignatelli, B. *Mutat. Res.* 1988, *202*, 307-324.
35. Tannenbaum, S.R., Moran, D., Falchuk, K.R., Correa, P. and Cuello, C. *Cancer Lett.* 1981, *14*, 131-136.
36. Yang, D., Tannenbaum, S.R., Büchi, G. and Lee, G.C.M. *Carcinogenesis* 1984, *5*, 1219-1224.
37. Pegg, A.E. *Adv. Cancer Res.* 1977, *25*, 195-267.
38. Mirvish, S.S. *Toxicol. App. Pharmacol.* 1975, *31*, 325-351.
39. Miwa, M., Stuehr, D.J., Marletta, M.A., Wishnok, J.S. and Tannenbaum, S.R. *Carcinogenesis* 1987, *8*, 955-958.
40. Challis, B.C. and Outram, J.R. *J. Chem. Soc. Chem. Comm.* 1978, 707-708.
41. Challis, B.C. and Kyrtopoulos, S.A. *Brit. J. Cancer* 1977, *35*, 693-696.
42. Challis, B.C.; Shuker, D.E.G.; Fine, D.H.; Goff, E.U. and Hoffman, G.A. In *N-nitroso Compounds: Occurrence and Biological Effects (IARC 41)*; Bartsch, H.; O'Neill, I.K.; Castegnaro, M. and Okada, M.; Eds.; IARC: Lyon, 1981, pp 11-20.
43. Challis, B.C. and Kyrtopoulos, S.A. *J. Chem. Soc. Perkin 1.* 1979, 299-304.
44. Challis, B.C. and Kyrtopoulos, S.A. *J. Chem. Soc. Chem. Comm.* 1976, *21*, 877-878.
45. Challis, B.C. and Kyrtopoulos, S.A. *J. Chem. Soc. Perkin 2.* 1978, *12*, 1296-1302.
46. Woods, L.F.J., Wood, M. and Gibbs, P.A. *J. Gen. Microbiol.* 1981, *125*, 399-406.
47. Stuehr, D.J. and Marletta, M.A. *Cancer Res.* 1987, *47*, 5590-5594.
48. Ralt, D., Wishnok, J.S., Fitts, R. and Tannenbaum, S.R. *J. Bacteriol.* 1988, *170*, 359-364.
49. Calmels, S., Ohshima, H., Rosenkranz, H., McCoy, E. and Bartsch, H. *Carcinogenesis* 1987, *8*, 1085-1087.
50. Liu, R.H., Baldwin, B., Tennant, B.C. and Hotchkiss, J.H. *Cancer Res.* 1991, *51*, 3925-3929.
51. Ehrenberg, L., Hiesche, K.D., Osterman-Golkar, S. and Wennberg, I. *Mutat. Res.* 1974, *24*, 83-103.
52. Leaf, C.D., Vecchio, A.J. and Hotchkiss, J.H. *Carcinogenesis* 1987, *8*, 791-795.
53. Leaf, C.D., Wishnok, J.S., Hurley, J.P., Rosenblad, W.D., Fox, J.G. and Tannenbaum, S.R. *Carcinogenesis* 1990, *11*, 855-858.

54.   Obiedzinski, M.W., Wishnok, J.S. and Tannenbaum, S.R. *Fd. Cosmet. Toxicol.* **1980**, *18*, 585-589.
55.   Geidushek, E.P. *Proc. Nat'l. Acad. Sci. USA* **1961**, *47*, 950-955.
56.   Kirchner, J.J. and Hopkins, P.B. *J. Amer. Chem. Soc.* **1991**, *113*, 4681-4682. 57. Shapiro, R., Dubelman, S., Feinberg, A.M., Crain, P.F. and McCloskey, J.A. *J. Amer. Chem. Soc.* **1977**, *99*, 302-303.
58.   Kirchner, J.J., Sigurdsson, S.T. and Hopkins, P.B. *J. Amer. Chem. Soc.* **1992**, *114*, 4021-4027.
59.   Wink, D.A., Kasprzak, K.S., Maragos, C.M., Elespuru, R.K., Misra, M., Dunams, T.M., Cebula, T.A., Koch, W.H., Andrews, A.W., Allen, J.S. and Keefer, L.K. *Science* **1991**, *254*, 1001-1003.
60.   Nguyen, T., Brunson, D., Crespi, C.L., Penman, B.W., Wishnok, J.S. and Tannenbaum, S.R. *Proc. Nat'l. Acad. Sci. USA* **1992**, *89*, 3030-3024.
61.   Beckman, J.S. *J. Devel. Physiol.* **1991**, *15*, 53-59.
62.   Fenton, H.J.H. *J. Chem. Soc.* **1894**, *65*, 899-910.
63.   Haber, F. and Weiss, J. *Naturwissenschaften* **1932**, *20*, 948-950.
64.   Haber, F. and Weiss, J. *Proc. Royal Soc. London* **1934**, *A147*, 332-351.
65.   Imlay, J.A., Chin, S.M. and Linn, S. *Science* **1988**, *240*, 640-642.
66.   Rush, J.D. and Koppenol, W.H. *J. Amer. Chem. Soc.* **1988**, *110*, 4957-4963. 67. Winterbourne, C.C. and Sutton, H.C. *Arch. Biochem. Biophys.* **1986**, *244*, 27-34.
68.   Beckman, J.S., Beckman, T.W., Chen, J., Marshall, P.A. and Freeman, B.A. *Proc. Nat'l. Acad. Sci. USA* **1990**, *87*, 1620-1624.
69.   Wink, D.A., Nims, R.W., Desrosiers, M.FG., Ford, P.C. and Keefer, L.K. *Chem. Res. Toxicol.* **1991**, *4*, 510-512.
70.   Koppenol, W.H., Pryor, W.A., Moreno, J.J., Ischiropoulos, H. and Beckman, J.S. *Chem. Res. Toxicol.* **1992**, *6*, 834-842.
71.   Ischiropoulos, H., Zhu, L. and Beckman, J.S. *Arch. Biochem. and Biophys.* **1992**, *298*, 446-451.
72.   Plumb, R.C. and Edwards, J.O. *J. Phys. Chem.* **1992**, *96*, 3245-3247.
73.   Wang, J.-F., Komarov, P., Sies, H. and De Groot, H. *Hepatol.* **1992**, *15*, 1112-1116.
74.   Dieter, P., Shultze-Specking, A. and Decker, K. *Europ. J. of Biochem.* **1988**, *177*, 61-67.
75.   Billiar, T.R., Curran, R.D., Ferrari, F.K., Williams, D.L. and Simmons, R.L. *J. Surg. Res.* **1990**, *48*, 349-353.
76.   McCall, T.B., Boughton-Smith, N.K., Palmer, R.M.J., Whittle, B.J.R. and Moncada, S. *Biochem. J.* **1989**, *261*, 293-296.
77.   Gryglewski, R.J., Palmer, R.M.J. and Moncada, S. *NATURE* **1986**, *320*, 454-456.
78.   Yang, G., Candy, T.E.G., Boaro, M., Wilkin, H.E., Jones, P., Nazhat, N.B., Saadalla-Nazhat, R.A. and Blake, D.R. *Free Rad. Biol. Med.* **1992**, *12*, 327-330.
79.   Hughes, M.N., Nicklin, H.G. and Sackrule, W.A.C.   *J. Chem. Soc. A* **1971**, 3722-3725.

80. Plumb, R.C., Edwards, J.O. and Herman, M.A. *J. Phys. Chem.* **1993**, (In Press)
81. Edwards, J.O. and Plumb, R.C. *J. Phys. Chem.* **1993**, (In Press)
82. Zhu, L., Gunn, C. and Beckman, J.S. *Arch. Biochem. Biophys.* **1992**, *298*, 452-457.
83. Beckman, J.S., Ischiropoulos, H., Zhu, L., Van der Woerd, M., Smith, C., Chen, J., Harrison, J., Martin, J.C. and Tsai, M. *Arch. Biochem. Biophys.* **1992**, *298*, 438-445.
84. Hollstein, M., Sidransky, D., Vogelstein, B. and Harris, C.C. *Science* **1991**, *253*, 49-53.

RECEIVED July 6, 1993

Chapter 11

Chemistry of the "NONOates"

Unusual N-Nitroso Compounds Formed by Reacting Nitric Oxide with Nucleophiles

Larry K. Keefer[1], Danae Christodoulou[1], Tambra M. Dunams[1], Joseph A. Hrabie[2], Chris M. Maragos[1], Joseph E. Saavedra[1], and David A. Wink[1]

[1]Chemistry Section, Laboratory of Comparative Carcinogenesis and [2]CSAL, PRI/DynCorp, Frederick Cancer Research and Development Center, National Cancer Institute, Frederick, MD 21702

Electrophilic attack by nitric oxide (NO) on certain nucleophiles (X⁻) produces isolable adducts of structure X-[N(O)NO]⁻ that have proven useful for the controlled biological delivery of NO, a newly-discovered bioregulatory agent. In this paper, selected contributions from the previous literature describing compounds containing the [N(O)NO]⁻ functional group are reviewed. Methods of synthesis are summarized, as are aspects of the compounds' physical properties and chemical reactivities. The results thus far suggest that such compounds should be useful probes for studies of NO's many bioeffector roles and may also provide a basis for drug design strategies.

Nitric oxide (NO) reacts with a surprising variety of nucleophiles to produce a class of unusual N-nitroso compounds containing the anionic structural element, X-[N(O)NO]⁻, where X is a nucleophile residue (equation 1) (1-5). The present essay will provide an overview of this chemistry: a brief history of nitric oxide's electrophilic reactions, including the origins of our interest in the area; how such compounds are synthesized; and a review of their structural characteristics, spectral properties, reactivity and biological applications.

$$X^- + 2\ NO \rightarrow X\text{-}[N(O)NO]^- \tag{1}$$

Amine/NO Adducts as Precursors for Carcinogenic Nitrosamines

We first became interested in the reactions of nucleophile/NO adducts in the 1970's when we noted an earlier series of papers by Russell Drago's research group describing the reaction of amines with nitric oxide. Of particular interest to us was the report that the adducts formed were prone to decomposition in air to form nitrosamines.

0097–6156/94/0553–0136$08.00/0

For example, the authors reported that bubbling nitric oxide through a cold ether solution of diethylamine produced a white powder (equation 2) (2) that, on standing in a watch glass overnight, was converted to a yellow liquid identified as diethylnitrosamine (equation 3) (5).

$$2\ Et_2NH\ +\ 2\ NO\ \rightarrow\ Et_2NH_2^+\ Et_2N\text{-}[N(O)NO]^- \tag{2}$$

$$Et_2NH_2^+\ Et_2N\text{-}[N(O)NO]^-\ \xrightarrow{[O]}\ Et_2N\text{-}NO \tag{3}$$

The authors speculated that the mechanism involved dissociation of the intermediate complex to amine and nitric oxide (equation 4), reaction of the NO with air to form NO_x (equation 5), and recombination of the NO_x (*i.e.* N_2O_3, N_2O_4, or other nitrosating agents produced in the autoxidation of NO) with the amine to form the nitrosamine (equation 6).

$$Et_2NH_2^+\ Et_2N\text{-}[N(O)NO]^-\ \rightarrow\ 2\ Et_2NH\ +\ 2\ NO \tag{4}$$

$$NO\ \xrightarrow{[O]}\ NO_x \tag{5}$$

$$Et_2NH\ +\ NO_x\ \rightarrow\ Et_2N\text{-}NO \tag{6}$$

We decided to reinvestigate this transformation as part of our continuing and intensive attempt to catalogue reactions by which carcinogenic N-nitroso compounds might form in the human environment. In particular, we suspected that a previously uncharacterized mechanism of nitrosamine formation might be operative in this case. Thus, direct attack of O_2 on the anionic $[N(O)NO]^-$ functional group might remove the elements of nitroxyl ion (NO⁻), forming the nitrosamine directly (equation 7). We reasoned that if this were the case, decomposition of the diethylamine/NO adduct in the presence of a different secondary amine would produce diethylnitrosamine as the only observable nitrosamine product. If, on the other hand, the mechanism postulated by the Drago group was correct (equations 4-6), the intermediate NO_x should be able to capture either amine, leading to a mixture of nitrosamine products.

$$Et_2N\text{-}[N(O)NO]^-\ +\ O_2\ \xrightarrow{?}\ Et_2N\text{-}NO\ +\ NO_3^- \tag{7}$$

We repeated the Drago/Paulik reaction (2), allowing the diethylamine/nitric oxide adduct to decompose in deuterium oxide solution containing excess pyrrolidine. Both diethylnitrosamine and N-nitrosopyrrolidine were produced in the ensuing decomposition reaction. The existence of extensive cross-over products in this reaction mixture demonstrated that the dissociative reaction mechanism proposed by Drago's group must be operative. Since NO_x was already known to everyone in the nitrosamine field as an excellent nitrosating agent, the mechanism of the reaction could not be considered very novel, so we published a short note on our observations (6) and shelved our work with the "Drago complexes." The mechanism of nitrosamine formation in this case, then, is as shown in equations 4-6.

Discovery of NO's Multifaceted Bioeffector Roles

For the same reasons that this chemistry was relatively uninteresting in 1982, however, it became extremely interesting a few years later. The turning point came when several research teams conclusively identified nitric oxide as the mediator of a variety of critical bioeffector roles. Initially, two regulatory functions were described: its production by cells of the immune system as an agent for arresting the growth of invading pathogens and tumor cells (*7,8*); and its identification as the "endothelium-derived relaxing factor", an agent produced by the layer of cells lining the blood vessels (*9,10*). The latter tissue is called the endothelium. It is responsible for signaling the underlying vascular smooth muscle to relax by bombarding it with NO, thereby dilating the vessels and causing the blood pressure to fall.

Since these discoveries in the mid-1980s, an astonishing variety of bioregulatory functions has been assigned to nitric oxide (*11,12*). It has been identified as a neurotransmitter, possibly the main one responsible for memory fixation (*13*). It has been shown to inhibit the aggregation and adhesion of platelets in the blood (*9,10*). It is produced naturally in male genitalia, causing the corpus cavernosum to relax and triggering penile erection (*14*). It has been implicated in bronchodilation (*15*), perception of pain (*16*), and regulation of gastrointestinal motility (*17*).

At first, we were very skeptical of these reports. We knew from two decades of research on carcinogenic N-nitroso compounds that nitric oxide is an abundant air pollutant (*18*), a harmful cigarette smoke constituent (*18*), and, as pointed out in the previous section, an important intermediate in nitrosamine formation (*19,20*). It thus seemed difficult to believe that anything so harmful and toxic could at the same time be such an important and widely distributed bioeffector species. Nevertheless, these reports quickly went from provocative when they first appeared to persuasive and compelling in subsequent years.

Of particular note was the recognition that drugs capable of releasing nitric oxide under controlled physiological conditions could be extremely useful for treating diseases arising from deficiencies of naturally produced nitric oxide in the body, including blood clotting disorders, respiratory distress, hypertension, and male impotence. With this incentive in mind, we decided to take a closer look at the diethylamine/NO adduct and related X-[N(O)NO]⁻ compounds to see if they might serve as useful sources of nitric oxide for pharmacological purposes.

A Literature Survey - NO as an Electrophile

Before beginning the laboratory portion of this investigation, it seemed prudent to learn what we could from the previous literature about the reactions of NO with nucleophiles. Our library search quickly turned up two noteworthy features. First of all, such reactions have been known for a very long time, the earliest example we found having been published nearly two centuries ago. Sir Humphrey Davy reported that in December of 1799 he had observed a reaction between "gaz nitreux" (presumably rich in NO) and aqueous potassium sulfite in the presence of a large excess of base (*21*). This reaction has since been widely studied by many groups, and it is now known to produce ⁻O₃S-[N(O)NO]⁻ (*1*). The structure of this ion was confirmed in 1948 by X-ray crystallography (*22*). Thus it is a bona fide example of a nucleophile/NO adduct, *i.e.* a compound of structure X-[N(O)NO]⁻, where X in this case is derived from sulfite ion.

The second noteworthy point to arise from our literature search was that, despite having been discovered a very long time ago, the reactions of nitric oxide as an electrophile remain rather obscure. In fact, we have been able to locate only one review on the subject. In 1962, Russell Drago provided an excellent summary of the earlier literature, including not only the sulfite reaction, but also those of amines (as in equation 2 above), phosphines, thiols, oximes, alkoxides (the Traube reaction), and carbon nucleophiles, including nitriles, nitro compounds, and carbonyl compounds containing hydrogen atoms α to the functional group (*1*).

Thus it is apparent that the reactivity of nitric oxide is not limited to metal complexation and oxidation, but also includes a rich chemistry as an electrophile. The fact that the X-[N(O)NO]⁻ ions thus produced are not nearly as extensively investigated as the metal nitrosyls and NO oxidation products led us to conclude that a consideration of these compounds' possible biological applications should be accompanied by a fuller exploration of their fundamental chemistry.

Chemistry of the X-[N(O)NO]⁻ Anion

Synthesis. Most of the nucleophile/NO adducts known to date have been prepared by the simple means of exposing nonhydroxylic solutions of the nucleophile to nitric oxide, either by bubbling NO through the cold solution or by placing five atmospheres thereof over the solution in a Parr bottle (*1*). We have used this method extensively in our own survey of the basic chemistry of compounds in this series. For example, we reacted a primary amine, a secondary amine, and a polyamine with nitric oxide to obtain three of the agents used in our initial examination of the chemical and pharmacological properties of these compounds (*23*). Two of the structurally diverse compounds we wished to examine, however, could not be prepared by this method.

One of these was Humphrey Davy's sulfite adduct. In this case, the anionic nucleophile is in the form of a salt that is insoluble in non-hydroxylic medium. Thus, the reaction could only be performed in water. Unlike most nucleophiles, sulfite reacted nicely with nitric oxide in aqueous solution (*24*). In fact, the reaction is so facile that, in the days before chemiluminescence detection of nitric oxide, it was used to quantify low concentrations of NO in polluted air; the recommended method involved exposing a certain volume of air to a highly alkaline solution of sulfite ion and measuring the ⁻O₃S[N(O)NO]⁻ stoichiometrically formed as a result (*25*).

One other such compound that cannot be made by the usual procedures is Na₂N₂O₃, commonly known as Angeli's salt (*26*). Instead of preparing it by reacting nitric oxide with the nucleophile, hydroxide ion, this agent is usually synthesized by nitration of hydroxylamine (*27*) as in equation 8.

$$H_2NOH \ + \ RONO_2 \ + \ 2\,NaOR \ \rightarrow \ Na_2N_2O_3 \ + \ 3\,ROH \qquad (8)$$

We thus prepared the five compounds needed for our basic chemistry investigation. These are shown in Figure 1, which also introduces the acronyms by which these compounds will be known later in this paper.

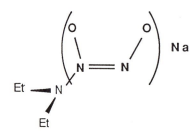

DEA/NO

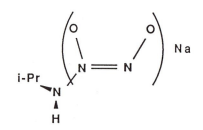

IPA/NO

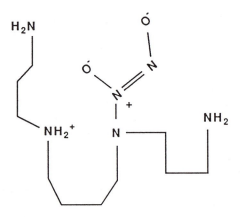

SPER/NO

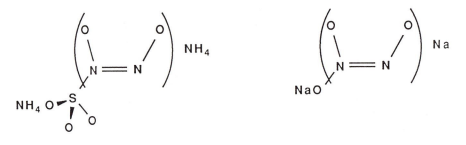

SULFI/NO

OXI/NO

Figure 1. Structures of five prototypical NONOates.

Structure and Spectral Characteristics. All X-[N(O)NO]⁻ compounds that we have been able to find in the literature whose structures have been characterized (*22,28,29*) appear to have the same arrangement of atoms in the [N(O)NO]⁻ functional group. The two nitrogens are always doubly bound, with the oxygens attached thereto lying in a cis arrangement.

In the electronic spectra, X-[N(O)NO]⁻ species (*23,29-31*) are like the dialkylnitrosamines (*32*) in showing a strong π→π* transition in the 230-250 nm region. The absorptivity ranges from 5000 to 9000 M^{-1} cm^{-1}. Unlike the more traditional N-nitroso compounds (*32*), however, n→π* transitions are not found in the anions' spectra (*23,29-31*).

The nuclear magnetic resonance spectra also show interesting contrasts with those of the dialkylnitrosamines. For example, there is no evidence for restricted rotation about the R_2N-N bond of an X-[N(O)NO]⁻ ion when X is a secondary amino group. Thus, the NMR spectra of the Et_2N-[N(O)NO]⁻ ion in chloroform-*d* showed no hint of magnetic nonequivalence between the ethyl groups even down to -50 °C (*6*). This finding is consistent with the ultraviolet spectral data in that both lines of reasoning point to an absence of significant resonance interaction between the substituent and the [N(O)NO]⁻ group. Also consistent with this conclusion is the single bond character of the Et_2N-N linkage found by X-ray crystallography.

The structural and spectral similarities among compounds containing the [N(O)NO]⁻ functional group, including all those shown in Figure 1, lead us to refer to them collectively as the "NONOates".

Decomposition to NO and/or N_2O

When exposed to acid, the NONOates display a very interesting variety of behaviors. When the NONOate group is attached directly to a carbon atom, the compounds appear to be stable under protonating conditions (*29*). When X is a heteroatom, however, the compounds are generally seen to undergo acid catalyzed decomposition (*23,31,33-36*). An example of this behavior is shown in Figure 2, which depicts the spectrum of DEA/NO as a function of time following its dissolution in aqueous buffer at pH 7.4. Under these conditions, the intense absorbance at 252 nm is shown to decrease with time. Semilog plots of the resulting data reveal cleanly first-order kinetics for the decomposition (*23*). This behavior, including clear isosbestic points, is seen in the decompositions of all NONOates we have examined to date.

But what are the products of this decomposition? Since confirmation that the reactions can produce nitric oxide, as suggested in the nitrosamine formation studies described above, was critical to the use of these agents in pharmacological applications, we followed the reaction course for the various NONOates shown in Figure 1 by the most selective method for nitric oxide analysis available today, chemiluminescence detection. A typical example is shown in Figure 3. In this case, OXI/NO was added to a reactor vessel at a high pH already shown to support little if any decomposition of this species. The reactor vessel containing the OXI/NO solution was purged for several minutes with helium to remove any remaining oxygen. At 6 min after introduction of OXI/NO into the reactor system, a large excess of pH 7.4 phosphate buffer was added to begin the decomposition at 37 °C. It is clear that nitric oxide was produced in significant quantities during this reaction. Qualitatively similar results are seen with the other NONOates of Figure 1 except for

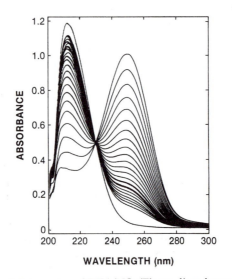

Figure 2. Ultraviolet spectra of DEA/NO (Figure 1) at intervals of 3 min (and at infinite time) after dissolution in pH 7.4 phosphate buffer (0.1 M) at room temperature.

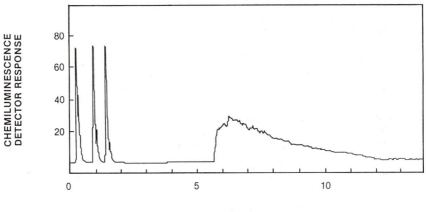

Figure 3. Chemiluminescence detector response to nitric oxide produced as a function of time after dissolution of OXI/NO (Figure 1) at 10 μM in pH 7.4 phosphate buffer at 37 °C. The three peaks at 0 - 2 min are replicates of the 0.361 nmol nitric oxide quantitative reference standard. OXI/NO was added at 5.5 min. (Reproduced courtesy of *J. Med. Chem.* from ref. 23.)

SULFI/NO, which has long since been shown to produce sulf*ate* and nitr*ous* oxide on decomposition (*37*) rather than reverting to the sulf*ite* and nitr*ic* oxide from which it was formed.

Values approaching the full theoretical two moles of nitric oxide per mole of NONOate decomposed have been seen in these reactions, especially among secondary amine adducts.

The OXI/NO reaction is particularly interesting also in that it produces both NO and N_2O, the ratio varying with conditions. We are currently probing the mechanism of this phenomenon in related decompositions.

Other reactivities of this interesting functional group will be described in due course, including the O-functionalization reactions akin to esterification of carboxylic acids (*38*) and the coordination of NONOates to metal centers, reactions described by Saavedra *et al.* and Christodoulou *et al.*, respectively.

Biomedical Applications of the NONOates

Since, except for SULFI/NO, these compounds release nitric oxide in a spontaneous first-order reaction simply on dissolution in buffers of physiological pH (*23*), it might be anticipated that they would be very useful in the pharmacological applications we had postulated for them. As predicted, the NONOates have displayed interesting biological properties in a number of investigations. For example, their ability to release nitric oxide, putatively the endothelium-derived relaxing factor, suggests that they should be able to dilate blood vessels. They did so when dissolved in the buffer bathing pieces of rabbit aorta, with a potency that was directly proportional to the amount of nitric oxide known from the physicochemical data to be released during the time course of the experiment (*23*). They have also shown cytostatic activity in human cancer cells in culture (*39*) as well as the ability to inhibit platelet aggregation (*40*), again more or less with the potency and time course anticipated from the data on their rate and extent of NO generation in solution.

One other type of biological activity is noteworthy from the point of view of both nitrosation chemistry and NONOate pharmacology. We had predicted (*41*) that the exocyclic NH_2 groups in DNA base residues might react with the NO_x produced during the aerobic dissociation of the Et_2N-$[N(O)NO]^-$ ion (equations 4 and 5), ultimately diazotizing and deaminating these bases as indicated in equations 9-12.

$$2\text{'-Deoxycytidine} + NO \xrightarrow{-e^-} 2\text{'-Deoxyuridine} + N_2 + H^+ \qquad (9)$$

$$5\text{-Methyl-2'-deoxycytidine} + NO \xrightarrow{-e^-} \text{Thymidine} + N_2 + H^+ \qquad (10)$$

$$2\text{'-Deoxyguanosine} + NO \xrightarrow{-e^-} 2\text{'-Deoxyxanthosine} + N_2 + H^+ \qquad (11)$$

$$2\text{'-Deoxyadenosine} + NO \xrightarrow{-e^-} 2\text{'-Deoxyinosine} + N_2 + H^+ \qquad (12)$$

To test these predictions, we mixed the model nucleoside, 2'-deoxycytidine, with Na^+ $Et_2N\text{-}[N(O)NO]^-$ in pH 7.4 phosphate buffer. As predicted, a portion of the base residues were deaminated to 2'-deoxyuridine (equation 9). We then dosed *Salmonella typhimurium*, Ames strain TA 1535, with SPER/NO to determine whether this chemistry could occur in the living cell. Again as predicted, the NO-releasing drug was mutagenic in the bacterium; the mutations observed were virtually all cytosine→thymine transitions, consistent with a deaminative mechanism of action (*41*). Thus, the NONOates have the potential for genotoxic action as well as the other pharmacological effects described above.

Conclusions

We have shown that many compounds containing the $X\text{-}[N(O)NO]^-$ functional group release nitric oxide spontaneously in neutral aqueous solution and that as a consequence they are useful for the controlled biological delivery of NO in a variety of pharmacological applications. In view of the potential medicinal utility of these compounds, we are intensively studying their fundamental physical and chemical properties. Their structural similarities suggest the name "NONOates" for the compound class, *i.e.* for ions containing the $[N(O)NO]^-$ functional group. A wide variety of such materials can be synthesized simply by exposing nucleophiles dissolved in suitable solvents to NO. While they are alike in their bonding arrangements, they diverge in their behavior on treatment with acid. Those in which the NONOate function is bound directly to carbon tend simply to protonate, while those attached via heteroatoms (N, O, S) decompose to NO and/or N_2O. Additional studies demonstrate that the NONOate group can be O-functionalized as well as coordinated to metal centers. Future investigations will focus on developing an increasingly refined understanding of their chemistry as a rational basis for designing drugs containing this structural element for a variety of biomedical applications.

Literature Cited

(1)   Drago, R. S. In *Free Radicals in Inorganic Chemistry*; American Chemical Society (Advances in Chemistry Series, Number 36): Washington, D. C., 1962; pp 143-149.

(2)   Drago, R. S.; Paulik, F. E. *J. Am. Chem. Soc.* **1960**, *82*, 96-98.

(3)   Drago, R. S.; Karstetter, B. R. *J. Am. Chem. Soc.* **1961**, *83*, 1819-1822.

(4)   Longhi, R.; Ragsdale, R. O.; Drago, R. S. *Inorg. Chem.* **1962**, *1*, 768-770.

(5)   Ragsdale, R. O.; Karstetter, B. R.; Drago, R. S. *Inorg. Chem.* **1965**, *4*, 420-422.

(6)   Hansen, T. J.; Croisy, A. F.; Keefer, L. K. In N-*Nitroso Compounds: Occurrence and Biological Effects*; Bartsch, H., O'Neill, I. K., Castegnaro, M., Okada, M., Eds.; International Agency for Research on Cancer (IARC Scientific Publications No. 41): Lyon, 1982; pp 21-29.

(7)   Marletta, M. A. *Chem. Res. Toxicol.* **1988**, *1*, 249-257.

(8)   Lancaster, J. R., Jr.; Hibbs, J. B., Jr. *Proc. Natl. Acad. Sci. USA* **1990**, *87*, 1223-1227.

(9) Ignarro, L. J. *Hypertension (Dallas)* **1990**, *16,* 477-483.
(10) Moncada, S.; Palmer, R. M. J.; Higgs, E. A. *Pharmacol. Rev.* **1991**, *43,* 109-142.
(11) *Nitric Oxide from L-Arginine: A Bioregulatory System*; Moncada, S. Higgs, E. A., Eds.; Elsevier Science Publishers (Excerpta Medica, International Congress Series 897): Amsterdam, 1990.
(12) *The Biology of Nitric Oxide*; Moncada, S., Marletta, M. A., Hibbs, J. B., Jr., Higgs, E. A., Eds.; Portland Press: London, 1992.
(13) Hoffman, M. *Science (Washington, D.C.)* **1991**, *252,* 1788.
(14) Rajfer, J.; Aronson, W. J.; Bush, P. A.; Dorey, F. J.; Ignarro, L. J. *N. Engl. J. Med.* **1992**, *326,* 90-94.
(15) Dupuy, P. M.; Shore, S. A.; Drazen, J. M.; Frostell, C.; Hill, W. A.; Zapol, W. M. *J. Clin. Invest.* **1992**, *90,* 421-428.
(16) Duarte, I. D. G.; Lorenzetti, B. B.; Ferreira, S. H. *Eur. J. Pharmacol.* **1990**, *186,* 289-293.
(17) Ozaki, H.; Blondfield, D. P.; Hori, M.; Publicover, N. G.; Kato, I.; Sanders, K. M. *J. Physiol. (London)* **1992**, *445,* 231-247.
(18) WHO Task Group on Environmental Health Criteria for Oxides of Nitrogen. *Oxides of Nitrogen*; World Health Organization (Environmental Health Criteria 4): Geneva, 1977.
(19) Challis, B. C.; Outram, J. R. *J. Chem. Soc., Chem. Commun.* **1978**, 707-708.
(20) Kosaka, H.; Wishnok, J. S.; Miwa, M.; Leaf, C. D.; Tannenbaum, S. R. *Carcinogenesis (London)* **1989**, *10,* 563-566.
(21) Davy, H. *Bibl. Brit (Serie Sci. Arts)* **1802**, 350-367.
(22) Cox, E. G.; Jeffrey, G. A.; Stadler, H. P. *Nature (London)* **1948**, *162,* 770.
(23) Maragos, C. M.; Morley D.; Wink, D. A.; Dunams, T. M.; Saavedra, J. E.; Hoffman, A.; Bove, A. A.; Isaac, L.; Hrabie, J. A.; Keefer, L. K. *J. Med. Chem.* **1991**, *34,* 3242-3247.
(24) Drago, R. S. In *Inorganic Syntheses*, Vol. V; Moeller, T., Ed.; McGraw-Hill Book Co., Inc.: New York, 1957; pp 120-122.
(25) Burns, E. A. In *The Analytical Chemistry of Nitrogen and its Compounds;* Streuli, C. A., and Averell, P. R., Eds.; Wiley-Interscience: New York, 1970; pp 85-144.
(26) Angeli, A.; Angelico, F. *Gazz. Chim. Ital.* **1903**, *33 (Part II),* 245-252.
(27) Akhtar, M. J.; Bonner, F. T.; Hughes, M. N.; Humphreys, E. J.; Lu, C.-S. *Inorg. Chem.* **1986**, *25,* 4635-4639.
(28) Hope, H.; Sequeira, M. R. *Inorg. Chem.* **1973**, *12,* 286-288.
(29) Hickmann, E.; Hädicke, E.; Reuther, W. *Tetrahedron Lett.* **1979**, 2457-2460.
(30) Addison, C. C.; Gamlen, G. A.; Thompson, R. *J. Chem. Soc.* **1952**, 338-345.
(31) Ackermann, M. N.; Powell, R. E. *Inorg. Chem.* **1967**, *6,* 1718-1720.
(32) Challis, B. C.; Challis, J. A. In *The Chemistry of Functional Groups. Supplement F. The Chemistry of Amino, Nitroso and Nitro Compounds and Their Derivatives*; Patai, S., Ed.; John Wiley & Sons: Chichester, 1982; p 1176.
(33) Bonner, F. T.; Ravid, B. *Inorg. Chem.* **1975**, *14,* 558-563.

(34) Hughes, M. N.; Wimbledon, P. E. *J. Chem. Soc., Dalton Trans.* **1976**: 703-707.
(35) Weitz, E.; Achterberg, F. *Ber. Dtsch. Chem. Ges. B* **1933**, *66*, 1718-1727.
(36) Clusius, K.; Schumacher, H. *Helv. Chim. Acta* **1957**, *40*, 1137-1144.
(37) Switkes, E. G.; Dasch, G. A.; Ackermann, M. N. *Inorg. Chem.* **1973**, *12*, 1120-1123.
(38) Saavedra, J. E.; Dunams, T. M.; Flippen-Anderson, J. L.; Keefer, L. K. *J. Org. Chem.* **1992**, *57*, 6134-6138.
(39) Maragos, C. M.; Wang, J. M.; Hrabie, J. A.; Oppenheim, J. J.; Keefer, L. K. *Cancer Res.* **1993**, *53*, 564-568.
(40) Diodati, J. G.; Quyyumi, A. A.; Hussain, N.; Keefer, L. K. Submitted.
(41) Wink, D. A.; Kasprzak, K. S.; Maragos, C. M.; Elespuru, R. K.; Misra, M.; Dunams, T. M.; Cebula, T. A.; Koch, W. H.; Andrews, A. W.; Allen, J. S.; Keefer, L. K. *Science (Washington, D.C.)* **1991**, *254*, 1001-1003.

RECEIVED June 29, 1993

Chapter 12

Nitric Oxide Production and Catalysis of *N*-Nitroso Compound Formation by Woodchucks *(Marmota monax)* and Woodchuck Hepatocytes in Culture

J. H. Hotchkiss, R. H. Liu, and T. J. Lillard

Institute for Environmental Toxicology, Stocking Hall, Cornell University, Ithaca, NY 14853

Humans are exposed to nitrosating agents from two primary sources: ingestion with water/diet (mostly in the form of nitrate which is reduced to nitrite in the mouth) and endogenous formation via the enzymic conversion of L-arginine to nitric oxide with subsequent reaction with oxygen to form nitrosating agents. Endogenously formed of nitric oxide can react directly with DNA resulting in inheritable mutations. Nitric oxide can also react with endogenous amines to form hepatocarcinogens. This is the source of endogenous nitrosation which occurs at locations other than the stomach. Nitric oxide and nitroso compound formation is elevated in woodchucks which are infected with woodchuck hepatitis virus. Primary cell cultures from infected woodchucks also form higher amounts of nitric oxide and N-nitroso compounds. This may explain the large increase in cancer risk associated with hepatitis B virus infection in humans.

Endogenous Formation of N-Nitroso Compounds

Several studies have investigated the endogenous formation of N-nitroso compounds and many have focused on the external and internal factors which qualitatively and quantitatively influence endogenous nitrosation (*1, 2*). Among the conclusions drawn from these studies are that humans and laboratory animals endogenously form N-nitrosamines and, perhaps, N-nitrosamides from endogenous and exogenous precursors and that for nonsmokers and individuals not occupationally exposed, exposure to N-nitroso compounds (NOC) formed within the body exceeds environmental exposure (*1*). It is clear that dietary and physiological factors qualitatively and quantitatively influence endogenous nitrosation. The primary dietary factor is the balance between ingestion of nitrosation precursors (e.g., nitrate) versus ingestion of nitrosation inhibitors (e.g., ascorbate) while the primary physiological factors are the level of oral nitrate reductase activity and the amount of oxides of nitrogen formed endogenously (*3-7*).

Sources of Endogenous Nitrosation. Endogenously formed NOC can be divided into two types based on source: those that are formed in the acidic environment of the stomach (i.e., intragastric) and those that are formed elsewhere in the body

0097–6156/94/0553–0147$08.00/0

(extragastric; 8). This conclusion is based on controlled studies in which the dietary N-nitrosamine contribution is monitored as well as nitrosamine excretion. For example, Perciballi et al. (9) monitored N-nitrosodimethylamine (NDMA) excretion in ferrets after consumption of nitrite-cured meats in order to determine if NDMA resulted from ingestion of nitrite-containing food. The food was analyzed for NDMA. In all but a few cases, excreted NDMA exceeded ingested. Labeling experiments suggested that the excreted excess nitrosamine was not formed in the animal's stomach but was formed at other unknown locations within the body.

Further evidence that NOC are formed at sites in addition to the stomach comes from human studies which investigated the effect of ascorbic acid on the endogenous formation of N-nitrosoproline (NPRO) which is a non-carcinogenic marker of nitrosation. Ascorbic acid is effective at inhibiting the endogenous formation of NPRO but only up to a point. Even very large doses of ascorbic acid can not reduce excretion to non-detectable levels. Leaf et al. (5) studied the dose-response relationship between endogenous NPRO formation and ascorbic acid intake. Ascorbate doses as large as 5.7 mmole asymptotically reduced NPRO formation to 24 nmol/day from 45 nmol/day but did not eliminate it. This ascorbate dose was estimated to be 20-fold greater than the amount of nitrite formed from the ingested nitrate. Analyses of the controlled diets precluded this as the source of the NPRO. Previous work by Wagner et al. (3) resulted in a similar finding leading them to conclude that there were two separate sources of endogenous nitrosation.

This unknown endogenous nitrosation occurs at a location(s) other than the gastric lumen and, therefore, must occur at pHs near neutrality. Nitrite does not appreciably nitrosate at neutral pH and is unlikely to be directly involved. However, Challis and Kyrtopoulos (10) have pointed out that oxides of nitrogen (NO_x) are potent nitrosating agents at physiological pH. This coupled with the discovery of Stuehr and Marletta (11) and Marletta et al. (12) that nitric oxide (NO) was produced in large amounts by macrophages *in vitro* suggests that NO production might be the source of the NOC endogenously formed at sites other than the stomach. The objective of our current work is to quantify the effects of nitric oxide synthesis on the formation of N-nitroso compounds both *in vitro* and *in vivo* in order to determine if this is the source of extra-gastric NOC formation previously observed.

Reactivity of Nitric Oxide

NO is an important bioregulatory molecule. However, NO is also a free radical which can react to form undesirable products, especially if NO were produced in excess of normal physiological amounts. Excessive NO synthesis could result in changes in macromolecules including DNA. This could occur in two ways: NO could react with endogenous amines or amides to produce genotoxic NOC which could alkylate DNA directly (nitrosamides) or after metabolism (nitrosamines). NO could also react directly with DNA by deaminating primary amine functional groups on nucleic acids resulting in mutations. Deamination of DNA bases by NO can cause G·C→A·T, A·T→G·C, G·C→T·A and A·T→T·A mutations. The latter has recently been demonstrated. Two groups have independently documented that NO is a direct mutagen and can deaminate nucleic acids. Wink et al. (13) showed that NO in the presence of oxygen, deaminates cytosine residues in intact DNA. As predicted, compounds which release NO were mutagenic in bacterial systems. Nguyen et al. (14) recently demonstrated similar deamination reactions and that NO is mutagenic in human cell lines.

Nitric Oxide Formation and Hepatitis

It is feasible that the source of extra-gastric NOC is related to NO synthesis, particularly in disease states in which there is stimulation of the immune system resulting in elevated NO production. One such situation might be chronic hepatitis. There is convincing epidemiological data that the risk of hepatocellular carcinoma (HCC) is associated with hepatitis B virus (HBV) infection in humans (*15, 16*). The HCC incidence in hepatitis surface antigen (HB_sAg) carriers is 223-fold higher than non-carriers (*16*). There is also a strong worldwide geographic correlation between the incidence of HCC and the prevalence of HB_sAg carriers (*15*). A large scale epidemiological survey conducted in China showed that HCC mortality was positively correlated with HB_sAg+ prevalence and unrelated to aflatoxin intake (*17*). However, the mechanisms by which hepatitis B virus influences liver cancer have not been elucidated.

   In addition to humans, some animals contract viral hepatitis and serve as laboratory models for the disease. The most studied is the woodchuck (*Marmota monax*) which contracts woodchuck hepatitis virus (WHV) in the wild and in the laboratory. Snyder (*18*) first reported that wild woodchucks were infected with chronic hepatitis and one-third of infected animals had primary hepatic neoplasms. The WHV DNA virus is morphologically similar and immunologically related to human HBV (*19*). Woodchucks chronically infected in the laboratory with WHV develop hepatocellular carcinoma (HCC) with 100% frequency within 17-36 months (*20*). Persistent antigenemia, active viral replication, and lesions of active, severe inflammation are observed in the liver during active hepatitis (*21, 22*). Woodchucks also develop HCC when exposed to diethylnitrosamine, and hepatic neoplasms are similar in size and appearance to those observed in WHV infected woodchucks (*23*).

   It is possible that NO would be locally available at high concentrations with stimulation of chronic inflammation such as occurs in hepatitis, gastritis, esophagitis and schistosomiasis infection. Some forms of cancer are epidemiologically associated with such chronic infections. For example, liver fluke infection results in an elevated risk for cholangiocarcinoma and patients have elevated urinary nitrate excretion (*24*). Tricker et al. (*25, 26*) reported that bilharzia patients with parasitic S. haematobium infection and patients with urinary diversions excreted more nitrite, nitrate and N-nitrosoamino acids than controls. However, they suggested that bacterial infection by nitrate-reducing bacteria and not immunostimulation were responsible for the higher urinary levels.

   Nitric oxide is formed from L-arginine and eventually oxidized *in vivo* to nitrate. In humans, approximately 50% of the nitrate is catabolized and 50% excreted unchanged. Thus, nitrate balance studies which measure the amount of nitrate formed endogenously are an indication of NO formation *in vivo* (*5*).

Nitrate and N-Nitrosodimethylamine in Woodchucks

Nitrate Balance. We recently undertook studies in which the objective was first to determine if WHV-infected animals excreted more nitrate than uninfected animals and second to determine if elevated NO synthesis resulting from WHV infection was a source of endogenously formed carcinogenic NOC. In these studies, urinary recovery of a bolus [^{15}N]-nitrate dose was 54 ± 13% in woodchucks (Table I), which is similar to the 53% reported in humans (*5*). There was no difference between infected and uninfected animals in nitrate recovery. However, when nitrate balance in 8 control and 8 WHV infected woodchucks were compared over a 5 day period and corrected for catabolism, infected animals formed 3-fold (p<0.01) more nitrate than control animals without added stimulus such as lipopolysaccharide (LPS; Table II; *27*). Previous studies have shown that

Table I. Urinary recovery in 24 hr. of a 33 μmole dose of [^{15}N]-nitrate in WHV-infected and control woodchucks

WHV Infected	49 ± 16% (n=6)
Control	58 ± 8.2% (n=6)
Combined	54 ± 13% (n=12)

Table II. Nitrate balance in control and WHV-infected woodchucks over 5 day period (μmol/24 h)

	Ingested	Excreted	Endogenous[a]
Control	2.73 ± 0.44	2.87[b] ± 0.77	2.58[c] ± 1.4
Infected[d]	2.25 ± 1.0	5.37[b] ± 2.0	7.71[c] ± 3.0

a. Endogenously synthesized = ([excreted/0.54] - ingested)
b. Statistically different (p<0.05)
c. Statistically different (p<0.01)
d. n=8 control; n=8 infected

Table III. Urinary [^{15}N]-nitrate excretion in woodchucks treated with ^{15}N$_2$]-L-Arg[a]

	Control	Infected
NO$^-_3$ excretion (μmol)	4.2	23
15N - NO$^-_3$ as a % of total NO$^-_3$ excreted	8.3	6.8
^{15}N - NO$^-_3$ (nmol)	350	1580
^{15}N - NO$^-_3$ as a % of ^{15}[N$_2$]-L-Arg dose (x10^2)	4.1	19

a. Treatment includes LPS, [^{15}N$_2$]-L-Arg, 4- Methylpyrazole and diuretic

Table IV. MS-SIM analysis of urine from woodchucks for [^{15}N]-N-Nitrosodimethylamine after treatment with [^{15}N$_2$]-L-Arg and LPS[a]. (M/Z 74, 75 = M and M + 1 for NDMA; M/Z 30, 31 = NO and NO + 1)

	M/Z			
	30	31	74	75
NDMA Standard	23	0.8	100	2.6
Spiked urine from uninfected animals	26	BT[b]	100	2.7
^{15}N-Arg treated infected animals	20	1.5	100	12

a. % of base peak
b. Below threshold of instrument

endogenous nitrate synthesis is elevated during immunostimulation in humans (28). Increased nitrate formation in WHV infected animals may be due to chronic hepatitis associated with persistent WHV infection (20, 21). There is also marked infiltration by other inflammatory cells into the liver in woodchucks chronically infected with WHV (20, 22). Hepatocytes, Kupffer cells and other macrophages have been shown in culture to metabolize L-arginine to citrulline and NO after LPS stimulation (29-32).

When infected and control woodchucks were treated with [$^{15}N_2$]L-arginine, a diuretic to increase nitrate output, 4-methylpyrazole which inhibits NOC metabolism so that NDMA can be detected in urine, and only glucose in water, infected animals excreted 5-fold more nitrate than control animals (Table III). The amount of ^{15}N enriched nitrate excreted was elevated in infected animals compared to control animals when both received [$^{15}N_2$]L-arginine but the level of enrichment as a percent of the total amount of nitrate excreted was similar in both control and infected animals. It has been reported (33) that L-arginine is the precursor to endogenously formed nitrate in other species as well. The guanidino nitrogen of L-arginine is the source of the nitrogen incorporated into NO, with the further oxidation of NO to nitrite and nitrate *in vitro* (32). Animals received only 10% glucose in order to decrease the dilution of [$^{15}N_2$]L-arginine by dietary L-arginine. The body pool of L-arginine is, however, large and likely responsible for the relatively low rate (approximately 8%) of [^{15}N]-nitrogen incorporation into nitrate.

N-Nitrosodimethylamine Excretion and Hepatitis. An increased abundance of [^{15}N]-nitrogen in the nitroso group of NDMA isolated from the urine of [$^{15}N_2$]L-arginine treated animals was apparent when the NDMA was analyzed by mass spectrometry. The 3-fold increase in the m+1 ion (i.e., m/z 75) and in the NO fragment ions (i.e., m/z 30, 31) in the mass spectrum of NDMA from infected animals compared to unlabelled NDMA extracted from control uninfected woodchuck urine indicates that a portion of the N-nitroso group in the NDMA contained [^{15}N]-nitrogen from [$^{15}N_2$]L-arginine (Table IV). When analyzed for NDMA content, livers acquired from infected animals at biopsy showed substantially higher amounts than uninfected animals (Table V). These data confirm that the L-arginine-NO pathway is a source of endogenous nitrosation and because the animals were not pre-treated with dimethylamine, that nitrosation of endogenous amines results in the formation of hepatocarcinogenic NOC.

These data from studies in woodchucks may explain work which found NDMA was excreted in larger amounts than ingested by humans. Garland et al. (34) developed a GC-high resolution MS method with high sensitivity to quantify levels of NDMA and NPRO in the urine of healthy volunteers. They reported a mean level of urinary NDMA at 38.2 ng/day and NPRO of 3.3 µg/day. Both nitrosamines were unaffected by ingestion of large amounts of ascorbic acid or α-tocopherol. Garland suggested that there might be an association between excretion of NDMA and concentration of oxides of nitrogen in the air but urinary excretion of carcinogenic NDMA was not correlated with the urinary excretion of non-carcinogenic NPRO. It is possible in our view that these NOC resulted from the NO-synthase pathway.

Cultured Hepatocytes

Macrophages in culture, when stimulated with LPS, produce nitrosating agents from L-arginine (12). Other cell types including endothelial cells (35), platelets (36), and neurons (37) have been reported to produce NO, nitrite and nitrate through the same pathway.

Addition of LPS to the woodchuck primary hepatocyte cultures resulted in

Table V. NDMA content of liver biopsies from WHV-
infected and control woodchucks (ng/50 g liver)

	Control	WHV-Infected
Sample Number	4	17
Number of Positive	1	12
Mean ± SE	3.8 ±3.8	16.5 ± 6.9
Range	ND* -- 15.3	ND -- 87.5

* ND = Not detected

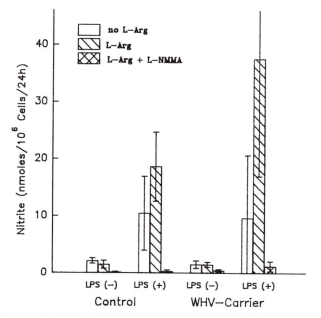

Figure 1. Nitrite formation by isolated primary woodchuck hepatocytes from
infected (WHV-carrier) and control animals. (Adapted from Ref. 42.)

the formation of large amounts of nitrite in the presence of L-arginine (Figure 1). LPS-treated hepatocytes from normal animals formed 12.5-fold more nitrite than unstimulated cells from the same animals. LPS-stimulated hepatocytes from infected animals formed 25-fold more nitrite than unstimulated hepatocytes from infected animals ($p < 0.01$). Hepatocytes from infected animals formed significantly ($p < 0.02$) more nitrite than hepatocytes from uninfected animals when both were treated with LPS. Both control and WHV carrier woodchuck hepatocytes produced small but detectable amounts of nitrite without LPS stimulation. Nitrite synthesis was effectively inhibited by N^G-monomethyl-L-arginine (L-NMMA), which is a selective inhibitor of NO-synthetase (*38*).

The amount of nitrite produced by woodchuck hepatocytes was directly related to the dose of LPS and L-arginine added to the medium (Figures 2 and 3). Nitrite synthesis was stimulated by as little as 0.01 μg/ml LPS and began to plateau at an LPS concentration of ≥ 1.0 μg/ml. Nitrite production by woodchuck hepatocytes treated with LPS increased with increasing L-arginine concentration and leveled off at an L-arginine concentration of 0.5 mM L-arginine.

Miwa et al. (*39*) have shown that immunostimulated macrophages nitrosate morpholine in vitro. Similarly, primary hepatocytes nitrosated morpholine to N-nitrosomorpholine (*42*). Nitrosation does not, in this case, appear to be acid catalyzed via nitrite (*29, 39*) but is mediated via nitric oxide which is an intermediate in the L-arginine to nitrate/nitrite pathway (*32*). Nitric oxide reacts rapidly with dissolved O_2 to yield NO_2, which exits in equilibrium with the effective nitrosating agents N_2O_3 and N_2O_4, both of which are capable of nitrosating dialkylamines in neutral aqueous solution to form nitrosamines or reacting with water to yield nitrate and nitrite. Assuming that NDMA is equally distributed in the body water which is approximately 55% of body weight (*40*), and that NDMA is not metabolized after administration of 4-methylpryazole, infected

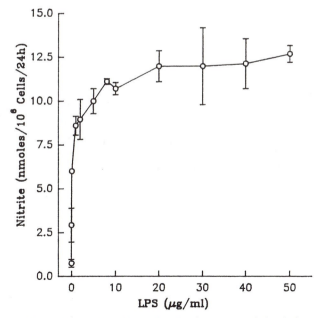

Figure 2. Nitrite formation by isolated primary woodchuck hepatocytes in response to increasing amounts of LPS in the growth medium. (Adapted from Ref. 42.)

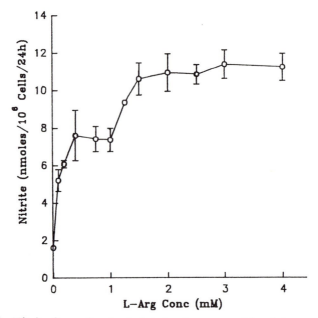

Figure 3. Nitrite formation by isolated primary woodchuck hepatocytes in response to increasing amounts of L-arginine in the growth medium. (Adapted from Ref. 42).

woodchucks endogenously formed 5.4 nmol NDMA/day/per animal while control animals formed 1.8 nmol/day/animal.

Summary

There is substantial evidence that nitrate is endogenously synthesized in mammals (*28, 5, 41*) and that the source of this nitrate is the formation of NO from L-arginine. This occurs in several cell types. Studies have shown that macrophages stimulated with LPS can nitrosate secondary amines in vitro (*39*). Recent data indicates that woodchucks chronically infected with hepatitis virus endogenously synthesize more nitrate than non-infected animals, and confirms that L-arginine is the precursor of nitrate. These data also show that stimulation of the L-arginine-NO pathway results in the endogenous formation of hepatocarcinogenic NDMA in greater amounts in infected animals compared to normal controls. This means that the process of forming NO in response to immunostimulation results in the formation of a potent hepatocarcinogen. NO also can react directly with DNA resulting in inheritable base substitutions. This provides an additional mechanism for chronic hepatitis to influence the risk of liver carcinoma.

Literature Cited

1.   Hotchkiss, J.H. In *Food Toxicology, A Perspective on Relative Risks*; Taylor, S.R. and Scanlan, R.A., Eds.; Marcel Dekker: New York, NY, **1989**; pp. 57-100.

2.   Bartsch, H.; Ohshima, H.; Pignatelli, B.; Calmels, S. *Cancer Surveys* **1989**, *8*, 335-362.

3.   Wagner, D.A.; Shuker, D.E.G.; Bilmazes, C.; Obiedzinski, M.; Baker, I.; Young, V.R.; Tannenbaum, S.R. *Cancer Res.* **1985**, *45*, 6519-6522.

4.   Mallet, A.K.; Rowland, I.R.; Walters, D.G.; Gangolli, S.D.; Cottrell, C.; Masser, R.C. *Carcinogenesis (London)* **1985**, *6*, 1585-1588.
5.   Leaf, C.D.; Vecchio, A.J.; Roe, D.A.; Hotchkiss, J.H. *Carcinogenesis (London)* **1987**, *8*, 791-795.
6.   Shapiro, K.B.; Hotchkiss, J.H.; Roe, D.A. *Fd. Chem. Toxicol.* **1991**, *29(11)*, 751-755.
7.   Helser, M.A.; Roe, D.; Hotchkiss, J.H. *J. Nutr. Biochem.* **1991**, *2(5)*, 268-273.
8.   Leaf, C.D.; Wishnok, J.S.; Tannenbaum, S.R. *Cancer Surveys* **1989**, *8*, 323-334.
9.   Perciballi, M.; Hotchkiss, J.H. *Carcinogenesis (London)* **1989**, *10(12)*, 2303-2309.
10. Challis, B.; Kryptopoulos, S.A. *J. Chem. Soc. Perkin 2*, **1978**, 1296.
11. Stuehr, D.J.; Marletta, M.A. *Proc. Nat. Acad. Sci. USA* **1985**, *82*, 7738-7742.
12. Marletta, M.A. *Chem. Res. Toxicol.* **1988**, *1*, 249-257.
13. Wink, D.A.; Kasprzak, K.S.; Maragos, C.M.; Elespuru, R.K.; Misra, M.; Dunams, T.J.; Cebula, T.A.; Koch, W.H.; Andrews, A.W.; Allen, J.S.; Keefer, L.K. *Science* **1991**, *254*, 1001-1003.
14. Nguyen, T.; Brunson, D.; Crespi, C.L.; Penman, B.W.; Wishnok, J.S.; Tannenbaum, S.R. *Proc. Natl. Acad. Sci. USA* **1992**, *89*, 3030-3034.
15. Szmuness, W. *Prog. Med. Virol.* **1978**, *24*, 40-69.
16. Beasley, R.P.; Hwang, L.Y.; Lin, C.C.; Chien, C.S. *Lancet* **1981**, *2*, 1129-1133.
17. Campbell, T.C.; Chen, J.; Liu, C.; Li, J.; Parpia, B. *Cancer Res.* **1990**, *50*, 6882-6893.
18. Snyder, R.L. *Am. J. Pathol.* **1968**, *52*, 32a-33a.
19. Snyder, R.L.; Tyler, G.; Sumers, J. *Am. J. Pathol.* **1982**, *107*, 422-425.
20. Popper, H.; Roth, L.; Purcell, R.H.; Tennant, B.C.; Gerin, J.L. *Proc. Natl. Acad. Sci. USA* **1987**, *84*, 866-870.
21. Popper, H.; Shih, J.W.-K.; Gerin, J.L.; Wong, D.C.; Hoyer, B.H.; London, W.T.; Sly, D.L.; Purcell, R.H. *Hepatology* **1981**, *1(2)*, 91-98.
22. Paronetto, F.; Tennant, B.C. In *Progress in Liver Disease*; Popper, H.; Schaffner, F., Eds.; **1990**, Vol. 9; pp. 463-483.
23. Baldwin, B.H.; Hornbuckle, W.E.; Hotchkiss, J.H.; Anderson, W.I.; Paronetto, F.; Popper, H.; Tennant, B.C. The 1990 International Symposium on Viral Hepatitis and Liver Disease (The 7th Triennial Congress), April, Westin Hotel, Houston, TX. *Scientific Program and Abstracts* **1990**, p. 206.
24. Srianujata, S.; Tonbuth, S.; Bunyaratvej, S.; Valyasevi, A.; Promvanit, N.; Chaivatsagul, W. In *Relevance of N-nitroso Compounds to Human Cancer: Exposures and Mechanisms*; Bartsch, H.; O'Neill, I.K.; Schulte-Hermann, Eds.; IARC Sci. Pub. No. 84; IARC: Lyon, 1987; pp. 544-546.
25. Tricker, A.R.; Mostafa, M.H.; Spiegelhalder, B.; Preussmann, R. *Carcinogenesis (London)* **1989**, *10(3)*, 547-552.
26. Tricker, A.R.; Kalble, T.; Preussmann, R. *Carcinogenesis* **1989**, *10(12)*, 2379-2382.
27. Liu, R.H.; Baldwin, B.; Tennant, B.C.; Hotchkiss, J.H. *Cancer Res.* **1991**, *51*, 3925-3929.
28. Wagner, D.A.; Tannenbaum, S.R. In *Banbury Report 12: N-nitrosamines and Human Cancer*; Magee, P.N., Ed.; Cold Spring Harbor Laboratory: Cold Spring Harbor, NY, 1982; pp. 437-443.
29. Iyengar, R.; Stuehr, D.J.; Marletta, M.A. *Proc. Natl. Acad. Sci. USA* **1987**, *84(18)*, 6369-6373.
30. Billiar, T.R.; Curran, R.D.; Stuehr, D.J.; West, M.A.; Bentz, B.G.; Simmons, R.L. *J. Exp. Med.* **1989**, *169*, 1467-1472.
31. Billiar, T.R.; Curran, R.D.; Ferrari, F.K.; Williams, D.L.; Simmons, R.L. *J. Surgical Res.* **1990**, *48*, 349-353.
32. Marletta, M.; Yoon, P.S.; Iyengar, R.; Leaf, C.D.; Wishnok, J.S. *Biochem.* **1988**, *27*, 8706-8711.

33. Leaf, C.D.; Wishnok, J.S.; Hurley, J.P.; Rosenblad, W.D.; Fox, J.G.; Tannenbaum, S.R. *Carcinogenesis (London)* **1990**, *11(5)*, 855-858.
34. Garland, W.A.; Kuenzig, W.; Rubio, F.; Kornychuk, H.; Norkus, E.; Conney, A. *Cancer Res.* **1986**, *46*, 5392-5400.
35. Palmer, R.M.J.; Ashton, D.S.; Moncada, S. *Nature* **1988**, *333*, 664-666.
36. Radomski, M.W.; Palmer, R.M.J.; Moncada, S. *Proc. Natl. Acad. Sci. USA* **1990**, *87*, 5193-5197.
37. Knowles, R.G.; Palacios, M.; Palmer, R.M.J.; Moncada, S. *Proc. Natl. Acad. Sci. USA* **1989**, *86*, 5159-5162.
38. Hibbs, J.B., Jr.; Vavrin, Z.; Taintor, R.R. *J. Immunol.* **1987**, *138*, 550-565.
39. Miwa, M.; Stuehr, D.J.; Marletta, M.A.; Wishnok, J.S.; Tannenbaum, S.R. *Carcinogenesis (London)* **1987**, *8*, 955-958.
40. Klaassen, C.D. In *Casarett and Doull's Toxicology, Third Edition*; Klaassen, C.D. et al., Eds.; Macmillan Publishing Company: New York, NY 1986; pp. 33-63.
41. Dull, B.J.; Hotchkiss, J.H. *Food Chem. Toxicol.* **1984**, *22*, 105-108.
42. Liu, R.H.; Jacob, J.R.; Tennant, B.C.; Hotchkiss, J.H. *Cancer Res.* **1992**, *52*, 4139-4143.

RECEIVED March 30, 1993

CHEMISTRY AND BIOCHEMISTRY
OF NITROSAMINE ACTIVATION
AND DETOXICATION

Chapter 13

Activation of *N*-Nitrosodialkylamines by Metalloporphyrin Models of Cytochrome P-450

M. Mochizuki, E. Okochi, K. Shimoda, and K. Ito

Kyoritsu College of Pharmacy, Shibakoen 1–5–30, Minato-ku, Tokyo 105, Japan

Chemical model oxidation systems, metalloporphyrins/oxidants, were used to investigate mechanism of metabolic activation of *N*-nitrosodialkylamines. Tetraphenylporphyrinatoiron(III) and -manganese(III) chloride were used in non-aqueous conditions, while tetrakis(4-sulfonatophenyl)porphyrinatoiron(III) was used in aqueous conditions. The model's efficiency was evaluated by the amounts of aldehydes formed from dealkylation through α-hydroxy nitrosamines, and by the mutagenicity derived from the intermediate α-hydroxy nitrosamines. In the model systems, *N*-nitroso-*N*-methylbutylamine and *N*-nitrosodialkylamines were all dealkylated and there is no selectivity of alkyl group in α-hydroxylation (alkyl = methyl, butyl and benzyl). Mutagenicity of activated *N*-nitrosodialkylamines by the models was assayed using *Salmonella typhimurium* TA1535 and *E. coli* WP2*hcr*⁻. *N*-Nitrosodibenzylamine and *N*-nitroso-*N*-methylbutylamine was activated to direct mutagen by the models. These results suggested that the chemical oxidation model systems are useful in elucidating the mechanism of activation of carcinogenic and mutagenic *N*-nitroso compounds.

N-Nitrosodialkylamines are environmental alkylating carcinogens which are metabolically activated through dealkylation or detoxified through denitrosation by cytochrome P-450 (*1*). The mechanism of metabolic interaction of *N*-nitrosodialkylamines with cytochrome P-450 has been extensively investigated. The major pathway for metabolic activation is considered to be α-hydroxylation. As a mechanism of the formation of α-hydroxy nitrosamines, a pathway through α-nitrosamino radical has been proposed (*2*). Stiborova et al. (*3*) showed metabolic activation of *N*-nitroso-*N*-methylaniline by a model system using peroxidase.

Recently, chemical model systems of cytochrome P-450 have been applied to study the mechanism of drug metabolism (*4*). Model systems have advantages over *in vivo* metabolism because it is easier to isolate products and determine their structures. Thus, chemical model oxidation systems are useful for investigation of the mechanism of the metabolic activation of *N*-nitroso compounds.

In 1979, Groves et al. (*5*) first employed model systems in epoxidation of olefins and hydroxylation of saturated hydrocarbons with iodosobenzene (PhIO) and

0097–6156/94/0553–0158$08.00/0
© 1994 American Chemical Society

tetraphenylporphyrinatoiron(III) chloride (FeTPPCl). Since then, metalloporphyrins such as iron- or manganese-porphyrins, coupled with oxidants such as PhIO, hydrogen peroxide (H_2O_2) or alkyl hydroperoxide as *tert*-butyl hydroperoxide (*t*-BuOOH) or cumene hydroperoxide (Cumene-OOH) (4), have been used as catalysts in the model reaction of cytochrome P-450. Various oxidation reactions similar to cytochrome P-450 have been reported for many compounds. Although there have been numerous studies of *N*-dealkylation of tertiary amines with models (6), very few reports have been published for *N*-nitroso compounds. Lindsay Smith et al. (7) reported oxidative dealkylation of *N*-nitrosodibenzylamine using FeTPPCl and MnTPPCl, and oxidants. Quite recently, Chauhan and Satapathy also reported dealkylation of *N*-nitrosodialkyl-amines with manganese-porphyrins containing phenyl or dichlorophenyl groups (8). However, both reports were not concerned with the detailed reaction mechanism. Such model systems also activated pro-carcinogens into direct mutagens. The FeTPPCl/PhIO model system activated benzo[*a*]pyrene and 2-aminofluorene in Ames assay (9). Aflatoxin B1 was also activated by a model system to mutagen in plasmid DNA of *E. coli* (10).

In this study, simplified chemical models were used to elucidate molecular mechanism of metabolic activation of *N*-nitrosodialkylamines. The formation of aldehydes through oxidative dealkylation via α-hydroxylation and bacterial mutation were used as markers to evaluate the efficiency of model systems for the metabolic activation of *N*-nitroso-dialkylamines.

Dealkylation by Non-aqueous Model System

Dealkylation of *N*-nitrosodialkylamine with an oxidant catalyzed by metalloporphyrins was investigated in non-aqueous conditions. Tetraphenylporphyrinatoiron(III) and -manganese(III) chloride (FeTPPCl and MnTPPCl) were the metalloporphyrins used, while *t*-BuOOH, Cumene-OOH and PhIO were the oxidants. After dissolving *N*-nitrosodialkylamine and metalloporphyrin in benzene, oxidant was added to the reaction mixture and the solution was incubated at 25°C. Aldehydes formed from dealkylation of *N*-nitrosodialkylamines were analyzed. Formaldehyde was determined colorimetrically by the method of Nash (11) and the other aldehydes were determined by reversed-phase HPLC after conversion to the corresponding 2,4-dinitrophenylhydrazones.

In dealkylation of *N*-nitrosodimethylamine (NDMA) by FeTPPCl and *t*-BuOOH, formation of formaldehyde was rapid and complete in the first 20 min (Figure 1). In the absence of either FeTPPCl or *t*-BuOOH, no formaldehyde was formed, demon-strating that the dealkylation of *N*-nitrosodialkylamine required both FeTPPCl and *t*-BuOOH. When using tetraphenylporphyrin without a central metal or when using an iron complex without a porphyrin ring (tris(acetylacetonato)iron(III)) no reaction occurred, demonstrating that the active catalyst for the dealkylation is a porphyrin ring with a central metal that has interacted with *t*-BuOOH. Although formation of formaldehyde also was observed in the absence of *N*-nitrosodialkylamines, this may result from interaction of *t*-BuOOH and metalloporphyrin (12).

The effect of an alkyl group in dealkylation was investigated. Three symmetric *N*-nitrosodialkylamines (alkyl = methyl, butyl and benzyl) afforded similar amounts of aldehydes (Figure 2A). When using *N*-nitroso-*N*-methylbutylamine (NMBA), there was no difference in the amount of formaldehyde and butyraldehyde formed (Figure 2B), which suggests that there is no selectivity of the alkyl group in dealkylation of *N*-nitrosodialkylamines by this model system.

Figure 3 shows the effect of an oxidant in dealkylation of *N*-nitrosodibenzylamine (NDBzA), where the formation of benzaldehyde was followed. Three oxidants, *t*-BuOOH, Cumene-OOH and PhIO, were all efficient for the model systems. The activity of *t*-BuOOH was the highest, followed by Cumene-OOH, and then PhIO. At higher concentration of oxidant no more increase in dealkylation was observed.

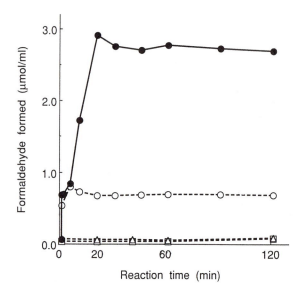

Figure 1.   The time course of the formation of formaldehyde in the dealkylation
of NDMA by FeTPPCl and *t*-BuOOH. NDMA (500µmol), FeTPPCl (0.5µmol)
and *t*-BuOOH (50µmol) were incubated in benzene 1 ml at 25°C. Complete
system(●), NDMA(–)(○), FeTPPCl(–)(□), *t*-BuOOH(–)(△).

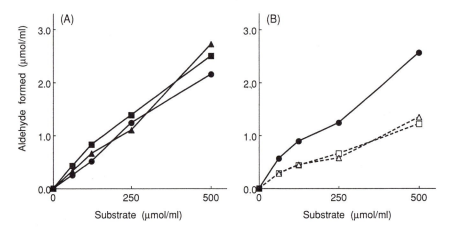

Figure 2.   The formation of aldehydes in the dealkylation of *N*-nitrosodi-
alkylamines by FeTPPCl and *t*-BuOOH. *N*-Nitrosodialkylamine (0-500µmol),
FeTPPCl (0.5µmol) and *t*-BuOOH (50µmol) were incubated in benzene 1 ml at
25°C for 20 min. (A) NDMA(●), NDBA(▲), NDBzA(■). (B) Formaldehyde(□),
butyraldehyde(△) and total aldehydes(●) from NMBA.

A central metal is necessary for the dealkylation of N-nitrosodialkylamines. The effect of metal in the complex was investigated using an iron- and manganese-porphyrin complex in the dealkylation of NDBzA. The formation of benzaldehyde was plotted against the concentration of metalloporphyrin (Figure 4). The time point for these determinations for the iron-porphyrin was 20 min and that of manganese-porphyrin was 3 h. The iron-porphyrin showed a higher efficiency and easily reached its maximum activity, while manganese-porphyrin showed a slower rate of dealkylation. Electronic properties such as the redox potential of the central metal may have profound effects on the catalytic activity of metalloporphyrins.

There are several reports that imidazole acts as a co-catalyst in the cytochrome P-450 model reaction (*13*). The effect of imidazole on dealkylation of N-nitrosodialkylamines was investigated in the FeTPPCl/t-BuOOH model system. In the presence of imidazole, formation of formaldehyde from dealkylation of NDMA increased about 60-130% over the original system (data not shown). Thus, imidazole showed a promoting effect on the dealkylation. Battioni et al. (*13*) reported that imidazole played at least two roles in metalloporphyrin/hydrogen peroxide model: (i) to promote heterolysis of hydrogen peroxide and enhance the formation of a high-valency imidazole-Mn(V)=O intermediate, and (ii) to favor H_2O_2 dismutaion as an acid-base catalyst. Such effects of nitrogen bases may suggest a promising method for the design of better models for the dealkylation of N-nitrosodialkylamines.

Dealkylation by Aqueous Model System

Aqueous models of cytochrome P-450 were used for dealkylation of N-nitrosodialkylamines. Tetrakis(4-sulfonatophenyl)porphyrinatoiron(III) (FeTPPS) and oxidants, t-BuOOH, Cumene-OOH or H_2O_2 were used. N-Nitrosodialkylamine and metalloporphyrin were dissolved in CH_3CN/H_2O (1:1) solution or 0.1M phosphate solution (pH 7.4) and oxidant was added. After reaction at 37°C, the aldehydes formed were analyzed. The corresponding 2,4-dinitrophenylhydrazones were extracted with hexane/CH_2Cl_2 (7:3) and quantified by reversed-phase HPLC (*14*).

In this model system in mixed solvent, NMBA, NDBzA and N-nitrosodibutylamine (NDBA) were dealkylated to the corresponding aldehydes. The dealkylation occurred quickly (Figure 5). However, the dealkylating activity of FeTPPS in aqueous systems was less than FeTPPCl in non-aqueous systems. In homogenous aqueous solution, 0.1M phosphate solution (pH 7.4) was used in FeTPPS/t-BuOOH model system. Acetaldehyde was formed from dealkylation of N-nitrosodiethylamine (NDEA) (Figure 6). This reaction did not occur in the absence of either FeTPPS or t-BuOOH.

The effect of oxidant (t-BuOOH, Cumene-OOH or H_2O_2) on dealkylation was examined in aqueous system using FeTPPS. Figure 7 shows that all oxidants were effective for the dealkylation. The efficiency was highest in t-BuOOH which was similar to the result in the non-aqueous model system. Figure 8 shows the effect of oxidant concentration in an aqueous model system. Dealkylation of NDEA was dependant on t-BuOOH concentration with the amount of acetaldehyde formed reaching a maximum at the highest concentration of t-BuOOH.

The effect of pH on dealkylation of NDEA by FeTPPS and t-BuOOH was investigated (Figure 9). The most rapid dealkylation occurred at pH 5, while in an alkaline solution, pH 9, a slower dealkylation was observed. El-Awady et al. studied the structures of porphyrin in aqueous solution (*15*). There is an equilibrium between monomer and μ-oxo dimer and, in alkaline solutions, μ-oxo dimer is the major structure. The dimer may react with oxidant after changing to the monomer. Hence, alkaline conditions may be unfavorable in this oxidation model.

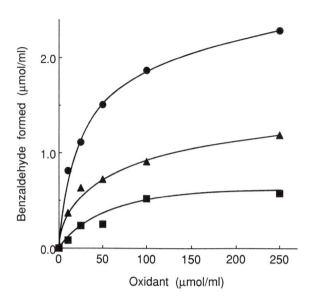

Figure 3.   The effect of oxidants on the dealkylation of NDBzA by FeTPPCl. NDBzA (500μmol), FeTPPCl (0.5μmol) and oxidant (0-250μmol) were incubated in benzene 1 ml at 25°C for 20 min. *t*-BuOOH(●), Cumene-OOH(▲), PhIO(■).

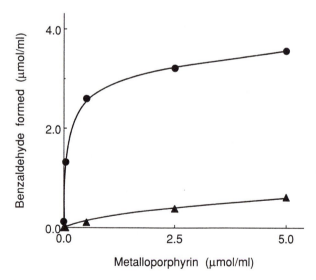

Figure 4.   The effect of metals on the dealkylation of NDBzA by metalloporphyrin and *t*-BuOOH. NDBzA (500μmol), metalloporphyrin (0-5μmol) and *t*-BuOOH (50μmol) were incubated in benzene 1 ml at 25°C for 20 min with FeTPPCl(●) or for 3 h with MnTPPCl(▲).

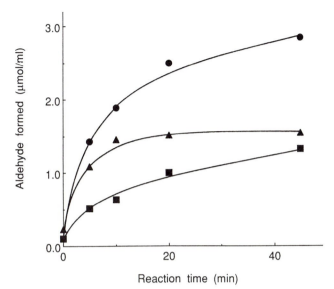

Figure 5.   The time course of formation of aldehydes in the dealkylation of
N-nitrosodialkylamines by FeTPPS and *t*-BuOOH. *N*-Nitrosodialkylamine (500
μmol), FeTPPS (0.5μmol) and *t*-BuOOH (50μmol) were incubated in CH₃CN/H₂O
(1:1) 1 ml at 37°C. NMBA(●), NDBzA(▲), NDBA(■).

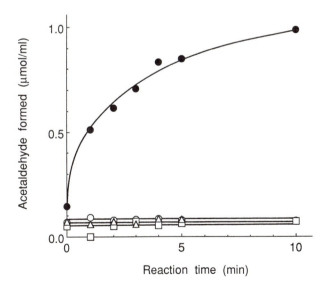

Figure 6.   The time course of formation of acetaldehyde in the dealkylation of
NDEA by FeTPPS and *t*-BuOOH. NDEA (500μmol), FeTPPS (0.5umol) and
t-BuOOH (250μmol) were incubated in 0.1M phosphate buffer (pH 7.4) 1 ml at
37°C. Complete system(●), NDEA(−)(○), FeTPPS(−)(□), *t*-BuOOH(−)(△).

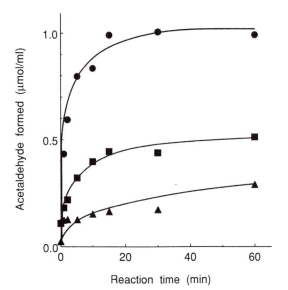

Figure 7.   The effect of oxidants on the dealkylation of NDEA by FeTPPS.
NDEA (500μmol), FeTPPS (0.5μmol) and oxidant (100μmol) were incubated in
0.1M phosphate buffer (pH 7.4) 1 ml at 37°C. *t*-BuOOH(●), Cumene-OOH(■),
H_2O_2(▲).

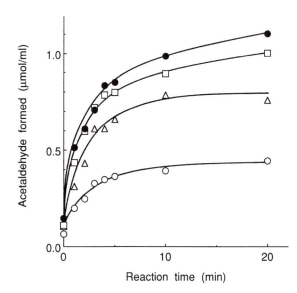

Figure 8.   The effect of *t*-BuOOH concentration on the dealkylation of NDEA
by FeTPPS. NDEA (500μmol), FeTPPS (0.5μmol) and *t*-BuOOH (0-250μmol)
were incubated in 0.1M phosphate buffer (pH 7.4) 1 ml at 37°C. *t*-BuOOH
250μmol(●), 100μmol(□), 50μmol(△), 10μmol(○).

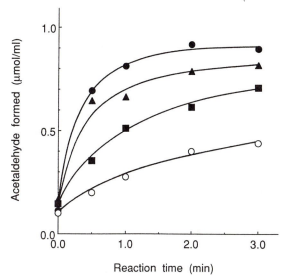

Figure 9. The effect of pH on the dealkylation of NDEA by FeTPPS and *t*-BuOOH. NDEA (500μmol), FeTPPS (0.5μmol) and *t*-BuOOH (250μmol) were incubated in 0.1M phosphate solutions 1 ml at 37°C. pH 5(●), pH 6(▲), pH 7.4(■) and pH 9(○).

Bacterial Mutation Assay

Because dealkylation is considered to be derived from α-hydroxy nitrosamine, one expects to see mutagenicity from these intermediate compounds. A bacterial mutation assay was employed to investigate the formation of alkylating species from *N*-nitrosodialkylamine in the model systems. The mutation test was based on the Ames method except a metalloporphyrin complex and oxidant were used rather than an S9 mix. The bacterial strains used were *Salmonella typhimurium* TA1535 and *E. coli* WP2*hcr*⁻. Figure 10 shows the mutation of NDBzA activated in a non-aqueous model system by FeTPPCl and *t*-BuOOH. Figure 11 shows the mutagenic activity of NDBA and NMBA in the presence of aqueous FeTPPS and *t*-BuOOH. In both *Salmonella typhimurium* and *E. coli*, the mutagenicity had a linear dose relationship to the nitrosamines dosage. In both the model system and the S9 mix from rat liver (*16*), the mutagenicity of NMBA in *Salmonella typhimurium* tended to be stronger than in *E. coli* (Figure 12).

The mutagenicity observed here is rather weak and the activity was sensitive to the assay conditions. Although, the formation of aldehydes was used as a chemical marker for the α-hydroxylation of *N*-nitrosodialkylamines, aldehydes are formed from both α-hydroxylation and denitrosation of *N*-nitrosodialkylamines (*1*). Besides, alcohols formed from the decomposition of α-hydroxy nitrosamines may be slowly oxidized to aldehydes by the models, and the aldehydes themselves may be slowly oxidized. Thus, a further investigation with more efficient models are required for elucidation of a precise mechanism of reactions.

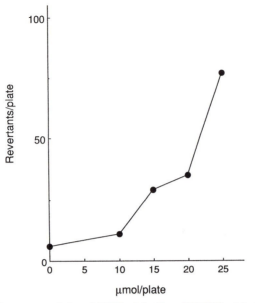

Figure 10.   The mutagenicity of NDBzA in *E. coli* WP2*hcr⁻* in the presence of
FeTPPCl and *t*-BuOOH. Conditions: FeTPPCl 0.1 nmol/plate, *t*-BuOOH 10
nmol/plate.

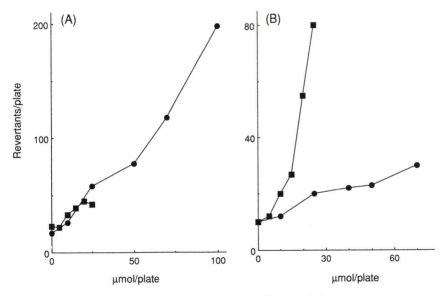

Figure 11.   The mutagenicity of NMBA and NDBA in *Salmonella typhimurium*
TA1535(A) and *E. coli* WP2*hcr⁻*(B) in the presence of FeTPPS and *t*-BuOOH.
Conditions: FeTPPS 0.1 nmol/plate, *t*-BuOOH 10 nmol/plate. NMBA(●) and
NDBA(■).

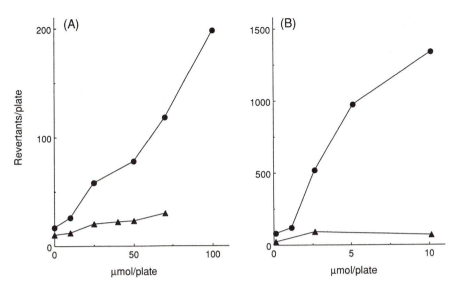

Figure 12.   The mutagenicity of NMBA in *Salmonella typhimurium* TA1535 and *Escherichia coli* WP2*hcr*⁻ in the presence of FeTPPS and *t*-BuOOH(A)[a] or S9 mix(B)[b]. *Salmonella typhimurium* TA1535(●) and *E. coli* WP2*hcr*⁻(▲). [a]Conditions: FeTPPS 0.1 nmol/plate, *t*-BuOOH 10 nmol/plate. [b]Suzuki et al.(*16*) S9 mix was prepared from liver of rats induced by phenobarbital.

Conclusion

N-Nitroso compounds are metabolically activated and inactivated by cytochrome P-450. In this study, the chemical model systems for cytochrome P-450 were used in non-aqueous and aqueous conditions to investigate dealkylation and mutagenicity due to the activation of *N*-nitrosodialkylamines. Dealkylation was observed by the formation of aldehydes with the model system using metalloporphyrin and oxidant. A direct mutation due to alkylating species from NMBA also occurred in *Salmonella typhimurium* TA1535 and *E. coli* WP2*hcr*⁻ in the model system. In the model system, the mutagenic activity was stronger in *Salmonella typhimurium* than in *E. coli*, which was similar to the mutagenicity in the presence of S9 mix from rat liver. In the model reaction, oxygen from the oxidants moved via a metal complex to *N*-nitrosodialkylamine to give α-hydroxy nitrosamine. From these results, one can conclude that oxidation by metalloporphyrins/oxidants can mimic the metabolic activation of *N*-nitroso-dialkyamine by cytochrome P-450. These investigations are useful in elucidating the mechanisms of alkylating carcinogens, and also in finding factors which inhibit activation and prevent human cancer.

Acknowledgments

This work was supported in part by a Grant-in-Aid for Cancer Research from the Ministry of Health and Welfare, Japan.

Literature Cited

1. Keefer, L. K.; Anjo, T.; Wade, D.; Wang, T.; Yang, C. S. *Cancer Res.* **1987**, *47*, 447-452.
2. Wade, D.; Yang, C. S.; Metral, C. J.; Roman, J. M.; Hrabie, J. A.; Riggs, C. W.; Anjo, T.; Keefer, L. K.; Mico, B. A. *Cancer Res.* **1987**, *47*, 3373-3377.
3. Stiborova, M.; Frei, E.; Schmeiser, H. H.; Wiessler, M.; Anzenbacher, P. *Cancer Lett.* **1992**, *63*, 53-59.
4. Mansuy, D.; Battioni, P.; Battioni, J. -P. *Eur. J. Biochem.* **1989**, *184*, 267-285, and the references cited therein.
5. Groves, J. T.; Nemo, T. E.; Myers, R. S. *J. Am. Chem. Soc.* **1979**, *101*, 1032-1033.
6. Lindsay Smith, J. R.; Mortimer, D. N. *J. Chem. Soc. Perkin Trans. II* **1986**, 1743-1749, and the references cited therein.
7. Lindsay Smith, J. R.; Nee, M. W.; Noar, J. B.; Bruice, T. C. *J. Chem. Soc. Perkin Trans. II* **1984**, 255-260.
8. Chauhan, S. M. S.; Satapathy, S. *Indian J. Chem.* **1991**, *30B*, 697-699.
9. Salmeen, I. T.; Aken, S. F. -V.; Ball, J. C. *Mutation Res.* **1988**, *207*, 111-115.
10. Wood, M. L.; Lindsay Smith, J. R.; Garner, R. C. *Mutation Res.* **1987**, *176*, 11-20.
11. Nash, T. *Biochem. J.* **1953**, *55*, 416-421.
12. Lindsay Smith, J. R.; Lower, R. J. *J. Chem. Soc. Perkin Trans. II* **1991**, 31-39.
13. Battioni, P.; Renaud, J. P.; Bartoli, J. F.; Reina-Artiles, M.; Fort, M.; Mansuy, D. *J. Am. Chem. Soc.* **1988**, *110*, 8462-8470, and the references cited therein.
14. Fung, K.; Grosjean, D. *Anal. Chem.* **1981**, *53*, 168-171.
15. El-Awady, A. A.; Wilkins, P. C.; Wilkins, R. G. *Inorg. Chem.* **1985**, *24*, 2053-2057.
16. Suzuki, E.; Mochizuki, M.; Takeda, K.; Okada, M. *Jpn. J. Cancer Res. (Gann)* **1985**, *76*, 184-191.

RECEIVED March 30, 1993

Chapter 14

Kinetics and Enzymes Involved in the Metabolism of Nitrosamines

Chung S. Yang, Theresa J. Smith, Jun-Yan Hong, and Shiqi Zhou

Laboratory for Cancer Research, College of Pharmacy, Rutgers University, Piscataway, NJ 08855-0789

The metabolic activation of nitrosodialkylamines is mainly catalyzed by cytochrome P450 enzymes. The catalytic activity of each P450 enzyme towards nitrosamines are determined by the structure of the alkyl groups. The distribution of the various P450 forms in different tissues are important in determining the tissue specificity of different nitrosamines. P450 2E1 has been shown to be a key enzyme in the activation of nitrosamines with low molecular weight, especially with methyl and ethyl groups. However, in the bioactivation of more complex nitrosamines, such as the tobacco carcinogen 4-(methylnitrosamino)-1-(3-pyridyl)-1-butanone, other enzymes may be more important. In tissues containing low levels of P450 enzymes, other enzymes may be important in the activation of nitrosamines.

Since the discovery of the carcinogenicity of nitrosamines, the organ and tissue specificity of these compounds have intrigued many investigators (*1, 2, 3*). In order to understand the molecular basis of these specificities or selectivities, we have extensively characterized the enzymes and mechanisms for the metabolism of many nitrosamines during the past ten years. In the present communication, we discuss the kinetics and enzymes involved in the metabolism of nitrosodialkylamines, from the simplest, *N*-nitrosodimethylamine (NDMA), to the more complex tobacco carcinogen, 4-(methylnitrosamino)-1-(3-pyridyl)-1-butanone (NNK), in their respective target tissues in mice, rats, and humans.

Mechanisms of Dealkylation and Denitrosation of Nitrosamines

It has long been recognized that α-hydroxylation is the key pathway leading to the formation of carcinogenic alkylating agents (*1*). The general pathways are illustrated in Figure 1. It is recognized by now that, in most of the cases studied, this pathway is due to an oxygenation of the α-carbon catalyzed by cytochrome P450 enzymes (*2*). The resulting hydroxylated intermediates (α-nitrosamino alcohol) which are not stable are readily rearranged to form an aldehyde and alkydiazohydroxide. In the specific example shown in Figure 1, the latter is a methylating agent which can react with DNA and other cellular molecules, leading to possible carcinogenesis and cytotoxicity. This is the well recognized dealkylation pathway for the activation of nitrosamines. The initial oxidation step is believed to form an α-nitrosamino radical and an ·OH at the P450 active site. These two radical species can recombine, in a process known as oxygen rebound, to produce the α-nitrosamino alcohol intermediate. Alternatively, the α-nitrosamino radical may fragment into nitric oxide and an imine which is readily hydrolyzed to an aldehyde and methylamine. The nitric oxide is oxidized to nitrite, possibly involving ·O_2 as an oxidizing agent (*4*). Since known alkylating agents are not produced, this denitrosation pathway is generally considered to be a detoxification pathway. The evidence generated in our laboratory indicated that the dealkylation pathway, sharing an α-nitrosamino radical intermediate with the dealkylation pathway, is an oxidative pathway. This is different from the views of other authors suggesting that denitrosation is a reductive pathway (*5*).

Via a similar mechanism, the other alkyl group (in this case a methyl group) can also undergo α-oxidation, leading to the demethylation and denitrosation pathways as described previously. The relative rates for the dealkylation and denitrosation reactions depend upon the structures of the nitrosamines and P450 enzymes for the catalysis. For example, with NDMA, the ratio of the rates' of demethylation (formaldehyde formation) to denitrosation is about 8 : 1. In the metabolism of *N*-nitrosomethylpentylamine, catalyzed by P450 2B1, the rates of the depentylation, demethylation, and denitrosation are at ratios of 100:50:25. In addition, the carbons of the pentyl moiety group can also be oxidized, forming 2-OH, 3-OH, 4-OH, and 5-OH derivatives at rate ratios (the same unit as above) of 0.6 : 1.5 : 26 : 2.4, much lower than the dealkylation rates (*6*).

Roles of Cytochrome P450 2E1 in the Metabolism of Nitrosamines

The vital role of P450 2E1 in the metabolism of NDMA is supported by several lines of evidence. Among all P450 enzymes isolated from rabbits, rats, and humans, P450 2E1 enzymes show the lowest K_m and highest k_{cat}/K_m in catalyzing the oxidation of NDMA (*7-11*). With purified P450 2E1 in the reconstituted system, a K_m value of 3.5 mM has been observed (*8*), much higher than the value of 20 μM observed in rat and human liver microsomes. This is probably due to the presence of glycerol and possibly other unknown inhibitors in the preparations of purified P450 2E1 and NADPH:P450 oxidoreductase. Recent work with heterologously expressed human

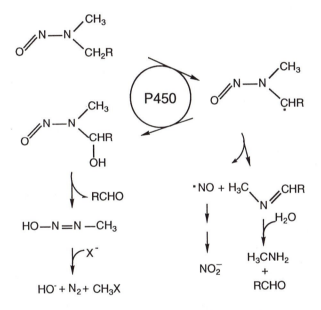

Figure 1. A general scheme for the metabolism of nitrosamines

P450 2E1 in Hep G2 cells indicates that the P450 2E1 containing microsomes displays a low K_m of 20 μM in the NDMA demethylase assay (*12*), confirming that P450 2E1 is responsible for the low K_m form of NDMA demethylase. In comparison to other P450 forms, P450 2E1 is also the form which most actively activates NDMA to a mutagen for the V79 cells (*13*). The predominant role of P450 2E1 in the metabolism of NDMA is suggested by the results that antibodies against P450 2E1 inhibit more than 80% of the NDMA demethylase and denitrosation activities in rat and human microsomes (*11, 14*).

P450 2E1 is also very effective in the metabolism of N-nitrosodiethylamine (NDEA). Antibodies against P450 2E1 inhibit NDEA deethylase in rat and human microsomes by 50 and 60%, respectively, suggesting that this enzyme also plays an important role in the metabolism of NDEA, but not as dominant a role as in the metabolism of NDMA (*11, 14*).

The enzyme specificity and alkyl group selectivity in the metabolism of nitrosamines have been investigated. It seems that the structures of the alkyl groups are more important than the nitroso group in determining the enzymes specificity in the metabolism of dialkylnitrosamines. Our tentative conclusion is that P450 2E1 is more effective in oxidizing the α-carbon of methyl and ethyl groups whereas other P450 forms, such as P450 2B1, are more effective in catalyzing the dealkylation of larger hydrocarbon chains (*15*). However, the alkyl group selectivity is an event which is dependent on the concentration of the substrate. When more sensitive methods become available for assaying the enzyme activity at lower substrate concentrations, this tentative conclusion may change.

Kinetic Parameters in the Metabolism of N-Nitrosomethylbenzylamine

N-Nitrosomethylbenzylamine (NMBzA) is a potent and rather specific esophageal carcinogen in rats. The molecular basis for the high specificity toward esophagus has attracted the attention of many investigators (*16, 17*). It is likely that NMBzA is very efficiently activated in the esophagus. Indeed, it was demonstrated recently that NMBzA is oxidized to benzaldehyde and benzoic acid, an activation pathway leading to the formation of a methylating agent, by an enzyme or enzymes displaying apparent K_m values of 3-5 μM in rat esophageal microsomes (*18*). This K_m value is about one tenth of the K_m value observed in liver microsomes (in the formation of benzaldehyde and benzyl alcohol); the V_{max} for the esophageal microsomes is approximately one fifth that of the liver microsomes. Again P450 2E1 is an important enzyme in the metabolism of micromolar concentrations of NMBzA in the liver microsomes; the activity is inducible by pretreatment of rats with acetone, an inducer of P450 2E1 (*18*). However, this enzyme is only present at low levels in the esophagus. The presence of P450 2E1 in esophagus is supported by immunoblot analysis, NDMA demethylase activity, and the presence of CYP 2E1 sequence in clones from a cDNA library of the rat esophagus. It appears that in rat esophagus, NMBzA is only partially (up to 20%) metabolized by P450 2E1; most of the activity and the low K_m form is probably due to other P450 enzymes. The identity of these enzymes remain to be established. NMBzA is also oxidized by human liver microsomes, with P450 2E1 again playing an important role (*11*). With human esophageal microsomes, NMBzA is metabolized but at a much lower rate than in rat esophageal microsomes. The possible existence of NMBzA in the human diet and its role in the causation of human esophageal cancer remains to be investigated.

Enzymology of the Metabolism of NNK

NNK is a potent tobacco-specific carcinogen and is believed to be a human carcinogen, especially in the causation of lung and oral cancers. The major pathways for NNK metabolism have been established (*19*). Again, the major activation pathways are through the α-oxidation of the methyl and methylene carbons, leading to the formation of a pyridyloxobutylating agent and a methylating agent, respectively (Figure 2). The detectable products for these pathways are keto alcohol [4-hydroxy-1-(3-pyridyl)-1-butanone] and keto aldehyde [4-oxo-1-(3-pyridyl)-1-butanone], respectively. In rat liver, lung, and nasal mucosa microsomes, mouse liver and lung microsomes, and human liver microsomes, NNK is mainly oxidized to these products by P450 enzymes (*20-25*). In rat liver microsomes, the reduction of NNK to NNAL [4-(methylnitrosamino)-1-(3-pyridyl)-1-butanol], catalyzed by carbonyl reductase, is the major route of metabolism whose rate is about 10 times higher than the oxidative pathway. Many P450 forms are capable of catalyzing the oxidization of NNK to different products displaying K_m values of 100-250 µM. The activity is inducible by commonly used P450 inducers such as phenobarbital, 3-methylcholanthrene, Aroclor 1254, pregnenolone 16α-carbonitrile, safrole, and isosafrole, but not by acetone (*25*). In a reconstituted system, purified P450 2B1 efficiently catalyzes the formation of keto aldehyde, keto alcohol, and NNK-*N*-oxide from NNK displaying K_m values of 130-300 µM (*22*). In the rat and mouse lung microsomes, however, the K_m values are much lower (3-30 µM). Antibodies against P450s 2A1 and 2B1 inhibit most of the activities for the formation of keto aldehyde. In addition, antibodies against P450 1A2 also significantly inhibit the oxidation of NNK to keto aldehyde in rat lung microsomes. The results suggest that these three P450 forms or immunochemically related forms are important in catalyzing the activation of NNK.

In comparison to other rodent samples studied, nasal mucosa microsomes possess the highest activity in catalyzing the oxidation of NNK (*24, 26*). With rat nasal mucosa microsomes, K_m values of approx. 10 µM and V_{max} values of approx. 3 nmol/min/mg protein have been observed for the formation of both keto aldehyde and keto alcohol (*24*). Immunoinhibition studies suggest the involvement of P450s 1A2, 2A1, and 3A in this pathway (*24*). With rats, microsomes from the olfactory nasal mucosa are 4- to 5-fold more active than those from the respiratory mucosa in catalyzing the oxidation of NNK to keto aldehyde and keto alcohol. With rabbit nasal microsomes, olfactory microsomes are also more active than respiratory microsomes but the difference is less than 2-fold (*27*). In addition, NNK-*N*-oxide is formed in rabbit nasal mucosa microsomes but not in rat nasal mucosa microsomes. Purified

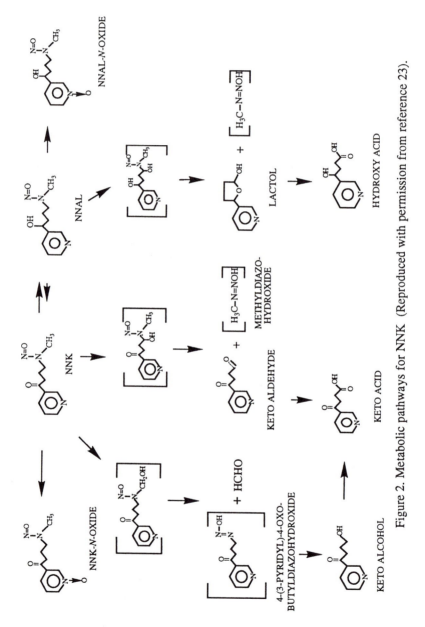

Figure 2. Metabolic pathways for NNK (Reproduced with permission from reference 23).

P450 NMa, a major constitutive P450 in rabbit nasal microsomes, shows low K_m values (9-15 μM) and high V_{max} values (1.3 nmol/min/nmol P450) in catalyzing the formation of keto aldehyde and keto alcohol. The olfactory specific P450 NMb has a higher K_m (180 μM) and low activity in catalyzing NNK oxidation (27). These studies indicate that in nonhepatic tissues, there are a few low K_m enzymes which catalyze the activation of NNK. In the liver, many P450 enzymes have the ability to catalyze the oxidation of NNK. In addition, the reduction of NNK is the major pathway in hepatic metabolism. The resulting NNAL can be oxidized in a manner similar to the oxidation of NNK, or perhaps more readily, be converted to a glucuronide conjugate and excreted. These metabolic properties may account for the lower susceptibility of the liver than the lung to NNK carcinogenesis.

Metabolism of NNK by Human P450s and Microsomes

Among the 12 human P450 forms expressed in Hep G2 cells, P450 1A2 has the highest activity, and P450s 2A6, 2B6, 2E1, 2F1, and 3A5 also have measurable activities in catalyzing the formation of keto alcohol. P450s 2C8, 2C9, 2D6, 3A3, 3A4, and 4B1 have no detectable activity (23). Keto aldehyde is not formed by all these P450 forms in Hep G2 cells. Low activity is also detected in P450 2D6 expressed in a B-lymphoblastoid cell line (28). The contribution of each of these forms in human tissues is dependent on the abundance and the activity of each enzyme. In one human liver microsome sample that we studied, keto alcohol was the major oxidative product formed. Immunoinhibition study suggests that P450s 1A2 and 2E1 each account for approximately 50% of the activity; whereas P450s 2C8, 2D1, 2D6, and 3A4 are not involved (23). With human liver microsomes, carbon monoxide inhibition study suggests that P450 enzymes contribute to 90% and 50% of the activities in the formation of keto alcohol and keto aldehyde, respectively. On the other hand, in human lung microsomes, keto aldehyde is the major oxidative metabolite and the activity was only inhibited 10-40% by carbon monoxide, dependent on the specific lung samples investigated (23). The results suggest the importance of P450-independent pathways in the activation of NNK in human lung. The identities of the enzymes in catalyzing these pathways is being investigated. Because of this significant species difference, caution has to be applied in extrapolating conclusions obtained from studies with animals to humans.

Inhibition of Nitrosamine Metabolism and Carcinogenesis by Dietary Chemicals

Many dietary constituents have been demonstrated to inhibit tumorigenesis and are considered as chemopreventive agents. Knowledge on the metabolism of nitrosamines provides a basis for us to study the mechanisms by which dietary chemicals affect nitrosamine bioactivation and carcinogenesis. For example, diallyl sulfide (DAS), a component derived from garlic, has been shown to affect P450 enzymes rather selectively. It inactivates P450 2E1 but induces P450 2B1 in the rat liver (29). Further studies indicate that DAS is metabolized to diallyl sulfoxide and then to diallyl sulfone which is a suicide inactivator of P450 2E1 (30). In addition, all three compounds are competitive inhibitors of P450 2E1-catalyzed reactions. These inhibitory actions are believed to be responsible for the protective effects of DAS against the toxicity or carcinogenicity caused by substrates of P450 2E1, such as NDMA, CCl_4, and 1,2-dimethylhydrazine (29, 31). Although the strong colon carcinogen 1,2-dimethylhydrazine is not a nitrosamine, its metabolites azoxymethane and methylazoxymethane are substrates of P450 2E1 and are metabolized similarly to NDMA to methyldiazonium hydroxide (31). In addition to the aforementioned substrates, P450 2E1 also catalyzes the oxidation of a variety of low molecular weight noncharged molecules, including ethanol and acetone (32). These substrates are also good competitive inhibitors for the metabolism of carcinogens such as NDMA. It is recognized that inhibition of the metabolic activation of NDMA in the liver would protect the liver against the deleterious effects of NDMA. However, the decrease in the first-pass removal of this compound would increase its concentration in nonhepatic tissues and thus may enhance its carcinogenicity in nonhepatic tissues (33, 34). It has been demonstrated that ethanol inhibited the metabolism of NMBzA in liver microsomes but not in esophageal microsomes (18). This result is consistent with the observation that alcohols enhanced DNA methylation by NMBzA (17) and suggests a mechanism for the enhancement of esophageal carcinogenesis by alcohol consumption.

The inhibition of NMBzA metabolism in rat esophagus by phenethyl isothiocyanate (PEITC), a compound derived from cruciferous vegetables, is likely to be the main mechanism for the inhibition of esophageal carcinogenesis by PEITC (35). Oral administration of PEITC to rats in acute or chronic feeding experiments markedly decreased the activity of the enzyme(s) which catalyzes the debenzylation of NMBzA in esophageal microsomes. With an acute dose, the effect in the esophagus was extensive (>95% inhibition) and lasted more than 24 hrs; whereas the effect in the liver was moderate (~50% inhibition) and vanished at 24 hrs. When added to the incubation with esophageal microsomes, PEITC was a very potent inhibitor for NMBzA debenzylation, showing a IC_{50} of 50 nM (18).

Both DAS and PEITC, when given orally, effectively decrease the activity for the oxidative metabolism of NNK in mouse lung microsomes, and this is believed to be the mechanism for their strong inhibitory effects against lung tumorigenesis. The mechanisms of PEITC action have been extensively investigated. It is a very potent competitive inhibitor for the low K_m enzyme(s) which catalyzes the activation of NNK in mouse lung, showing K_i values of 51-93 nM (31, 36). Through its chemical reactivity, PEITC also inactivates enzymes involved in NNK bioactivation. Again, the inhibitory effect in the lung is more extensive and long lasting than in the liver. This property is advantageous for the purpose of chemoprevention because it causes inhibition of the metabolic activation of carcinogens in specific target organs without significantly affecting the xenobiotic metabolism in the liver. These examples illustrate that investigation on the modulation of nitrosamine metabolism by dietary compounds not only helps us to understand the mechanisms of inhibition of carcinogenesis but may also be useful for predicting anti-carcinogenic activities of dietary chemicals for screening chemopreventive agents.

References

1. Magee, P. N. and Barnes, J. M. *Adv. Cancer Res.* **1967**, 10, 163-246.
2. Yang, C. S.; Smith, T.; Ishizaki, H. and Hong, J.-Y. In *Relevance to Human Cancer of N-Nitroso Compounds, Tobacco Smoke and Mycotoxins* O'Neill, I. K.; Chen, J. C. and Bartsch, H. Eds., IARC Publications, Lyon, France, **1990**, pp 291-299.
3. Lijinsky, W. In *Relevance to Human Cancer of N-Nitroso Compounds, Tobacco Smoke and Mycotoxins* O'Neill, I. K.; Chen, J. C. and Bartsch, H. Eds., IARC Publications, Lyon, France, **1990**, pp 305-310.
4. Wade, D.; Yang, C. S.; Metral, C. J.; Roman, J. M.; Hrabie, J. A.; Riggs, C. W.; Anjo, T.; Keefer, L. K. and Mico, B. A. *Cancer Res.* **1987**, 47, 3373-3377.
5. Appel, K. E.; Ruhl, C. S. and Hildebrandt, A. G. *Chem. Biol. Interactions* **1985**, 53, 69-76.
6. Ji, C.; Mirvish, S. S.; Nickols, J.; Ishizaki, H.; Lee, M. J. and Yang, C. S. *Cancer Res.* **1989**, 49, 5299-5304.
7. Yang, C. S.; Tu, Y. Y.; Koop, D. R. and Coon, M. J. *Cancer Res.* **1985**, 45, 1140-1145.
8. Patten, C. J.; Ning, S. M.; Lu, A. Y. H. and Yang, C. S. *Arch. Biochem. Biophys.* **1986**, 251, 629-638.
9. Levin, W.; Thomas, P. E.; Oldfield, N. and Ryan, D. E. *Arch. Biochem. Biophys.* **1986**, 248, 158-165.
10. Thomas, P. E.; Bandiera, S.; Maines, S. L.; Ryan, D. E. and Levin, W. *Biochemistry* **1987**, 26, 2280-2289.
11. Ishizaki, H.; Yoo, J.-S. H.; Guengerich, F. P. and Yang, C. S. **1992**, (In preparation).
12. Patten, C.; Aoyama, T.; Lee, M. J.; Ning, S. M.; Gonzalez, F. and Yang, C. S. *Arch. Biochem. Biophys.* **1992**, 299, 163-171.
13. Yoo, J.-S. H. and Yang, C. S. *Cancer Res.* **1985**, 45, 5569-5574.
14. Yoo, J.-S. H.; Ishizaki, H. and Yang, C. S. *Carcinogenesis (Lond.)* **1990**, 11, 2239-2243.
15. Lee, M.; Ishizaki, H.; Brady, J. F. and Yang, C. S. *Cancer Res.* **1989**, 49, 1470-1474.
16. Mehta, R.; Labuc, G. E. and Archer, M. C. *J. Natl. Cancer Inst.* **1984**, 72, 1443-1447.
17. Ludeke, B.; Meier, T. and Kleihues, P. In *Relevance to Human Cancer of N-Nitroso Compounds, Tobacco Smoke and Mycotoxins* O'Neill, I. K.; Chen, J. C. and Bartsch, H. Eds., IARC Publications, Lyon, France, **1990**, 2. pp 286-293.
18. Zhou, S.; Ishizaki, H. and Yang, C. S. (in preparation).

19. Hecht, S. S.; Young, R. and Chen, C.-H. B. *Cancer Res.* **1980**, 40, 4144-4150.
20. Smith, T. J.; Guo, Z.-Y.; Thomas, P. E.; Chung, F.-L.; Morse, M. A.; Eklind, K. and Yang, C. S. *Cancer Res.* **1990**, 50, 6817-6822.
21. Guo, Z.; Smith, T. J.; Thomas, P. E. and Yang, C. S. *Cancer Res.* **1991**, 51, 4798-4803.
22. Guo, Z.; Smith, T. J.; Ishizaki, H. and Yang, C. S. *Carcinogenesis (Lond.)* **1991**, 12, 2277-2282.
23. Smith, T. J.; Guo, Z.; Gonzalez, F. J.; Guengerich, F. P.; Stoner, G. D. and Yang, C. S. *Cancer Res.* **1992**, 52, 1757-1763.
24. Smith, T. J.; Guo, Z.-Y.; Hong, J.-Y.; Ning, S. M.; Thomas, P. E. and Yang, C. S. *Carcinogenesis (Lond.)* **1992**, 13, 1409-1414.
25. Guo, Z.; Smith, T. J.; Thomas, P. E. and Yang, C. S. *Arch. Biochem. Biophys.* **1992**, 298, 279-286.
26. Hong, J.-Y.; Smith, T.; Lee, M.-J.; Li, W.; Ma, B.-L.; Ning, S.-M.; Brady, J. F.; Thomas, P. E. and Yang, C. S. *Cancer Res.* **1991**, 51, 1509-1514.
27. Hong, J.-Y.; Ding, X.; Smith, T. J.; Coon, M. J. and Yang, C. S. *Carcinogenesis (Lond.)* **1992**, 13, 2141-2144.
28. Penman, B. W.; Reece, J.; Smith, T.; Yang, C. S.; Gelboin, H. V.; Gonzalez, F. J. and Crespi, C. L. *Pharmacogenetics* **1992**, (in press).
29. Brady, J. F.; Wang, M.-H.; Hong, J.-Y.; Xiao, F.; Li, Y.; Yoo, J.-S. H.; Ning, S. M.; Fukuto, J. M.; Gapac, J. M. and Yang, C. S. *Toxicol. Appl. Pharmacol.* **1991**, 108, 342-354.
30. Brady, J. F.; Ishizaki, H.; Fukuto, J. M.; Lin, M. C.; Fadel, A.; Gapac, J. M. and Yang, C. S. *Chem. Res. Toxicol.* **1991**, 4, 642-647.
31. Sohn, O. S.; Ishizaki, H.; Yang, C. S. and Fiala, E. S. *Carcinogenesis (Lond.)* **1991**, 12, 127-131.
32. Swanson, B. A.; Dutton, D. R.; Lunetta, J. M.; Yang, C. S. and Ortiz de Montellano, P. R. *J. Biol. Chem.* **1992**, 266, 19258-19264.
33. Griciute, L.; Castegnaro, M. and Bereziat, J.-C. *Cancer Lett.* **1981**, 13, 345-352.
34. Anderson, L. M. *Carcinogenesis (Lond.)* **1988**, 9, 1717-1719.
35. Stoner, G. D.; Morrissey, D. T.; Heur, Y.-H.; Daniel, E. M.; Galati, A. J. and Wagner, S. A. *Cancer Res.* **1991**, 51, 2063-2068.
36. Smith, T. J.; Guo, Z.; Li, C.; Thomas, P. E. and Yang, C. S. (submitted for publication).

RECEIVED November 1, 1993

Chapter 15

Potential Mechanism of Action of Nitrosamines with Hydroxy, Oxo, or Carboxy Groups

G. Eisenbrand and C. Janzowski

Department of Chemistry, Food Chemistry and Environmental
Toxicology, University of Kaiserslautern, 6750 Kaiserslautern, Germany

Potential activation mechanisms for nitrosamines with OH-, oxo- and carboxy groups include Cytochrome P-450 (P-450) mediated α-oxidation, ß-oxidation by alcohol dehydrogenase (ADH) or by enzymes of the mito-chondrial fatty acid oxidation machinery and sulfate conjugation. Dietha-nolnitrosamine (NDELA) induces dose-dependent O^6-deoxyguanosine (dG)-hydroxyethylation of mouse and rat liver DNA. O^6-dG-hydroxyethylation and DNA single strand break (SSB) induction in rat liver can be abolished by applying an ADH-inhibitor. However, nitroso-2-hydroxymorpholine, the putative ADH-metabolite of NDELA, although inducing DNA SSB in rat liver, does not hydroxyethylate O^6-dG. Although sulfotransferase inhibitors alleviate DNA SSB induction in rats, 3′phosphoadenosine-5′-phosphosulfate (PAPS)-deficient (brachymorphic) mice show neither significantly reduced DNA SSB nor O^6-hydroxyethylation. Thus, although NDELA clearly is a hydroxyethylating agent, the underlying mechanism of activation is not yet clear.
  Butyl-3-carboxypropylnitrosamine (BCPN) is the proximate meta-bolite of dibutylnitrosamine (DBN) relevant for induction of urinary bladder tumors. Its methyl analog, methyl-3-carboxypropylnitrosamine (MCPN) is a metabolite in common to even chain alkyl-methyl-nitrosamines with carcinogenic effectiveness to the bladder. BCPN and MCPN are activated by α-C-hydroxylation at low rates; they also undergo ATP-dependent mitochondrial ß-oxidation to butyl-2-oxopropylnitrosamine (BOPN) and MOPN, respectively. The latter are readily activated by α-oxidation. BOPN is a strong DNA-damaging agent and a potent bacterial mutagen whereas BCPN is not; in cytogenetic assays BOPN is more potent than BCPN. In calf thymus DNA incubated with rat liver microsomal fraction MOPN formed more O^6-methyl dG than MCPN. Mitochondrial ß-oxidation and/or P-450 mediated α-oxidation thus appears to play a major role in activating those nitrosamines to bladder carcinogens.

0097–6156/94/0553–0179$08.00/0
© 1994 American Chemical Society

Some environmentally important nitrosamines appear to be activated by biotransformations other than cytochrome P-450 (P-450) mediated α-hydroxylation. An example is diethanolnitrosamine (NDELA), an extremely water soluble carcinogen that, according to a great number of experimental data, has not been found to be activated by P-450.

Biotransformation into polar metabolites does not always result in deactivation but might rather be responsible for organ-specific tumor induction via ultimate activation of a polar metabolite into a genotoxic agent. This applies, for instance, to di-n-butylnitrosamine (DBN) and butyl-4-butanolnitrosamine (BHBN) that are metabolized to butyl-3-carboxypropylnitrosamine (BCPN). Certain even chained alkylmethylnitrosamines are enzymatically degraded to methyl-3-carboxypropylnitrosamine (MCPN). BCPN and MCPN are known to be responsible for specific tumor induction in the urinary bladder by the parent nitrosamines. Their mechanism of action, however, is not clear yet.

In this paper, data on various metabolic pathways potentially relevant to activation of NDELA and of bladder carcinogenic nitrosamines are discussed, together with biological effects of parent compounds and metabolites. DNA strand breaking potential (SSB), formation of O^6-dD adducts in DNA and bacterial/mammalian mutagenicity are evaluated as a consequence of specific biotransformation.

Diethanolnitrosamine

NDELA is a relatively potent carcinogen, causing tumors of liver and nasal cavity in rats at a dosage as low as 1.5 mg/kg/day (*1*). It is, however, at best, a poor substrate for P-450 of rat liver, displaying binding characteristics in rat liver microsomal preparations according to type II (*2*) which is not related to monooxygenation. A similar binding to P-450 is observed with ethanol (*3*). Accordingly, most mutagenicity assays involving rat liver microsomal or S-9 fractions have been negative (*3-5*).

NDELA has been found to be a substrate for liver alcohol dehydrogenase (ADH) (*6,2*). Its activation to a mutagen in Salmonella typhimurium TA 100 and TA 98 with NAD/ ADH from horse liver has been demonstrated (*7*). In-vivo, NDELA is metabolized only to a minor percentage of a given dose, with up to 80% being excreted apparently unchanged in the urine. The only urinary metabolite identified so far, N-nitroso-2-hydroxyethylglycine (NHEG) is formed at a rate of less than 10% of a given dose, presumably by ADH/ aldehyde dehydrogenase (*8*) (Figure 1). N-Nitroso-2-hydroxymorpholine (NHMOR) was identified as a metabolite in vitro (*8-11*). After oral application of NHMOR to rats about 70% of the dose (1.5 mmol/kg) appeared as NHEG in urine and about 10% as NDELA (*12*). Efficient systemic metabolism of NHMOR therefore might prevent NHMOR from being detected as a metabolite formed in vivo from NDELA.

Alcohol dehydrogenase-activation. NHMOR has been found to be a direct acting mutagen (*10, 13*). When reacted with deoxyguanosine (dG) in-vitro at elevated temperature, it forms an $1,N^2$-glyoxal adduct (*14*). Because of its direct mutagenic effectiveness and since NHMOR has been found to induce DNA-damage in rat liver (> 0.2 mmol/kg b.w.) and in primary rat hepatocytes (12.5 mM) (*13, 15*), it has been considered a candidate for a proximate/ultimate carcinogen. However, studies in Fischer rats revealed rather weak carcinogenic activity of NHMOR after oral application at relatively low dosage (*16, 17*). It is unclear whether deactivating biotransformation of NHMOR at low dosage prevents it from reaching relevant target organs or whether other explanations might apply.

O^6-dG Hydroxyethylation. Formation of O^6-hydroxyethyl dG (O^6-HdG) adducts in rat liver DNA was measured 4 h after oral application of equimolar doses of NDELA and NHMOR. An immuno-slot-blot technique, using a primary rabbit antibody raised against O^6-HdG, was used for adduct determination (*18*). For confirmation, adducts were also determined by HPLC after acid hydrolysis of DNA. Dose dependent formation of O^6- HEdG was observed in rat liver DNA 4h after p.o. application of NDELA (0.15 - 2.25 mmol/kg). With NHMOR, (0.375 - 1.5 mmol/kg) O^6-HdG formation was barely detectable and dose dependence was not observed (*12, 18*) (Table I). When the noncompetitive ADH-inhibitor n-butylthiolane-1-oxide (*19*) was given simultaneously with NDELA, the extent of O^6-HdG formation dropped to almost background level (*12*). A similarly strong inhibitory effect of n-butylthiolane-1-oxide was also observed on DNA single strand break (SSB) induction in rat liver by NDELA. However, no inhibition was seen for SSB induced by NHMOR (*20*).

These results lend support to the hypothesis that ADH plays a role for in-vivo activation of NDELA to a hydroxyethylating agent. It cannot be excluded, however, that the ADH-inhibitor n-butylthiolane-1-oxide might also influence other enzymes relevant for activation. NHMOR-induced strand breaks in rat liver DNA are not paralleled by O^6-dG hydroxyethylation. Therefore, O^6-dG hydroxyethylation by NDELA does not appear to be caused via formation of NHMOR.

Potential Relevance of Sulfation. Sulfate conjugation of hydroxyethylnitrosamines as potentially activating mechanism was first proposed by Michejda et al. (*21*). Our group found a complete inhibition of NDELA induced DNA strand breaking in rat liver with the phenol sulfotransferase inhibitor 2,4-dichloronitrophenol, orally applied in 1,2-propanediol (propylene glycol, PG) (*22*). It was, however, found later by Michejda's group (*24*) that the observed abrogation of DNA-damage (*12*) can practically completely be attributed to PG, a potent competitive inhibitor of alcohol sulfotransferase (*23*).

Further evidence for O-sulfation as activating process for hydroxyethylnitrosamines was provided by the finding that PG inhibited methyl-2-hydroethylnitrosamine-induced O^6-G-methylation, 7-G-methylation and O^6-G-hydroxyethylation of rat liver DNA but not 7-G-hydroxyethylation (*24*). In-vitro, 7-G-methylation of calf thymus DNA by rat liver cytosol occurred in the presence of the sulfotransferase cofactor 3'-phosphoadenosine-5'-phosphosulfate (PAPS) but was not detectable in its absence. Furthermore, in rat primary hepatocytes, 7-G-methylation was found to be dependent on inorganic sulfate whereas O^6 -G-hydroxyethylation was not (*24*).

The extent of DNA O^6-G-hydroxyethylation by NDELA in normal mice (57

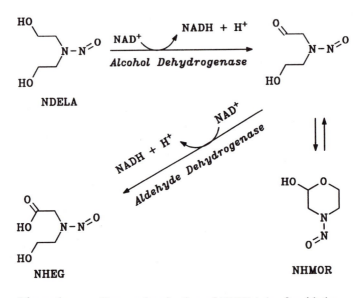

Figure 1:   Proposed activation of NDELA by β-oxidation

**Table I. Hydroxyethylation of Liver DNA after
Application of NDELA and NHMOR to Rats**

Compound	Dose (mmol/kg b.w.)	μmol O-^{6}HdG/mol dG[a]
NDELA	0.15	3.5
	0.375	6.7
	0.75	10.1
	1.5	24.2
	2.25	53.3
NHMOR	0.375	2.0 / 2.1
	1.5	5.9 / 5.1

[a] obtained from two pooled livers of identically treated wistar rats

BL) has been compared to that in PAPS deficient brachymorphic mice (bm/bm) (*12*). Only minor differences were seen in both groups (Table II). A similar experiment with methyl-2-hydroxyethylnitrosamine likewise did not show significant differences in DNA adduct formation in brachymorphic mice as compared to heterozygous siblings, except for 7-G-hydroxyethylation (*24*).

Different activation mechanisms appear to be operative for DNA strand breaking and O^6-G-hydroxyethylation by NDELA. Presumably this also applies to 7-G-adduct formation, as already proposed for methylhydroxyethylnitrosamine (*24*).

In summary, NDELA is a carcinogen that induces O^6-G-hydroxyethylation of DNA in rat and mouse liver. It is, however, not yet clear how activation to a hydroxyethylating agent proceeds. Some experimental evidence using enzyme inhibitors supports the relevance of alcohol dehydrogenase and of sulfotransferase. Other results are not in accordance, especially those monitoring O^6-G-hydroxyethylation in PAPS deficient mice. NDELA potentially acts as a substrate for various activating enzymes. If one specific pathway becomes compromised, alternative routes of activation might become relevant. Moreover, DNA-interactions other than hydroxyethylation might also be important for genotoxic efficacy of NDELA. In fact, they might even be more relevant and therefore urgently need to be elucidated, in view of the environmental importance of NDELA.

Dibutylnitrosamine

DBN is a carcinogen found in the air at specific working places and in rubber articles (*25, 26*). DBN and its ω-oxidized metabolites butyl-4-hydroxybutylnitrosamine (BHBN) and butyl-3-carboxypropylnitrosamine (BCPN) are potent urinary bladder carcinogens in rodents and other laboratory animals after oral application (*27, 28*). DBN ω-oxidation is initiated by P-450 dependent monooxygenases, followed by ADH and aldehyde dehydrogenase oxidation to generate BCPN (*29, 30, 31*). This biotransformation occurs at high rates in the liver. Structure/activity investigations with DBN and its metabolites revealed that the specific tumor induction in the urinary bladder is related to the formation of BCPN which is excreted at high rates in urine (*32*). BHBN and BCPN are selective urinary bladder carcinogens. The relevance of ω-oxidation was further underlined by experiments with fluorinated analogs of DBN in which ω-oxidation is bloked by trifluorination (*33*). Bis(4,4,4-trifluorobutyl)nitrosamine did not induce urinary bladder tumors in rats (*34*).

One possible activation mechanism of BCPN is α-hydroxylation (see Figure 2), even though its polarity should not favor monooxygenation. Alternatively, BCPN might be subject to ß-oxidation: after application of BHBN to rats, small amounts of butyl-2-oxopropylnitrosamine (BOPN) were found in urine (*35, 36*). Microsomal monooxygenases and ADH or mitochondrial (peroxisomal) enzymes of the fatty acid degradation machinery might be responsible for ß-oxidation of BCPN.

In vitro Metabolism of BCPN. Microsomal fractions from rat liver and pig urinary bladder epithelium were found to α-oxidize BCPN (Table III), both alkyl chains being dealkylated at low rates (*37*). The monooxygenase inhibitor SKF 525A (1 mM) effected a 50% decrease of debutylation (*37*). DBN and BHBN were dealkylated at

Table II. Hydroxyethylation of Liver DNA
After Application of NDELA to Mice

Dose (mmol/kg b.w.)	Mouse Strain	$\mu mol\ O\text{-}^6HdG/mol\ dG$[a]
0	normal	1.2 ± 0.2
	brachymorph	5.2 ± 0.7
0.375	normal	36.2 ± 4.0
	brachymorph	33.1 ± 0.9
0.75	normal	48.9 ± 2.4
	brachymorph	45.2 ± 2.2

[a] $n = 2$

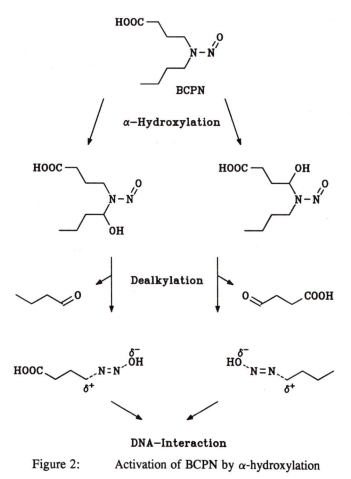

DNA–Interaction

Figure 2: Activation of BCPN by α-hydroxylation

**Table III.   Incubation of DBN, BHBN, BCPN
and BOPN with Microsomal Fraction**

Origin of microsomal fraction	Compound	generated aldehydes $[nmol \times mg\ prot^{-1} \times 60\ min^{-1}]$			
		butyraldehyde mean (SD; n)		4-oxobutyric acid mean (SD; n)	
PB induced rat liver	DBN	300	($\pm$ 22; 4)		
	BHBN	68	($\pm$ 4; 5)		
	BCPN	1.5	($\pm$ 0.3; 7)	2.8	($\pm$ 0.7; 4)
	BOPN	100	($\pm$ 11; 8)		
uninduced rat liver	DBN	60	($\pm$ 9; 6)		
	BHBN	34	($\pm$ 8; 5)		
	BCPN	1	($\pm$ 0.3; 2)	2.0	($\pm$ 0.6; 6)
	BOPN	17	($\pm$ 1.6; 4)		
pig urinary bladder	DBN	n.d	(- ; 2)		
	BHBN	1.0	($\pm$ .5; 2)		
	BCPN	0.4	($\pm$ .01; 2)	0.8	($\pm$ 0.2; 5)
	BOPN	0.8	($\pm$ 0.3; 2)		

Substrate concentration: 10 mM, Incubation 1h, 37°C
Determination of aldehyde-2.4-DNPH by HPLC/VIS$_{336}$

n.d. = not detectable [< 0.7]
PB = phenobarbital

substantial rates in liver microsomes, whereas urinary bladder microsomes produced low, but significant, α-oxidation rates. In the case of DBN, high analytical background compromised quantification of butyraldehyde.

Potential ß-oxidation of BCPN, mediated by mitochondrial enzymes in analogy to the ß-oxidation pathway of fatty acids (Figure 3) was studied in rat liver mitochondrial fractions (38). The 7000xg pellet was incubated with BCPN (0.2 mM) in the presence or absence of ATP (1 h, 37°C, protein: 0.7-1.7 mg/ml). BOPN formation, indicative for ß-oxidation, was found to be time and dose dependent (Table IV). Without addition of ATP, metabolic rates were drastically reduced. Low residual rates of BOPN formation observed without ATP might be explained by small intrinsic amounts of ATP in the mitochondrial preparations. Addition of the cofactors NAD, CoASH or carnitine did not exert significant effects on the ß-oxidation rate. BOPN formation was inhibited in the presence of octanoic acid, supporting involvement of fatty acid metabolizing enzymes. BOPN formation did not occur with 7000xg supernatant. Mitochondrial preparations from rat kidney also metabolized BCPN to BOPN.

The 7000xg pellet (rat liver) was subfractionated by percoll gradient centrifugation into mitochondria and peroxisomes. Marker enzyme activities were monitored: cytochrome c oxidase for mitochondria, catalase for peroxisomes (39). BOPN formation was almost exclusively localized in mitochondrial fraction.

In summary BCPN obviously is metabolized by α- and ß-oxidation. BCPN α-oxidation also has been observed by Airoldi et al. (40, 41) in isolated rat urinary bladder, in urinary bladder homogenate and in rat liver S-9 fraction. ß-Oxidation takes place in mitochondria, most probably following the degradation pathway of medium chain fatty acids (42, 43). As an alternative, BOPN might also be generated from BCPN by 2-monooxygenation/dehydrogenation in the intact cell. However, the putative precursor metabolite butyl-2-hydroxy-3-carboxypropylnitrosamine was not converted to BOPN by horse liver ADH in-vitro (44).

Biological Activity of BCPN and BOPN.   Low rates observed for in-vitro BCPN-oxidation may explain negative results in various short term tests with and without metabolic activation, such as bacterial mutagenicity, induction of DNA single strand breaks and induction of HPRT mutations in mammalian cells (45, 37, 44). More sensitive cytogenetic assays, however, revealed BCPN to induce micronuclei, sister chromatid exchanges, chromosomal aberrations in tumor cell lines and in primary rabbit urinary bladder epithelial cells (46-48). Micronuclei induction was enhanced in the presence of phenobarbital(PB)-induced rat liver microsomal mix and inhibited by the monooxygenase inhibitor SKF 525A. The non-genotoxic analogue butyl-tert.-butylnitrosamine was negative under these conditions. The results suggest α-oxidation by microsomal enzymes to be responsible for BCPN-activation to a mutagen.

However, ß-oxidation might be more important for biological effects since the water soluble metabolite BCPN is converted into the lipophilic metabolite BOPN. The latter is a much better substrate for α-oxidation and is debutylated at a substantial rate by P-450 monooxygenases from rat liver (49). Since it also is a better substrate than BCPN for urinary bladder microsomes, it might well represent a relevant proximate carcinogen, prone to local activation processes. BOPN is a potent mutagen (44, 48)

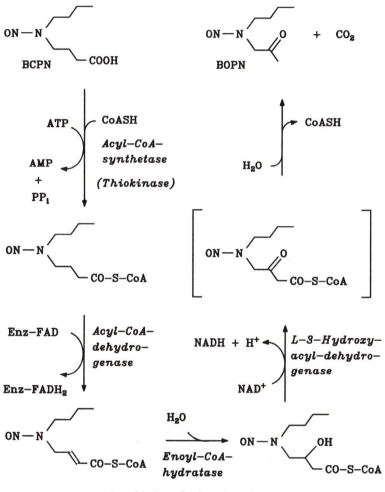

Figure 3:   β-Oxidation of BCPN by mitochondrial enzymes

and carcinogen that has been found to induce liver tumors after oral application to rats (*50*).

Biological Activity of Model Compounds (Nitrosoureas). Butylnitrosourea (BNU) and 3-carboxypropylnitrosourea (CPNU) were chosen as model compounds for α-activated BCPN, 2-oxopropylnitrosourea (OPNU) as a model compound for α -activated BOPN.

While BNU is a potent mutagen and carcinogen (*51, 52*), CPNU is a weak mutagen in a repair deficient E. coli strain (*53*). OPNU was found a relatively weak

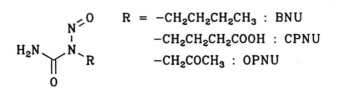

$$R = -CH_2CH_2CH_2CH_3 \; : \; BNU$$
$$-CH_2CH_2CH_2COOH \; : \; CPNU$$
$$-CH_2COCH_3 \; : \; OPNU$$

carcinogen in rats and hamsters, acting as methylating agent (*54, 55*).

BNU, CPNU and OPNU were tested for their ability to induce DNA-SSB in tumor cell lines. In bovine urinary bladder epithelial cells (Figure 4) highest genotoxic potency was observed with OPNU, CPNU being negative up to a concentration of 12 mM (*44*). In Namalva cells, however, CPNU exerted weak SSB-induction (*46*). A similar ranking of effectiveness was seen in mutagenicity experiments in Salmonella typhimurium TA 1535 (*46, 51*).

CPNU has been found to decompose in aqueous solution mainly into γ-butyrolactone (73%). Therefore, intramolecular quenching of alkylating activity might explain the low genotoxic potency of CPNU. 4-Hydroxybutyric acid (17%), vinylacetic acid (5%) and 3-hydroxybutyric (4%) were detected as minor products of aqueous decomposition.

In summary, BCPN α- **and** ß- oxidation- generates ultimate electrophiles of varying genotoxic and mutagenic potency. Highest genotoxic potency obviously is afforded α-C-hydroxylation of BOPN at the butyl chain. The same biotransformation leads, however, only to a marginally potent genotoxic agent in the case of BCPN. Activation of BCPN and BOPN can occur in liver and/or urinary bladder. For induction of urinary bladder tumors by DBN and its ω-oxidized metabolites, formation of ultimate electrophiles in the target organ is regarded as decisive.

Methylalkylnitrosamines

Certain methylalkylnitrosamines with long even-numbered alkyl chains (nC ≥ 8) show a distinct organotropic effect to the urinary bladder, whereas analogs with shorter alkyl groups induce tumors in liver, lung and esophagus (*56, 57, 58*). Some of these methylalkylnitrosamines (alkyl = dodecyl to octadecyl) have been detected as contaminants in cosmetics, dishwashing liquids and household cleaning preparations (*59-61*). The presence of methyl-3-carboxypropylnitrosamine (MCPN) in the urine of rats

Table IV. ß-Oxidation of BCPN by Incubation with Rat Mito-chondrial Fraction

origin of mitocho ndria	ATP [mM]	octanoic acid [mM]	generated BOPN [nmol x mg protein $^{-1}$ x h^{-1}] mean	(n; range)
liver	1.5	-	0.80	(8; ± 0.1)
	-	-	0.15	(8; ± 0.04)
	1.5	1	n.d.[a]	(2)
	1.5	0.2	0.12	(2; ± 0.04)
kidney	1.5	-	0.15	(3; ± 0.05)
	-	-	0.10	(2; ± 0.01)

[a] not detectable (< 0.02)
Incubation: 1 h, 37 °C
BCPN concentration: 0.2 mM
Protein concentration: 0.7 - 1.7 mg/ml

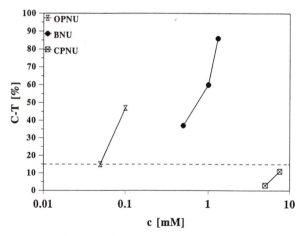

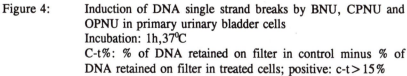

Figure 4: Induction of DNA single strand breaks by BNU, CPNU and OPNU in primary urinary bladder cells
Incubation: 1h,37⁰C
C-t%: % of DNA retained on filter in control minus % of DNA retained on filter in treated cells; positive: c-t > 15%

treated with even numbered alkylmethylnitrosamines suggested oxidative degradation via ω-oxidation and subsequent ß-oxidation (*62, 63*). Similar to BCPN, MCPN induces urinary bladder tumors after oral application to rats and is regarded as the proximal metabolite relevant for local tumor induction (*27, 64*).

MCPN might be activated by α- or ß-oxidation as already described for BCPN. Methyl-2-oxopropylnitrosamine (MOPN), the metabolite resulting from ß-oxidation, was detected in trace amounts in urine of rats treated with even numbered alkylmethylamines (*63*). MOPN is mutagenic and induces tumors of liver and esophagus when orally applied to rats (*65, 54, 64*). Upon intravesicular instillation, however, urinary bladder tumors were observed (*66*).

ß-Oxidation of Methyl-ω-carboxyalkylnitrosamines.   MCPN and methyl-5-carboxypentylnitrosamine (MCPeN) were incubated with rat liver mitochondrial fractions (7000xg pellet) as described for BCPN. With both carboxylated nitrosamines, ATP-dependent ß-oxidation was observed (incubation: 1h, 37°C; MCPN: 0.21 nmol MOPN x mg protein^{-1} ; MCPeN: 0.28 nmol MOPN x mg protein^{-1}). Although MCPeN needs two rounds of ß-oxidation to generate MOPN, it is metabolized more rapidly than MCPN. Carboxylated nitrosamines with longer alkyl chains therefore appear to be better substrates for acylcoA-synthetase.

α-Oxidation of MCPN and MOPN.   MCPN and MOPN were incubated with calf thymus DNA in the presence of PB-induced rat liver microsomal fractions. Formation of O ^{6}-MethyldG was determined by ^{32}P-postlabelling according to Wilson et al. (*67*) with modifications   (Figure 5) (*68*). Both compounds exhibited   O^6-G methylation, a higher rate was, however, observed with MOPN. As discussed for BOPN the higher methylating potential of MOPN can be ascribed to its higher lipophilicity, making it a better substrate for P-450. Moreover, both substituents generate methylating agents on α-activation. In summary, methylalkylnitrosamines are transformed into MCPN by mitochondrial metabolism of the primary ω-carboxylated metabolite. MCPN is activated to methylating agents by α- and/or ß-oxidation.

Conclusion

Nitrosamines carcinogenic to the bladder have intermediate 3-carboxypropyl metabolites in common that are rather weak genotoxic agents and are subject to P-450 mediated α-oxidation to a minor extent only. On the other hand, they can be converted by ATP-dependent mitochondrial ß-oxidation/decarboxylation into lipophilic oxopropyl metabolites. These are much better substrates for P-450 enzymes and are potent DNA damaging mutagens, generating methylating electrophils from α-oxidation of the oxopropyl chain. The sequence of metabolic events leading ultimately to oxopropylating/methylating agents therefore appears causally related to bladder cancer induction.

Thus, in contrast to dialkylnitrosamines known to be subject mainly to P-450 mediated α-oxidation, nitrosamines with hydroxy, oxo- or carboxy groups are subject to metabolic transformations by a variety of enzymes that are of great influence for their activation into genotoxic agents.

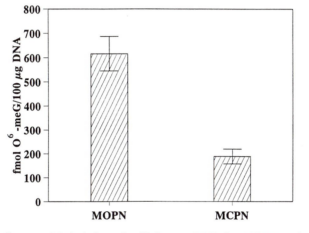

Figure 5:   Methylation of calf thymus DNA by MOPN and MCPN
Incubation: 1h,37°C
nitrosamine concentration:10 mM
DNA concentration: 1mg/ml
protein concentration: 2 mg/ml

Literature Cited

1   Preussmann, R.; Habs, M.; Habs, H.; Schmähl, D. *Cancer Res.* **1982**, *42*, 6167
2   Denkel, E. Thesis **1986**, University of Kaiserslautern
3   Gilbert, P.; Rollmann, B.; Rondelet, J., Mercier, M.; Poncelet, F. *Arch. Int. Physiol. Biochim.* **1979**, *87*, 813
4   Lee, S.Y.; Guttenplan, J.B. *Carcinogenesis* **1981**, *2*, 1339
5   Lijinsky, W.; Andrews, A.W. *Mut. Res.* **1983**, *111*, 135
6   Loeppky, R.N.; Tomasik, W.; Kovacs, D.A.; Outram, J.R.; Byington, K. H. In *N-nitroso compounds: occurrence, biological effects and relevance to human cancer*; O'Neill, I.K.; von Borstel, R.C.; Miller, C.T.; Bartsch, H. Eds, IARC Sci. Publ. No 57, International Agency For Research on Cancer: Lyon, France, 1984, pp 429-436
7   Eisenbrand, G; Denkel, E; Pool, B. *J. Cancer Res. Clin. Oncol.* **1984**, *108*, 76
8   Airoldi, L.; Bonfanti, M.; Fanelli, R; Bove, B.; Benfenati, E.; Gariboldi, P.; *Chem.- Biol. Interact.* **1984**, *51*, 103
9   Manson, D.; Cox, P.J.; Jarman, M. *Chem.-Biol. Interactions* **1978**, *20*, 341
10 Hecht, S.S. *Carcinogenesis* **1984**, *5*, 1745
11 Airoldi, L.; Macri, A.; Bonfanti, M.; Bonati, M.; Fanelli, R. *Fd. Chem. Toxicol.* **1984**, 133
12 Scherer, G. Thesis **1989**, University of Kaiserslautern
13 Denkel, E.; Pool, B.L.; Schlehofer, J.R.; Eisenbrand, G. *J. Cancer Res. Clin. Oncol.* **1986**, *111*, 149
14 Chung, F.-L.; Hecht, S.S. *Carcinogenesis* **1985**, *6*, 1671
15 Denkel., E; Sterzel, W.; Eisenbrand, G. In *The relevance of N-nitroso compounds to human cancer: Exposures and mechanisms*; Bartsch, H.; O'Neill, I.K.; Schulte-Herrmann,R., Eds; IARC Sci. Publ. No 84; International Agency for Research on Cancer: Lyon, France, 1987, pp 83-90
16 Hecht, S.S.; Lijinsky, W.; Kovatch, R.M.; Chung, F-.L.; Saveedra, J.E. *Carcinogenesis*, **1989**, *10*, 1475-1477
17 Lijinsky, W.; Saaveedra, J.E.; Kovatch, R.M. *In Vivo* **1991**, *5*, 85
18 Scherer, G; L; Ludeke, B; Kleihus, P.; Loeppky, R.N.; Eisenbrand, G. In *Relevance to human cancer of N-nitroso compounds, tobacco smoke and mycotoxins*; O'Neill, I.K.; Chen, J.; Bartsch, H. Eds.; IARC Sci. Publ. No 105, International Agency for Research on Cancer: Lyon, France, 1991, pp 339-342
19 Chada et al, *J. Med. Chem.* **1985**, *28*, 36
20 Eisenbrand, G.; Schlemmer, K.-H.; Janzowski, C. In *Combination effects in chemical carcinogenesis*; Schmähl, D. Ed.; VCH Verlagsgesellschaft, Weinheim, Germany, 1988, pp 1-29
21 Michejda, C.J.; Andrews, A.W.; Köpke, S.R. *Mut. Res.* **1979**, *67*, 301
22 Sterzel W.; Eisenbrand, G. *J. Cancer Res. Clin. Oncol.* **1986**, *111*, 20
23 Spencer, B. *Biochem. J.* **1960**, *77*, 294

24   Kroeger-Koepke, M.B.; Koepke, S.R.; Hernandez, L.; Michejda, C.J. *Cancer Res.* **1992**, *52*, 3300

25   Spiegelhalder, B.; Preussmann, R. In *N-Nitroso compounds: Occurrence and biological effects*; Bartsch, H.; Castegnaro, M; O'Neill, I.K.; Okada, M., Eds.; IARC Sci. Publ. No 41; International Agency for Research on Cancer: Lyon, France, 1982, pp 231-244

26   Fajen, J.M.; Rounbehler, D.P.; Fine, D.H. In *N-Nitroso Compounds: Occurrence and biological effects*; Bartsch, H.; Castegnaro, M; O'Neill, I.K.; Okada, M., Eds.; IARC Sci. Publ. No 41; International Agency for Research on Cancer: Lyon, France, 1982, pp 223-229

27   Okada, M; Suzuki, E; Mochizuki, M. *Gann* **1976**, *67*, 771

28   Preussmann, R; Stewart, B.W. In *Chemical carcinogens*; Searle, C.E., Ed.; ACS Monograph 182; American Chemical Society: Washington D.C., 1984, Vol. 2; pp 742-868

29   Irving, C.C.; Daniel, D.S. *Biochem. Pharmacol.* **1988**, *37*, 1642

30   Irving, C.C.; *Carcinogenesis* **1988**, *9*, 2109

31   Pastorelli, R.; Ancidei, A.; Benfenati, E.; Fanelli, R; Airoldi, L. *Toxicology* **1988**, *48*, 71

32   Okada,M.; Suzuki, E.; Aoki, J.; Iiyoshi, M.; Hashimoto, Y. In *Gann Monograph on Cancer Research*; Odashima, S; Takayama, S; Sato, H. Eds., University Park Press, 1975, No 17, pp 61-176

33   Janzowski, C.; Gottfried, J.; Eisenbrand, G.; Preussmann, R. *Carcinogenesis* **1982**, *3*, 777

34   Preussmann, R.; Habs, M.; Habs, H.; Stummeyer, D. *Carcinogenesis* **1982**, *3*, 1219

35   Suzuki, E.; Okada, M., *Gann* **1980**, *71*, 856

36   Irving, C.C.; Daniel, D.S. *Carcinogenesis* **1987**, *8*, 1309

37   Janzowski, C.; Jacob, D.; Henn, I.; Zankl, H.; Pool-Zobel, B.L.; Eisenbrand, G. *Toxicology* **1989**, *59*, 195

38   Rickwood, D.; Wilson, M.T.; Darley-Usmar, V.M. In *Mitochondria: a practical approach*; Darley-Usmar, V.M.; Rickwood, D.; Wilson, M.T., Eds.; The Practical Approach Series; IRL Press: Oxford, England, 1987; pp1-16

39   Kolvraa, S.; Gregersen, N. *Biochim. Biophys. Acta* **1986**, *876*, 515

40   Airoldi, L.; Magagnotti, C.; Bonfanti, M.; Fanelli, R. *Carinogenesis* **1990**, *11*, 1437

41   Airoldi, L.; Magagnotti, C.; de Gregorio, G.; Moret, M.; Fanelli, R. *Chem.-Biol. Interactions* **1992**, *82*, 231

42   Aas, M. *Biochim. Biophys. Acta* **1971**, *231*, 32

43   Waku, K. *Biochim. Biophys. Acta* **1992**, *1124*, 101

44   Jacob, D. Thesis **1991**, University of Kaiserslautern

45   Pool, B.L.; Gottfried-Anacker, J.; Eisenbrand, G.; Janzowski, C. *Mut. Res.* **1988**, *209*, 79

46   Janzowski, C.; Jacob, D.; Henn, I.; Zankl, H.; Pool-Zobel, B.L.; Wiessler, M.; Eisenbrand, G. In *Relevance to human cancer of N-nitroso compounds, tobacco smoke and mycotoxins*; O'Neill, I.K.; Chen,J.; Bartsch, H. Eds.; IARC Sci. Publ. No 105, International Agency for Research on Cancer: Lyon, France, 1991, 332-337

47	Henn, I.; Janzowski, C.; Eisenbrand, G; Zankl, H. *Mutagenesis* **1990**, *5*, 621
48	Henn-Christmann, I. Thesis **1992**, University of Kaiserslautern
49	Janzowski, C.; Jacob, D.; Goelzer, P.; Henn, I.; Zankl, H.; Eisenbrand G. *J. Cancer Res. Clin. Oncol.* **1991**I.; *117*, 16
50	Okada, M.; Hashimoto, Y. *Gann* **1974**, *65*, 13
51	Lijinsky, W.; Elespuru, R.K.; Andrews, A.W. *Mut. Res.* **1987**, *178*, 157
52	Druckrey, H.; Preussmann, R; Ivancovic, S.; Schmähl, D. *Zeitschr. Krebsforsch.* **1967**, *69*, 103
53	Kohda , K.H.; Ninomiya, S.; Washizu, K.; Shiraki, K.; Ebie, M.; Kawazoe, Y. *Mut. Res.* **1987**, *177*, 219
54	Lijinsky, W. *Cancer Lett.* **1991**, *60*, 121
55	Leung , K.H.; Archer, M.C. *Chem.-Biol. Interactions* **1984**, *48*, 169
56	von Hofe, E.; Schmerold, I.; Lijinsky, W.; Jeltsch, W; Kleihus, P. *Carcinogenesis* **1987**, *8*, 1337
57	Lijinsky, W., Reuber, M.D. *Cancer Lett.* **1984**, *22*, 83
58	Lijinsky, W.; Saveedra, J.E., Reuber, M.D. *Cancer Res.* **1981**, *41*, 1288
59	Kamp, E; Eisenbrand G. *Fd. Chem. Toxic.* **1991**, *29*, 203
60	Hecht, S.S.; Morrison, J.B.; Wenninger, J.A. *Fd. Chem. Toxic.* **1982**, *20*, 165
61	Morrison, J.B.; Hecht, S.S.; Wenninger, J.A. *Fd. Chem. Toxic.* **1983** , *21*, 69
62	Suzuki, E.; Iiyyoshi, M.; Okada, M. *Gann* **1981**, *72*, 113
63	Singer, G.M.; Lijinsky, W.; Buettner, L.; McClusky, G.A. *Cancer Res.* **1981**, *41*, 4942
64	Lijinsky, W.; Reuber, M.D.; Saveedra, J.E.; Singer, G.M. *J. Natl. Cancer Inst.* **1983**, *70*, 959
65	Langenbach, R,; Gingell, R,: Kuszynski, C., Walker, B., Nagel, D., Pour, P., *Cancer Res.* **1980**, *40*, 3463
66	Thomas, B.J., Kovatch, R.M., Lijinsky, W. *Jpn. J. Cancer Res. (Gann)* **1988**, *79*, 309
67	Wilson, V.L.; Basu, A.K.; Essigmann, J.M.; Smith, R.A.; Harris, C.C. *Cancer Res.* **1988**, *48*, 2156
68	Kamp, E. Thesis **1991**, University of Kaiserslautern

RECEIVED October 11, 1993

Chapter 16

Activation of β-Hydroxyalkylnitrosamines
Evidence for Involvement of a Sulfotransferase

Christopher J. Michejda, Steven R. Koepke, Marilyn B. Kroeger Koepke,
and Lidia Hernandez

Molecular Aspects of Drug Design Section, Macromolecular Structure
Laboratory, ABL-Basic Research Program, Frederick Cancer Research
and Development Center, National Cancer Institute,
Frederick, MD 21702

Activation of most dialkylnitrosamines involves oxidation of an α-carbon by a cytochrome P-450 monooxygenase. However β-hydroxyalkylnitrosamines are not good substrates for these enzymes. Oxidation of the β-hydroxylated carbon to the corresponding carbonyl derivative by alcohol dehydrogenases has been shown to activate some β-hydroxyalkylnitrosamines to mutagens. However, it was postulated that these nitrosamines could also be activated by alcohol sulfotransferase-catalyzed sulfation. Experiments on N-nitrosomethyl-2-(hydroxyethyl)amine (NMHEA) showed that chemical sulfation of the hydroxyl group transformed the nitrosamine into a potent alkylating agent, capable of methylating DNA. Evidence is presented here that NMHEA can be activated by sulfation to a DNA methylating agent in intact animals, in primary hepatocytes, in cultured liver cells, and in post-mitochondrial cytosol prepared from weanling rats. However, the data also indicate that there are pathways other than sulfation which can transform NMHEA to a DNA damaging agent. Implications of these findings to the metabolism of hydroxylated nitrosamines are discussed.

The predominant pathway of metabolic activation of most simple dialkyl and heterocyclic nitrosamines involves the cytochrome P450-catalyzed hydroxylation of the α-carbon. The resulting α-hydroxynitrosamines spontaneously break down to aldehydes (derived from the group bearing the α-hydroxyl substituent) and alkyldiazonium ions from the other side. The electrophilic diazonium ions are the putative ultimate carcinogens from those nitrosamines (1). There are some nitrosamines, however, which are poor substrates for microsomal monooxygenases. The most important are those which are already oxidized on the side chains. This is consistent with the generalization that most cytochrome-P450 monooxygenases prefer lipophilic substrates.

N-Nitrosodiethanolamine (NDELA), which is one of the most

0097–6156/94/0553–0195$08.00/0

environmentally significant nitrosamines, does not appear to be a substrate for any of the cytochrome-P450 isozymes, as was shown by Farrelly *et al.* (*2*). Consistent with its hydrophilic nature, 70-90% of NDELA is excreted in the urine of rats, depending on the initially applied dose (*3*). However, there is some metabolism of NDELA *in vivo*, which has been partially delineated. Airoldi and co-workers (*4*) showed that N-nitroso-N-(2-hydroxyethyl)glycine was a urinary metabolite of NDELA in the rat. Eisenbrand *et al.* (*5*) demonstrated that NDELA, which is not a bacterial mutagen either alone or when activated by rat liver microsomes, became a potent mutagen in Salmonella typhimurium TA98 and TA100 strains when activated by alcohol dehydrogenase. The product of this oxidation was the cyclic hemiacetal of the initially formed aldehyde, N-nitroso-2-hydroxymorpholine (NHMOR) (*6*). Thus, it is clear that while β-hydroxyalkylnitrosamines may resist cytochrome P450 activation, they can be substrates for alcohol dehydrogenases in a reaction which can constitute an activation process.

Chemical Studies

Previously we postulated that some β-hydroxyalkylnitrosamines, especially N-nitrosomethyl-(2-hydroxyethyl)amine (NMHEA), undergo activation by conjugation of the hydroxyl group by a good leaving group, which we hypothesized to be sulfate (*7*). This postulate was based on purely chemical evidence. We observed that the p-toluenesolfonate (tosylate) ester of NMHEA solvolyzed very rapidly in glacial acetic acid with the formation of the corresponding acetate. We explained the very large rate acceleration in terms of anchimeric assistance by the N-nitroso group (*8*). The anchimeric assistance suggested that the reaction proceeded via the intermediacy of a cyclic intermediate, 3-methyl-1,2,3-oxadiazolinium tosylate. Indeed, the cyclic intermediate hypothesis was strengthened considerably by the finding that the acetolysis of *S*-N-nitrosomethyl-(2-hydroxypropyl)amine resulted in the formation of the corresponding *S*-acetate. The retention of configuration in a nucleophilic displacement reaction demanded a double inversion, first in the formation of the cyclic intermediate, followed by the opening of the intermediate to the final product (*9*). Finally, the 3-methyloxadiazolinium ion itself could be isolated readily by warming a solution of NMHEA tosylate in a non-nucleophilic solvent, such as methylene chloride. These reactions are summarized in Figure 1. The oxadiazolinium ion also formed readily when NMHEA was treated with sulfur trioxide in pyridine.

   NMHEA is not a mutagen in the Ames assay in Salmonella typhimurium with or without rat liver postmitochondrial fraction (S9) activation. It is, however, activated to a mutagen by hamster liver S9 (Koepke and Andrews, unpublished data). Conversion of NMHEA to the tosylate, or cyclization of the tosylate to the oxadiazolinium ion, produced potent, directly acting mutagens in the TA1535 strain of Salmonella typhimurium. The dose-response curves for the tosylate and the oxadiazolinium ion derived from it were essentially congruent, suggesting that the tosylate cyclized prior to the reaction with the bacterial genome (7). Interestingly, the next higher homolog of NMHEA, N-nitrosomethyl-(3-hydroxypropyl)amine, was not activated to a mutagen by conversion to the tosylate, a result consistent with the solvolytic behavior of that compound, which behaved much more like a simple primary tosylate.

The chemistry of the oxadiazolinium ion was briefly explored in our laboratory. It is conceivable that the ion has three sites for nucleophilic attack, as shown in Figure 2.   The solvolysis results showed that the attack of acetate on the ion produced exclusively the ring-opened acetate ester. This indicated that reaction at site **b** was the most important for that nucleophile. The reaction of one equivalent of the oxadiazolinium tosylate with one equivalent of aniline in methylene chloride solution for two hours at room temperature produced the β-anilino derivative of NMHEA in 82% yield. This again suggested that site **b** was favored for this nucleophile.   However, the reaction of the oxadiazolinium ion with 3,4-dichlorothiophenol in methylene chloride under nitrogen for two hours produced a 90% yield of 3,4-dichlorophenylmethylsulfide,   indicating that site **a** also could be attacked by an appropriate nucleophile.   Methylation of guanine (0.3 mM) by the oxadiazolinium ion (0.3mM) was also observed in dry dimethylsulfoxide. Interestingly, the results were the same if equivalent amounts of the uncyclized tosylate was used in the reaction. The reaction was carried out for 24 hours. The products were separated from the solution with dry ether. The oily residue was extracted with methylene chloride to remove unreacted starting materials. The resulting yellow solid was separated by column chromatography on the acidic form of Dowex 50 resin using an HC1 gradient.   The yields of the methylated guanines were: 7-methylguanine 38%, 3-methylguanine 25%, and 1-methylguanine 28%.   The identity of the products was confirmed by comparison with authentic standards, and supported by high resolution mass spectrometry.

A similar reaction of the oxadiazolinium ion (or the tosylate) with guanosine, (the reaction carried out in DMSO with the products separated by chromatography on Dowex 50), showed that 7-methylguanosine was a major product. Its identity, however, was established only by co-chromatography with authentic 7-methylguanosine.

It thus appears that **a** and **b** are the most likely sites for nucleophilic attack on the oxadiazolinium ion. No evidence for the reaction at site **c** was detected in our studies. It was also significant that within the resolution of the experiments cited above, there were no mixed reactions. Nucleophilic attacks at the methyl group of the oxadiazolinium ion or at carbon 5 are predicted by *ab initio* molecular orbital calculations on the ion carried out using the Gaussian 90 program at the 3-21G basis set level, since these atoms have the highest concentration of positive charge. These calculations also predict that carbon 5 ought to be more susceptible to attack by "soft"nucleophiles, since the charge on the methyl group is less diffuse. Conversely, the calculation also shows that site **c** is not a good site for nucleophilic attack.

The reaction of the oxadiazolinium ion at site **a** causes methylation of the nucleophile and the formation of an apparently unstable product, 4,5-

dihydro-2,3-oxadiazoline (Figure 3). This compound is a cyclic diazotic acid ether, a molecule which would be predicted to fragment readily. Molecular orbital calculations at both the semi-empirical PM3 level and at the *ab initio* 3-21G basis set level (*10*) show that the molecule is capable of existence, but our calculation at this and higher basis set levels indicate that it undergoes fragmentation with relative ease, just as intuitive predictions suggested. The computational data predict that the most favorable pathway for fragmentation leads to the formation of diazomethane and formaldehyde (Figure 4), (Kroeger Koepke and Michejda, unpublished data).

This result, while somewhat surprising, is supported by some experimental data (see below). We do not yet know, however, how the oxadiazoline behaves in water. It is entirely possible that it might be subject to other transformations, such as acid-catalyzed ring opening to the hydroxyethyldiazonium ion. In fact, *ab initio* SCF calculations indicate that attachment of a proton to the ether oxygen causes a synchronous opening to the 2-hydroxyethyldiazonium ion (Kroeger Koepke and Michejda, unpublished data). This result is also consistent with the protonation behavior of the isoelectronic molecule, 4,5-dihydro-1,2,3-triazole (*11*).

The chemical experiments described above suggested strongly that β-hydroxynitrosamines such as NMHEA could be activated metabolically to electrophiles capable of alkylating cellular targets simply by conjugation of the hydroxyl group with a good leaving group, such as sulfate. The following discussion presents evidence that such an activation pathway may be important in the case of NMHEA.

Activation by Sulfotransferase

NMHEA is a carcinogen in rats (*12*). In contrast to most nitrosamines, which show little or no sex preference in carcinogenesis, NMHEA was a potent liver carcinogen in female F344 rats but produced only a few liver tumors in males. However, while it caused nasal cavity tumors in males, such tumors were completely absent in the females. NMHEA tosylate was a weaker carcinogen on a molar basis than NMHEA but produced the same spectrum of tumors. The chemicals were administered to the rats by gavage in corn oil on a twice-weekly schedule (26 μmol per application). The median survival of the females from start of treatment was about nine months, while that of the males was twelve months (*12*). NMHEA tosylate was at least partly hydrolyzed to NMHEA because it was possible to detect NMHEA in blood plasma of NMHEA tosylate-treated rats. We postulated that NMHEA tosylate, which probably cyclized to the oxadiazolinium ion in the rat stomach, was not able to pass through the gastrointestinal mucosa. The liver tumors, which were observed in lower frequency in rats treated with the tosylate as compared to NMHEA, were probably caused by NMHEA which was formed by the hydrolysis of the tosylate.

It is interesting to note that the next higher homolog to NMHEA, N-nitrosomethyl-(3-hydroxypropyl)amine (NMHPA) was a much weaker

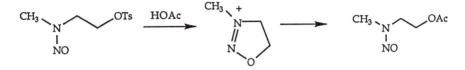

Figure 1. Solvolysis of NMHEA tosylate and formation of 4,5-dihydro-3-methyl-oxadiazolinium tosylate $Ts=-SO_2C_6H_4CH_3$.

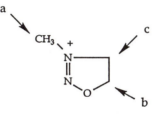

Figure 2. Sites for attack by nucleophilic agents on 4,5-dihydro-3-methyloxadiazolinium ion.

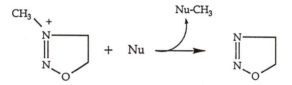

Figure 3. Formation of 4,5-dihydro-2,3-oxadiazoline.

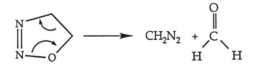

Figure 4. Fragmentation of 4,5-dihydro-2,3-oxadiazoline.

carcinogen on an equimolar basis than NMHEA. Nevertheless, several of the animals treated with NMHPA had adenocarcinomas of the lung (10/19 males and 2/20 females) as well as neoplastic nodules in the liver. The median time of death was at about twenty three months after start of treatment. These data are consistent with the chemical evidence which showed that conjugation of NMHPA with a tosyl group failed to enhance its reactivity or mutagenicity. There is no evidence that NMHPA is conjugated by sulfation and is probably activated to the ultimate carcinogenic form by a different mechanism from NMHEA.

The ability of NMHEA to modify genomic DNA was examined *in vivo* in F344 rats (*13*). In one series of experiments, groups of five male and five female rats were treated by gavage by varying concentrations of NMHEA dissolved in corn oil. After four hours, the animals were sacrificed, and the livers, the lungs and the kidneys were removed. The DNA from each of these tissues was isolated, purified and subjected to neutral thermal, and acid catalyzed hydrolysis. The hydrolysates were separated by high pressure liquid chromatography (hplc), and the modified guanines were quantitated by fluorescence detection (*14*). NMHEA gives rise to four fluorescent alkylated guanines, 7-and O^6-methylguanines, and 7-and O^6-hydroxyethylguanines. Dose-response data were obtained for doses ranging from 10-50 mg/kg at 10 mg/kg intervals. Although the levels of alkylation at 10 and 20 mg/kg were the same for both sexes, at higher doses the levels of all adducts were higher in females than in males.

The higher levels of alkylation observed in the liver for all adducts in female rats were dramatically evident in an experiment where a group of rats of both sexes was treated with 5.3 mg/week of NMHEA for a period of six weeks, either by gavage or in drinking water (*15*). Animals were sacrificed every week up to six weeks and the adducts in liver DNA were determined. In general, the level of all adducts increased with time for the entire period for the rats treated with NMHEA in drinking water and by week six, the female rats showed an approximately twofold higher level of alkylation than the males. It should be pointed out that female rats were ~30% lighter than the males throughout our study and thus received a slightly higher dose per unit weight than the males. In the gavage treated animals, the level of all adducts was also about twice as high in females as in males but no accumulation with time was noted (i.e. the levels at week one were about the same as at week six). These data suggest that the greater hepatocarcinogenicity of NMHEA in female F344 rats can be directly correlated with the generally higher levels of alkylation at all sites, as compared to the male rats. There are only minor sex differences in the *in vivo* persistence of adducts, suggesting that there are no significant differences in the rates of repair of the DNA lesions between sexes (*13*). These data also suggest that the sex-linked difference in carcinogenicity of NMHEA must be at the level of carcinogen activation, rather than at some post-metabolic step of carcinogenesis. The amounts of alkylation in the lung were much lower than in the liver and negligible in the kidneys.

It has been known for some time that the levels of the various sulfotransferase isozymes vary with species and sex though the picture is still somewhat confused (*16*, *17*). However, Singer *et al.* (*18*) reported that adult female F344 rats had levels of cortisol sulfotransferase which were 6-8-fold higher than in males. Cortisol sulfotransferase is capable of sulfating simple alcohols. It has also been known that sulfotransferase activity could be inhibited by competitive inhibitors. Thus, substances which are substrates for the enzyme can act as inhibitors for other possible substrates. Phenolic sulfotransferases are inhibited strongly *in vitro* and *in vivo* by various halogenated phenols such as 2,6-dichloro-4-nitrophenol (DCNP) (*19*), while propylene glycol (PG), a common vehicle for many drugs, has been found to be an inhibitor for alcohol sulfotransferase (*20*). Table I shows the effect of these two inhibitors on DNA guanine alkylation *in vivo*. We used N-nitrosodimethylamine (NDMA) as a negative control substrate because alkylation by that substance would not be expected to be influenced greatly by inhibitors of sulfotransferases, at least for short exposure times.

The methylation of DNA by NDMA was not greatly affected by either DCNP or by PG. The somewhat depressed levels of methylation in the PG-treated males may have been due to some inhibition of mixed function monooxygenases by PG, since it is known that various alcohols, including the closely related ethylene glycol, do inhibit NDMA demethylase (*21*). Alkylation by NMHEA is almost completely blocked by PG, except for the formation of 7-(hydroxyethyl)guanine (Table I). That adduct is only slightly inhibited. This result shows that PG, which is presumably a sulfotransferase inhibitor, blocks most, but not all of the alkylation. This implies that if sulfotransferase activation is operative in the case of NMHEA, it is not the only activation pathway. Clearly the 7-hydroxethylguanine adduct must be formed largely by a different pathway. Further, this result suggests that since only the hydroxyethylation of the 7-position of guanine was not inhibited by PG, the reactive agent had to be similar to ethylene oxide, which reacts by an SN_2 mechanism and would not be expected to give rise to any O^6-guanine adducts. When administered in corn oil, DCNP, the phenolic sulfotransferase inhibitor, failed to significantly inhibit any of the adducts except 7-hydroxyethylguanine in female rats. In fact, it dramatically increased methylation but only in male rats. The origin of this enhancement is unknown, but the result suggests that DCNP causes a metabolic switch, which favors methylation in male rats. This activity is unlikely to be connected with DCNP's ability to inhibit at least one class of sulfotransferases. A combination of PG and DCNP was most effective in inhibiting alkylation at all sites and in both sexes. Interestingly, the presence of PG completely suppressed the methylation enhancement in male rats induced by DCNP.

The *in vivo* inhibition data, such as those shown in Table I, are very difficult to interpret because any inhibitor, unless targeted at a specific enzyme, will have multiple modes of action and, as a consequence, will have the capacity for affecting more than one biochemical pathway. Thus, the data in Table I support the hypothesis that NMHEA is activated to an alkylating agent by an alcohol sulfotransferase, but they do not constitute a proof. Inhibitors

Table I. Yields of guanine adducts in rat liver DNA following treatment with NMHEA, NDMA, and various inhibitors

Treatment [a]	Sex	Adducts (μmol/mol guanine) (% control)				
		7-MeG	O^6-MeG	7-EtOHG	O^6-EtOHG	
NDMA	M	2798(100)	319(100)			
	F	3399(100)	462(100)			
NDMA,PG	M	1936(69)	191(60)			
	F	3219(95)	383(83)			
NDMA,DCNP in corn oil	M	2375(85)	266(84)			
	F	2861(84)	304(66)			
NMHEA	M	1372(100)	206(100)	693(100)	75(100)	
	F	1656(100)	248(100)	742(100)	91(100)	
NMHEA,PG	M	167(12)	15(7)	478(69)	3(4)	
	F	307(19)	25(10)	625(84)	7(8)	
NMHEA,DCNP in corn oil	M	2620(264)	535(260)	513(74)	80(88)	
	F	1566(95)	244(99)	215(29)		
NMHEA,DCNP in PG	M	269(20)	14(7)	72(10)	3(4)	
	F	465(28)	26(10)	71(10)	6(6)	

[a] Rats (16 weeks old) were treated with 5 mg/kg of NDMA or 25 mg/kg NMHEA by gavage in corn oil (0.2 ml). PG (1.0 ml/kg, 13.6 mmol) and DCNP (26 μmol/kg) were administered by i.p. injection 2 h prior to nitrosamine treatment. The animals were sacrificed 4 h after nitrosamine treatment.

have been used by other investigators to test the role of sulfotransferases. Dehydroepiandrosterone, which is a potent inhibitor of hydrocortisone sulfation, inhibited the sulfation of the carcinogen, 7,12-dihydroxymethylbenz[a]anthracene, but the phenolic sulfotransferase inhibitors, DCNP and pentachlorophenol, were ineffective (22). Sterzel and Eisenbrand (23) found that DCNP in PG as carrier inhibited the development of strand breaks in rat liver DNA in animals treated with NDELA and with N-nitrosoethyl-(2-hydroxyethyl)amine.   The effect of DCNP and PG on strand breaks produced by N-nitrosodiethylamine was slight.

It was desirable to obtain more direct evidence for the involvement of alcohol sulfotransferase in the activation of NMHEA, since the *in vivo* inhibition data were equivocal.   Sulfotransferase enzymes all utilize 3'-phosphoadenosine 5'-phosphosulfate (PAPS) as a cofactor (16).   The biosynthesis of PAPS ultimately depends on inorganic sulfate. Thus, one of the possible ways of providing strong evidence for involvement of sulfotransferases in the activation process would be to examine the effect of sulfate concentration on alkylation. However, it is difficult to deplete an intact animal of inorganic sulfate because the ion is formed by catabolism of sulfur-containing amino acids.   Likewise, it is difficult to increase the amount of circulating sulfate because the organism tends to maintain sulfate homeostasis.   Thus, manipulation of sulfate levels *in vivo* does not produce unequivocal results.

We attempted to obtain direct *in vivo* evidence for sulfotransferase activation by measuring the extent of DNA alkylation by NMHEA in the livers of B6C3F2 brachymorphic mice (*bm/bm*) and in their heterozygous litter mates (+/?) (15). These mice carry a mutation which makes them deficient in their ability to synthesize PAPS. There were, however, no significant differences between alkylation levels in the brachymorphic and in the heterozygous mice. These data, however, do not disprove the sulfotransferase hypothesis because it appears that in the case of this nitrosamine, and perhaps in the case of other similar nitrosamines, closing down one pathway shunts the metabolism to other pathways.

We therefore elected to determine whether the level of DNA modification by NMHEA could be a function of inorganic sulfate concentration in intact cells *in vitro*. Figure 5 shows the changes in DNA-guanine methylation in primary rat liver hepatocytes with changes in the concentration of inorganic sulfate dissolved in the culture medium (15). The medium also contained exogenous DNA so that the observed 7-methylguanine was mainly derived from the extracellular DNA, whose concentration was much higher (5mM) than that of the genomic DNA. This experiment clearly showed that the methylation of DNA *in vitro* by NMHEA was sulfate-dependent up to about 3mM sulfate but higher concentrations, in combination with NMHEA, appeared to be toxic to the hepatocytes.   Figure 6 shows a related experiment, wherein DNA methylation was measured in cultured CL9 rat liver cells, as a function of added sulfate and as a function of NMHEA concentration.   The CL9 cells appeared to be more sensitive to higher sulfate concentrations than were the primary hepatocytes.   However, these data show a clear dependence of

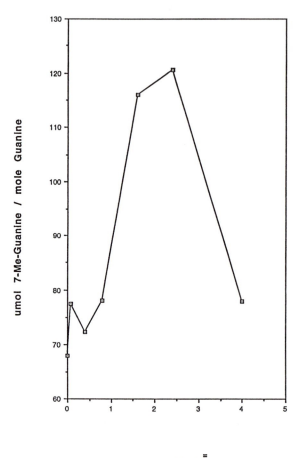

mM SO$_4^=$

Figure 5. Dose-response for methylation of guanine in exogenous DNA in presence of primary hepatocytes by NMHEA as a function of inorganic sulfate. Typically, reactions, consisting of 4.4 x 10^6 cells, 5 mg of calf-thymus DNA, 10 mM NMHEA, were incubated for 2 hrs. DNA was extracted, hydrolyzed and the guanines were analyzed by HPLC with fluorescence detection. Each point represents two experiments. (Reproduced with permission from reference 15. Copyright 1992 Cancer Research.)

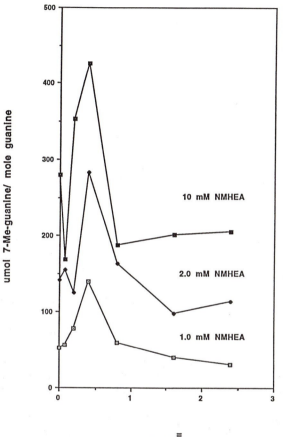

Figure 6. Effect of inorganic sulfate on 7-methylguanine levels in DNA from CL9 cells in the presence of varying amounts of NMHEA. Experiments were performed in cultures approximately 80% confluent, in sulfate-free phosphate-buffered saline with 10% fetal bovine serum, glucose, $CaCl_2$, $MgCl_2$, insulin, phenol red, and varying amounts of sodium sulfate. Cells were incubated for 6 h, and DNA was extracted from scraped monolayers with phenol, hydrolyzed, and analyzed by HPLC coupled with fluorescence detection. Points are representative of two typical experiments.

methylation on sulfate, and on the NMHEA concentration. In this case there was no exogenous DNA, so the observed methylation was that which was produced in the genomic DNA.

Finally, we examined the methylation of DNA guanine in an incubation mixture containing [methyl-^{14}C]NMHEA, calf thymus DNA, cytosol from the rat liver, and the sulfotransferase cofactor, PAPS. The relevant chromatograms are shown in Figure 7. The detection of 7-methylguanine in incubations containing PAPS, and the absence of the adduct in absence of PAPS, suggested that only the cytosolic activity which caused methylation by NMHEA was a PAPS-dependent sulfotransferase. Interestingly, treatment of calf-thymus DNA with [hydroxyethyl-^{14}C]NMHEA in presence of rat liver cytosol and in presence of PAPS also caused the methylation of guanine, which was not found in the absence of PAPS. This curious result is consistent with methylation of DNA by diazomethane produced by the decomposition of the byproduct of the oxadiazolinium ion, the oxdiazoline. It may be recalled that this result was predicted by theoretical calculations (see above).

Conclusions

It could be argued that each of the foregoing experiments could be explained without invoking the necessity of sulfotransferase activation. However, the combination of experiments and ample literature precedent strongly support the hypothesis that NMHEA is activated, at least in part, by conversion to sulfate. Accumulated evidence implicates sulfate conjugation in the activation of diverse carcinogens, including N-2-fluorenylacetamide (24), 1'-hydroxysafarole (25), 4-aminobenzene (26), and 1-hydroxymethylpyrene (27). In an experiment particularly relevant to the present work, Kokkinakis et al. (28) found significant quantities of the sulfate ester of N-nitroso-(2-hydroxypropyl) (2-oxopropyl)amine in the urine of hamsters treated with the nitrosamine. These workers also detected a low level of sulfation of β-hydroxyalkylnitrosamines in rats (29).

An assessment of the combined experiments on NMHEA suggests a rather complex metabolic picture, which is depicted in Scheme 1. Experiments utilizing ^{15}N-doubly labeled NMHEA in metabolism of the chemical by freshly isolated rat hepatocytes indicated that there was very little, if any, metabolism of the chemical by oxidation either on the methyl group, or on the α-carbon of the hydroxyethyl group since very little doubly labeled molecular nitrogen ($>2.3\%$ of the total metabolism) was evolved during the reaction (30). While it is possible that oxidation of the α-carbons of NMHEA may be more efficient in vivo, it is unlikely that this oxidative reaction is an important pathway of metabolism of the chemical. The formation of 7-hydroxyethylguanine in DNA, which is not inhibited by PG, continues to be unexplained. It cannot be formed from any pathway derived from the methyl oxadiazolinium ion, or from P450-catalyzed oxidation of the methyl group. A most likely reaction would involve its formation from ethylene oxide, but how that chemical is formed from NMHEA remains elusive.

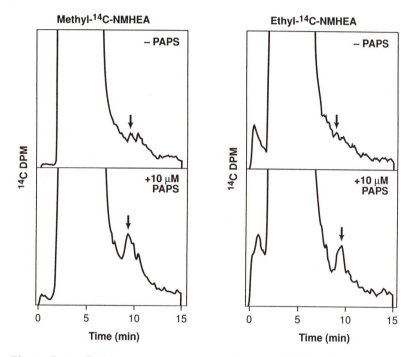

Figure 7. Radiochromatograms of hydrolysates of DNA which had been treated with [*methyl*-¹⁴C]NMHEA and [*hydroxyethyl*-¹⁴C]NMHEA in the presence of cytosol from rat liver. Cytosol was the post-100,000g supernatant from liver homogenates of weanling (12 day) female F344 rats. Experiments combined cytosol, 0.15 μM ¹⁴C-NMHEA, 1 mM MgCl$_2$, 10 mM Tris-HCl (pH 7.4), 2 mg. calf thymus DNA and either 10 μM PAPS, or no PAPS. The mixture was incubated for 2 hrs at 37 °C. The DNA was extracted with buffered phenol-chloroform, hydrolyzed by boiling and analyzed by cation-exchange HPLC with radiometric detection. Arrows indicate the 7-[¹⁴C-methyl]guanine peak; the large peak at ~5 min. is due to the residual unmetabolized, labeled NMHEA.

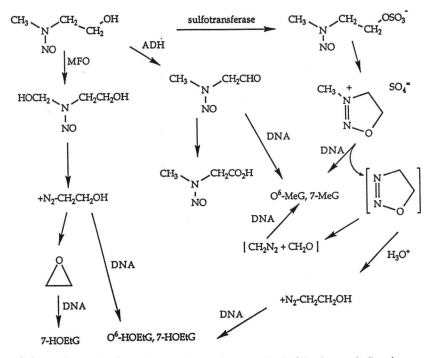

Scheme 1. Metabolic pathways for N-nitrosomethyl (2-hydroxyethyl)amine. Parts of this scheme are still unproven.

In summary, it is likely that NMHEA is activated by the sulfate conjugation pathway. It is clear, however, that other paths of metabolism become dominant if that pathway is blocked. This suggests that there are several, almost equally probable, metabolic activation pathways for NMHEA and probably other β-hydroxyalkylnitrosamines. This gives rise to possibilities of metabolic switching, resulting in difficulties of blocking metabolic activation of this and related nitrosamines. This conclusion may have profound effects on attempts to block carcinogenesis by metabolic inhibitors, and has implications for the design of chemoprevention strategies.

Acknowledgments

Research sponsored by the National Cancer Institute, DHHS, under contract no. NO1-CO-74101 with ABL. The contents of this publication do not necessarily reflect the views or policies of the Department of Health and Human Services, nor does mention of trade names, commercial products, or organizations imply endorsement by the U.S. Government. By acceptance of this article, the publisher or recipient acknowledges the right of the U.S. Government and its agents and contractors to retain a nonexclusive royalty-free license in and to any copyright covering the article.

Literature Cited

1. Preussmann, R.; Stewart, B.W. In *Chemical Carcinogens*; Searle, C., Ed.; American Chemical Society: Washington, D.C., 1984, Vol. 2; pp. 643-828.
2. Farrelly, J.G.; Stewart, M.L.; Lijinsky, W. *Carcinogenesis* (Lond.) **1984**, *5*, 1015-1019.
3. Preussman, R.; Spiegelhalder, B.; Eisenbrand, G.; Würtele, G.; Hoffmann, I. *Cancer Letters* **1981**, *4*, 207.
4. Airoldi, L.; Bonfanti, M.; Benfenati, E.; Tavecchia, P.; Fanelli, R. *Biomed. Mass Spectrum.* **1983**, *10*, 334-337.
5. Eisenbrand, G.; Denkel, E.; Pool, B. *J. Cancer Res. Clin. Oncol.* **1984**, *108*, 76-80.
6. Hecht, S.S. *Carcinogenesis* (Lond.) **1984**, *5*, 1745-1747.
7. Michejda, C.J.; Andrews, A.W.; Koepke, S.R. *Mutat. Res.* **1979**, *67*, 301-308.
8. Michejda, C.J.; Koepke, S.R. *J. Am. Chem. Soc.* **1978**, *100*, 1959-1960.
9. Koepke, S.R.; Kupper, R.; Michejda, C.J. *J. Org. Chem.* **1979**, *44*, 2718-2722.
10. Sapse, A.-M.; Allen, E.B.; Lown, J.W. *J. Am. Chem. Soc.* **1988**, *110*, 5671-5675.
11. Wladkowski, B.D.; Smith, R.H., Jr.; Michejda, C.J. *J. Am. Chem. Soc.* **1991**, *113*, 7893-7897.
12. Koepke, S.R.; Creasia, D.R.; Knutsen, G.L.; Michejda, C.J. *Cancer Res.* **1988**, *48*, 1533-1536.
13. Koepke, S.R.; Kroeger Koepke, M.B.; Bosan, W.; Thomas, B.J.; Alvord, W.G.; Michejda, C.J. *Cancer Res.* **1988**, *48*, 1537-1542.

14. Herron, D.C.; Shank, R.C. *Anal. Biochem.* **1979**, *100*, 58-63.
15. Kroeger Koepke, M.B.; Koepke, S.R.; Hernandez, L.; Michejda, C.J. *Cancer Res.* **1992** *52*, 3300-3305.
16. Mulder, G.J. In *Sulfation of Drugs and Related Substances*; Mulder, G.J., Ed.; CRC Press: Boca Raton, FL, 1981, pp. 131-185.
17. Mulder, G.J. *Chem.-Biol. Interactions* **1986**, *57*, 1-15.
18. Singer, S.; Moshtaghie, A.; Lee, A.; Kutzer, T. *Biochem. Pharmacol.* **1980**, *29*, 3181-3188.
19. Mulder, G.J.; Scholtens, E. *Biochem. J.* **1977**, *165*, 533-559.
20. Spencer, B. *Biochem. J.* **1960**, *77*, 294-304.
21. Yoo, J.-S.H.; Cheung, R.J.; Patten, C.J.; Wade, D.; Yang, C.S. *Cancer Res.* **1987**, *47*, 3378-3383.
22. Watabe, T.; Hiratsuka, A.; Ogura, K. *Carcinogenesis* (Lond.) **1987**, *8*, 445-453.
23. Sterzel, W.; Eisenbrand, G. *J. Cancer Res. Clin. Oncol.* **1986**, *111*, 20-24.
24. Lai, C.-C.; Miller, J.A.; Miller, E.C.; Liem, A. *Carcinogenesis* (Lond.) **1985**, *6*, 1037-1045.
25. Boberg, E.W.; Miller, E.C.; Miller, J.A.; Poland, A.; Liem, A. *Cancer Res.* **1983**, *43*, 5163-5173.
26. Delclos, K.B.; Miller, E.C.; Miller, J.A.; Liem, A. *Carcinogenesis* (Lond.) **1986**, *7*, 227-237.
27. Surh, Y.-J.; Blomquist, J.C.; Liem, A.; Miller, J.A. *Carcinogenesis* (Lond.) **1990**, *11*, 1451-1460.
28. Kokkinakis, D.M.; Hollenberg, P.F.; Scarpelli, D.G. *Cancer Res.* **1985**, *45*, 3586-3592.
29. Kokkinakis, D.M.; Scarpelli, D.G.; Subbarao, V.; Hollenberg, P.F. *Carcinogenesis* (Lond.) **1987**, *8*, 295-303.
30. Michejda, C.J.; Koepke, S.R.; Kroeger Koepke, M.B.; Bosan, W.S. In *Relevance of N-Nitroso Compounds to Human Cancer: Exposures and Mechanisms*; Bartsch, H.; O'Neill, I.K.; Schulte-Hermann, R., Eds. IARC Scientific Publications No. 84, International Agency for Research in Cancer: Lyon, France, **1987**, pp. 77-82.

RECEIVED July 2, 1993

Chapter 17

Hemoglobin Adducts, DNA Adducts, and Urinary Metabolites of Tobacco-Specific Nitrosamines

As Biochemical Markers of Their Uptake and Metabolic Activation in Humans

Stephen S. Hecht, Neil Trushin, and Steven G. Carmella

Division of Chemical Carcinogenesis, American Health Foundation, 1 Dana Road, Valhalla, NY 10595

Methods have been developed for quantitation in humans of hemoglobin and DNA adducts resulting from metabolic activation of 4-(methylnitrosamino)-1-(3-pyridyl)-1-butanone (NNK) and N'-nitrosonornicotine (NNN), two carcinogenic tobacco specific nitrosamines, and for quantitation in human urine of two NNK metabolites, 4-(methylnitrosamino)-1-(3-pyridyl)-1-butanol (NNAL) and its glucuronide. The hemoglobin and DNA adducts are formed by α-hydroxylation of NNK and NNN, and release 4-hydroxy-1-(3-pyridyl)-1-butanone (HPB) upon mild hydrolysis. The released HPB is derivatized and analyzed by gas chromatography-mass spectrometry. Hemoglobin adduct levels are elevated above background in 15-20% of smokers and in most snuff-dippers. DNA adduct levels are higher in lung tissue from smokers than non-smokers. Diastereomeric NNAL glucuronides have been characterized as major urinary metabolites of NNK in the patas monkey. This led to the development of a gas chromatography-Thermal Energy Analysis method for detection of NNAL and its glucuronides in smokers' urine. These NNK metabolites have been detected in all smokers but not in non-smokers. The results of this research are providing new insights on the metabolic activation and detoxification of carcinogenic nitrosamines in humans.

Tobacco alkaloids are nitrosated during the curing and processing of tobacco, producing a group of nitrosamines called tobacco-specific nitrosamines (1,2). Two of these compounds, 4-(methylnitrosamino)-1-(3-pyridyl)-1-butanone (NNK) and N'-nitrosonornicotine (NNN), have well documented carcinogenic effects in laboratory animals. NNK is a potent pulmonary carcinogen, inducing predominantly adenocarcinoma in mice, rats, and hamsters independent of the route of administration (1-5). Extensive dose response studies in rats have demonstrated that the total doses of NNK required to produce lung tumors are similar to the total doses to which life-long smokers would be exposed, based on the amounts of NNK in mainstream cigarette smoke (6). NNK and its metabolite NNAL also induce pancreatic tumors in rats (4). They are the only tobacco smoke constituents known to induce acinar and ductal tumors of the pancreas in laboratory animals. NNN, which produces esophageal tumors in rats, is the most prevalent esophageal carcinogen in tobacco smoke (*1,2*). A mixture of NNK and NNN has been shown to cause oral cavity tumors in rats (*7*). These carcinogenicity data are discussed in more detail in the chapters by Hoffmann et al, and Murphy. Based on these data, we have proposed that NNK and/or NNN are possible causative agents for cancers of the lung, pancreas, esophagus, and oral cavity observed in humans who use tobacco products (*8*).

In this chapter, we review our work on the development and application of methods to assess human uptake and metabolic activation of these nitrosamines. Our goal is to understand their mechanisms of metabolic activation and detoxification in humans. NNK and NNN are suitable substrates for achieving this goal because human exposure to these compounds is extensive. Our hypothesis is that the probability of tumor development in an exposed person will be at least partially determined by that individual's ability to metabolically activate or detoxify these nitrosamines. Two types of approaches will be discussed. In one, we are quantifying hemoglobin adducts and DNA adducts of NNK and NNN in order to provide an estimate of the dose of metabolically activated substrate which reaches cells. In the other, we are assessing levels of urinary metabolites of NNK to obtain an estimate of its uptake, and eventually, a profile of its metabolic activation and detoxification.

Hemoglobin and DNA Adducts of NNK and NNN.

Figure 1 summarizes pathways of NNK and NNN metabolism, based on studies carried out in laboratory animals (*1,2,9,10*). The important reactions leading to hemoglobin and DNA adduct formation are the α-hydroxylation pathways (*11-14*). α-Methylene hydroxylation of NNK gives intermediate 9, which spontaneously decomposes to methane diazohydroxide (16). This electrophile alkylates DNA and hemoglobin. α-Methyl hydroxylation of NNK produces intermediate 10, which upon loss of formaldehyde generates the pyridyloxobutane diazohydroxide 18. This diazohydroxide reacts with aspartate or glutamate in hemoglobin, with the formation of ester adducts 22 (*15*). Hydrolysis of these adducts with mild base releases 4-hydroxy-1-(3-pyridyl)-1-butanone (HPB, 26). DNA adducts are also produced by 18. Acid hydrolysis of these adducts gives HPB. The formation of HPB-releasing hemoglobin and DNA adducts by NNK in rats has been investigated in some detail (*12-20*). The pathway leading to pyridyloxobutylation of globin and DNA also occurs in animals treated with NNN, via α-hydroxylation at the 2'-position to 12. α-Hydroxylation of NNN at the 5'-position is a known metabolic pathway but DNA and globin adducts arising in this way have not been characterized.

Quantitation of hemoglobin or DNA adducts of NNK and NNN will provide data on the extents to which these metabolic activation pathways may occur in humans. The use of hemoglobin adducts as dosimeters of carcinogen exposure was suggested by Ehrenberg and co-workers (*21*). Several groups have developed methodology which can now be applied to assess human exposure to, and/or metabolic activation of, a number of carcinogens including ethylene oxide, aromatic amines, and polynuclear aromatic hydrocarbons (*22*). Advantages of hemoglobin adducts as dosimeters include the relatively long lifetime of the erythrocyte in humans (approximately 120 days), which permits integration of dose over a somewhat extended period, and the relative ease with which ample quantities of hemoglobin can be obtained. Disadvantages include the probable lack of relevance of the hemoglobin adducts to the carcinogenic process, and the necessity to establish a predictable relationship between hemoglobin adduct levels and the biologically relevant DNA adducts. A more direct approach is the quantitation of DNA adducts, which has been attempted in numerous studies using techniques such as immunoassays, ^{32}P-post-labelling, fluorescence spectroscopy, and gas chromatography-mass spectrometry (GC-MS)(*23*). The clear advantage of this approach is the relevance of some DNA adducts to carcinogenesis. However, DNA is difficult to obtain from potential target tissues in quantities sufficient for analysis and the interpretation of adduct measurements can be confounded by repair and other mechanisms of removal. In spite of some of these limitations, quantitation of hemoglobin or DNA adducts can potentially provide important information on human exposure to important DNA damaging intermediates produced by carcinogen metabolism.

We have chosen HPB, released by hydrolysis of hemoglobin or DNA, as a dosimeter of NNK and NNN metabolic activation. HPB releasing hemoglobin and DNA adducts can only be formed from NNK, its metabolite NNAL, and NNN, as far as we are aware. Thus, detection of these adducts can be traced specifically to these compounds or certainly to tobacco-derived exposures. In contrast, methylation of hemoglobin and DNA by NNK would not be a specific dosimeter, because methylation has many sources, including endogenous ones, leading to high backgrounds and potentially confounding interpretation of the data. Both the methylation and pyridyloxobutylation pathways of NNKmetabolism have biological significance with respect to its carcinogenic activity; pyridyloxobutylation is essential for NNN carcinogenicity (*24*).

It should be noted that hemoglobin and DNA adducts, such as those releasing HPB, give a measure of both carcinogen uptake and metabolic activation in an individual. This measurement of "internal dose" is distinct from biomarkers of overall exposure to tobacco smoke, such as urinary cotinine.

The methodology employed to quantify HPB, released by base hydrolysis of human hemoglobin, has been described (*25,26*). In this method, an HPB-enriched fraction is prepared by a series of partition steps. The HPB is derivatized as its pentafluorobenzoate and, after an HPLC cleanup step, the derivative is separated and detected by capillary column gas chromatography-negative ion chemical ionization mass spectrometry, with selected ion monitoring (GC-NICI-MS-SIM). Deuterated HPB is used as an internal standard. The sensitivity of this method is excellent, with a detection limit of approximately 0.1 fmol HPB-pentafluorobenzoate. A trace obtained from a smoker's hemoglobin is presented in Figure 2.

Data obtained from several completed or ongoing studies are presented in Figure 3 (*24-26*). In non-smokers, we have generally not observed levels of HPB releasing hemoglobin adducts which are significantly above background levels of the method (approximately 200 fmol/g Hb). This is consistent with the fact that NNK and NNN are tobacco-specific nitrosamines. In three cases, however, elevated levels were observed suggesting exposure of these individuals to environmental tobacco smoke. This requires further investigation. In smokers, we have detected HPB-releasing hemoglobin adducts above background in approximately 15-20% of the individuals tested. Hemoglobin adduct levels have not been found to correlate with plasma cotinine or numbers of cigarettes smoked. Thus, the elevated levels may relate to more efficient metabolic activation of NNK or NNN in these smokers than in the others. Limited studies carried out to date on snuff-dippers indicate that these individuals have generally higher levels of HPB releasing adducts than do smokers, possibly due in part to higher levels of exposure to NNK and NNN. Hemoglobin adduct levels in F344 rats treated chronically with NNK, and determined by the same GC-MS method, are also illustrated in Figure 3. Most of the hemoglobin adduct values observed in humans fall within the same range as those quantified in rats that had been exposed to expected human doses of NNK and NNN, based on the amounts of these nitrosamines in tobacco products. However, some are well above this range. The factors governing higher hemoglobin adduct levels in some individuals require further research.

Based on these data, and on our experience with this assay, we can conclude that the pathway illustrated in Figure 1, leading to pyridyloxobutylation of hemoglobin via the intermediate 18 (or a related electrophile), exists in humans exposed to NNK and NNN. Our studies to date have demonstrated that the published GC-NICI-MS-SIM methodology is accurate and reproducible, giving a reliable quantitation of HPB in samples of base treated hemoglobin. However, there are many questions to be

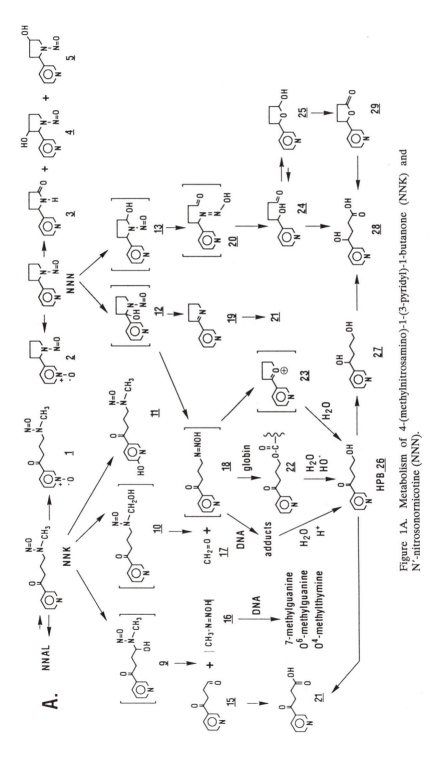

Figure 1A. Metabolism of 4-(methylnitrosamino)-1-(3-pyridyl)-1-butanone (NNK) and N'-nitrosonornicotine (NNN).

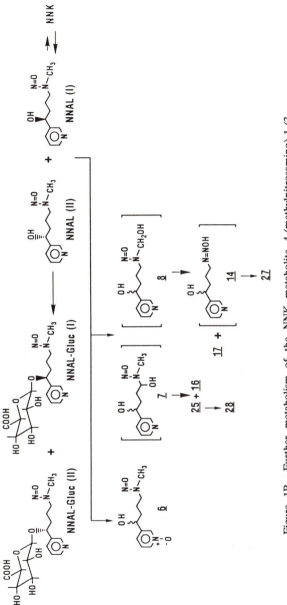

Figure 1B. Further metabolism of the NNK metabolite 4-(methylnitrosamino)-1-(3-pyridyl)-1-butanol (NNAL).

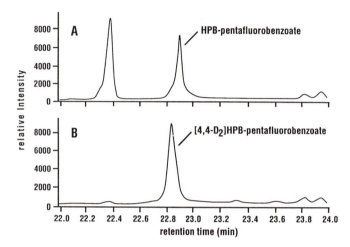

Figure 2.   GC-NICI-MS-SIM chromatograms obtained upon analysis of a
smokers' hemoglobin; A) SIM at m/z 359 (molecular ion of HPB-
pentafluorobenzoate and B) SIM at m/z 361 (molecular ion of internal standard,
[4,4-D$_2$]HPB-pentafluorobenzoate.

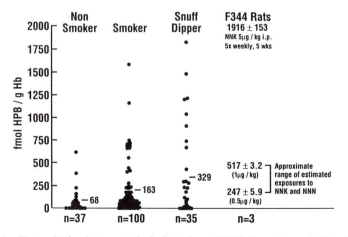

Figure 3.   Data obtained upon analysis for released HPB from hemoglobin of non-
smokers, smokers, snuff-dippers, and F-344 rats treated with NNK.   Data are from
ongoing studies conducted in our laboratory by S.E. Murphy, P.G. Foiles, and S.
Akerkar, as well as from published studies (ref. 25).

answered. The mechanism of adduct formation in hemoglobin will be investigated to better define the origin and significance of the adducts. The kinetics of formation and removal of the adduct in humans will be determined to supplement what is already known in rats. The relationship between hemoglobin adduct levels and DNA adduct levels in a given individual will be assessed in future studies. In rats, a predictable but non-linear relationship between hemoglobin adducts and DNA adducts in lung and liver has been observed in studies to date (*17*). Data from these studies will facilitate interpretation of adduct levels determined in humans and may provide insights on the parameters which lead to high adduct levels in only certain individuals. Properly designed epidemiologic studies of tobacco use and cancer induction, which incorporate HPB releasing hemoglobin adduct data, may provide insight on the possible relationship of adduct levels to the probability of cancer development.

The methodology for determination of HPB-releasing DNA adducts is similar to that employed for hemoglobin adducts, except that the first step requires acid hydrolysis to release HPB (*27*). This method has not been applied as widely as the hemoglobin assay in studies to date, principally because DNA is more difficult to obtain in the required quantities. In one study, DNA adduct levels ranging from 3-49 fmol/mg DNA were found in smokers' periperal lung and trachea, obtained at immediate autopsy. The amounts of HPB-releasing DNA adducts were higher than in non-smokers (*27*). As in rats, levels of HPB-releasing adducts were higher in DNA than in hemoglobin, when expressed per weight of macromolecule (*17*). These studies support the hypothesis that NNK or NNN are metabolically activated to intermediates which pyridyloxobutylate DNA in human lung.

Metabolism of NNK in the Patas Monkey

Although the metabolism of NNK has been well characterized in rodents, limited data are available in primates (*28,29*). Since one of our goals is to identify and quantify metabolites of NNK in human urine, we initiated a study of NNK in the patas monkey. These studies were carried out in collaboration with Lucy M. Anderson and Jerry M. Rice of the National Cancer Institute. Female monkeys were given i.v. injections of 0.1 µg/kg-4.9 mg/kg [5-^{3}H]NNK, labelled with tritium at the 5-position of the pyridine ring. Blood and urine were collected and analyzed by HPLC. The time course of NNK and its metabolites in serum is illustrated in Figure 4. NNK disappeared rapidly from serum, while NNAL and its glucuronides persisted for longer times. The results for NNK and NNAL are consistent with previous observations in baboons. An important observation was the rapid and extensive formation of keto acid 21 and hydroxy acid 28. These are products of α-hydroxylation which entail the formation of adducts, as discussed above.

HPLC analysis of serum and urine demonstrated the presence of a major peak which did not coelute with any of the known metabolites of NNK. Hydrolysis experiments with ß-glucuronidase indicated that this metabolite was a glucuronide of NNAL. However, it had a different chromatographic retention time from NNAL-Gluc that had been previously characterized in rat and mouse urine (arbitrarily assigned as NNAL-Gluc(I)-Figure 1-since the absolute configuration is unknown)(*30*). This peak was collected and its ^{1}H- and ^{13}C-NMR spectra were determined. They were quite similar to those of NNAL-Gluc(I) indicating that the new metabolite was the diastereomer, NNAL-Gluc(II). This was confirmed by HPLC isolation of NNAL-Gluc(I) and (II) from monkey urine, followed by hydrolysis to the enantiomeric NNAL(I) and NNAL(II), which were converted to the corresponding diastereomeric carbamates by reaction with R-(+)-α-methylbenzyl isocyanate. Significantly, the levels of NNAL-Gluc(I) and (II) in monkey urine accounted for up to 23% of the urinary metabolites after a dose of 0.1 µg/kg NNK, approximately equivalent to a smokers' exposure to NNK. These results encouraged us to analyze human urine for NNAL-Gluc.

Metabolites of NNK in Human Urine

The analytical method which we developed for analysis of NNAL and NNAL-Gluc in human urine is summarized in Figure 5. Fraction 1 contained unconjugated NNAL. The aqueous portion of the urine was treated with ß-glucuronidase, and the released NNAL was further purified, silylated, and analyzed by GC-TEA. Traces from 5 smokers and 5 non-smokers are illustrated in Figure 6. In each case, the peak marked with the asterisk, which corresponds in retention time to the trimethylsilyl ether of NNAL, was detected in Fractions 1 and 2 from smokers. Of the seven non-smokers examined, only small amounts of NNAL-Gluc were detected in one; all other non-smoker urines were negative. The following evidence supports the identity of NNAL and its glucuronides in smokers' urine: (1) the GC retention time of silylated NNAL was identical to that of a standard but well resolved from that of silylated 4-(methyl-nitrosamino)-4-(3-pyridyl)-1-butanol (iso-NNAL), which was added to samples to determine silylation efficiency (see Figure 6); (2) the material which was collected from HPLC, prior to silylation, had the same retention time as NNAL; and (3) the GC retention time of NNAL from these samples was the same as that of a standard, in analyses that were performed without silylation. Some experiments were carried out without ß-glucuronidase, or with sulfatase instead of ß-glucuronidase, or with ß-gluc-uronidase in the presence of saccharic acid 1,4-lactone, an inhibitor of ß-glucuronidase activity. No NNAL was detected in Fraction 2 of these samples. Other experiments, involving the addition of nitrite or monitor amines to urine samples, demonstrated that artefactual formation of NNAL and NNAL-Gluc under our conditions was minimal.

The results of analyses of urine from 11 smokers are summarized in Table I. Six samples were also analyzed for NNK, by collecting the appropriate fraction from HPLC, reducing with NaCNBH$_3$, and analyzing as above. NNK was not detected.

Table I. NNAL-Gluc and NNAL in Smokers' Urine

Subject	Sex	Cigarettes/Day	NNAL (μg/24 h)		
			glucuronide[a]	free	total
1	M	36	2.4	0.46	2.9
2	F	29	3.0	0.37	3.4
3	F	20	3.5	0.35	3.9
4[b]	M	25	0.92	0.43	1.4
5	F	20	2.1	0.69	2.8
6[b]	F	29	2.1	1.0	3.1
7	F	25	2.3	0.23	2.5
8	F	25	0.31	0.70	1.0
9[b]	F	12	0.76	0.87	1.6
10[b]	F	20	1.8	0.29	2.1
11	F	15	1.5	0.28	1.8

a. Expressed as NNAL equivalents. For conversion to μg NNAL-Gluc, multiply by 1.85
b. Levels of NNK in mainstream smoke of the cigarettes used by these volunteers were available. These were multiplied by cigarettes per day to obtain estimated daily dose of NNK, as follows (in μg) #4, 3.4; #6, 4.7; #9, 1.6; #10, 2.7

These data clearly demonstrate the presence in smokers' urine of NNAL and NNAL-Gluc. Their amounts are in the range expected based on the levels of NNK in the mainstream smoke of cigarettes used by four of the subjects. These data confirm the uptake of NNK by smokers in quantities which are comparable to those which induce lung tumors in laboratory animals, as previously estimated based on NNK levels in mainstream cigarette smoke (6). It is possible that some of the NNAL and

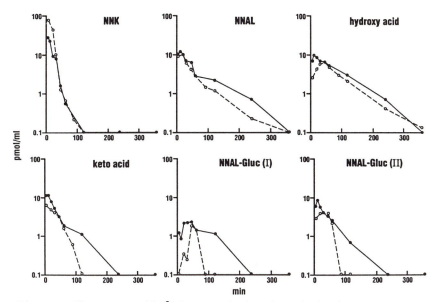

Figure 4. Time course of [5-³H]NNK and selected metabolites in patas monkey serum after i.v. injection of [5-³H]NNK. Hydroxy acid and keto acid are compounds 28 and 21 of Figure 1. Open and closed symbols represent experiments with two different monkeys.

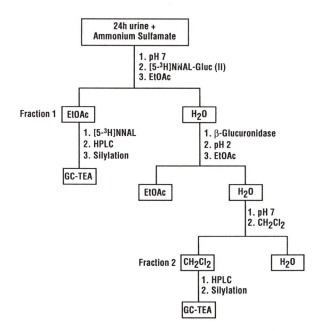

Figure 5. Scheme for analysis of NNAL and NNAL-Gluc in human urine.

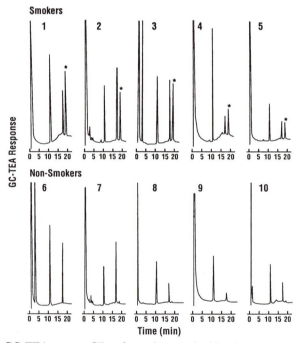

Figure 6. GC-TEA traces of Fraction 2 (see method in Figure 5) from smokers
(1-5) and non-smokers (6-10). The peaks eluting at approximately 10 and 17 min
in all traces are internal standards, nitrosoguvacoline (injection standard) and iso-
NNAL (silylation standard). The peak marked with the asterisk is the
trimethylsilyl ether of NNAL.

NNAL-Gluc in urine may arise from NNAL in cigarette smoke, but its presence in smoke has not been reported. The presence of NNAL and NNAL-Gluc in smokers' urine, but apparent absence of NNK, is consistent with the studies described above which showed that NNAL-Gluc(I) and (II) were major constituents of patas monkey urine, in contrast to NNK (*10*). The ratios of NNAL-Gluc to NNAL in human urine are potentially interesting. They vary from 0.44 to 10. Most smokers had ratios greater than 1, indicating preferential excretion of NNAL as its glucuronide conjugate. It is possible that smokers who conjugate NNAL poorly may α-hydroxylate it more extensively, potentially leading to higher levels of hemoglobin and DNA adducts.

Perspectives

Although our knowledge of carcinogen metabolism and mechanisms of carcinogenesis in laboratory animals has greatly increased over the past four decades, our understanding of these processes in humans has lagged behind. The development of reliable techniques to quantify carcinogen metabolic activation, detoxification, and macromolecular binding in humans should lead to advances in understanding causes of human cancer. The techniques of molecular biology have led to the detection of mutations in critical cellular genes associated with cancer such as *ras* and p53 (*31,32*). However, the origin of these changes with respect to particular carcinogen exposures is still speculative.

Human exposure to tobacco-specific nitrosamines through use of tobacco products is widespread in spite of the known hazards of smoking. The methodology reported here has already produced new insights on their metabolic activation and detoxification in humans. The detection of elevated levels of tobacco-specific nitrosamine hemoglobin adducts in only 15-20% of smokers was a surprising finding although it should be noted that further refinements in the GC-MS assay may lead to inclusion of a greater percentage of subjects above background levels. Nevertheless, the reasons for these relatively elevated adduct levels need to be determined since they may indicate potential susceptibility factors. While the interpretation of data obtained from analysis of hemoglobin adducts and DNA adducts is somewhat complex because of multiple factors which may influence their formation and removal from cells, interpretation of urinary metabolite data is more straightforward and can be useful with respect to determination of overall uptake and individual profiles of metabolic activation and detoxification reactions. The two metabolites analyzed here can arise only from NNK or NNAL, while other metabolites of NNK and NNN may also be formed from nicotine. Nevertheless, stereochemical differences in the formation of some of these metabolites from NNK and NNN versus from nicotine may allow us to confidently assign their origin. By combining data from adduct determinations with those from analysis of urinary metabolites, we expect to achieve a more complete understanding of NNK and NNN metabolic activation and detoxification in humans.

Literature Cited

1. Hoffmann, D; and Hecht, SS. Cancer Res., **45**, 935(1985).
2. Hecht, SS; and Hoffmann, D. Carcinogenesis, **9**, 875(1988).
3. Belinsky, SA; Foley, JF; White, CM; Anderson, MW; and Maronpot, RR. Cancer Res., **50**, 3772(1990).
4. Rivenson, A; Hoffmann, D; Prokopczyk, B; Amin, S; and Hecht, SS. Cancer Res., **48**, 6912(1988).
5. Lijinsky, W; Thomas, BJ; and Kovatch, RM. Jpn. J. Cancer Res., **82**, 980(1991).

6.   Hecht, SS; and Hoffmann, D. In: "The Origins of Human Cancer: A Comprehensive Review", Cold Spring Harbor, NY, Cold Spring Harbor Laboratory Press, 1991, 745.
7.   Hecht, SS; Rivenson, A; Braley, J; DiBello, J; Adams, JD; and Hoffmann, D. Cancer Res., **46**, 4162(1986).
8.   Hecht, SS; and Hoffmann, D. Cancer Surv., **8**, 273(1989).
9.   Hecht, SS; Castonguay, A; Rivenson, A; Mu, B; and Hoffmann, D. J. Environ. Health Sci. CI, **CI 1**, 1(1983).
10. Hecht, SS; Trushin, N; Reid-Quinn, C; Burak, E; Jones, AB; Southers, J; Gombar, C; Carmella, SG; Anderson, LM; and Rice, JM. Carcinogenesis, in press (1992).
11. Hecht, SS; Trushin, N; Castonguay, A; and Rivenson, A. Cancer Res., **46**, 498(1986).
12. Carmella, SG; and Hecht, SS. Cancer Res., **47**, 2626(1987).
13. Hecht, SS; Spratt, TE; and Trushin, N. Carcinogenesis, **9**, 161(1988).
14. Hecht, SS; and Trushin, N. Carcinogenesis, **9**, 1665(1988).
15. Carmella, SG; Kagan, SS; and Hecht, SS. Chem. Res. Toxicol., **5**, 76(1992).
16. Spratt, TE; Trushin, N; Lin, D; and Hecht, SS. Chem. Res. Toxicol., **2**, 169(1989).
17. Murphy, SE; Palomino, A; Hecht, SS; and Hoffmann, D. Cancer Res., **50**, 5446(1990).
18. Carmella, SG; Kagan, SS; Spratt, TE; and Hecht, SS. Cancer Res., **50**, 5453(1990).
19. Peterson, LA; Carmella, SG; and Hecht, SS. Carcinogenesis, **11**, 1329(1990).
20. Peterson, LA; Mathew, R; Murphy, SE; Trushin, N; and Hecht, SS. Carcinogenesis, **12**, 2069(1991).
21. Ehrenberg, L; and Osterman-Golkar, S. Teratog., Carcinog., Mutag., **1**, 105(1976).
22. Skipper, PL; and Tannenbaum, SR. Carcinogenesis, **11**, 507(1990).
23. Santella, RM. Environ. Carcino. & Ecotox. Revs., **C9(1)**, 57(1991).
24. Hecht, SS; Carmella, SG; Foiles, PG; Murphy, SE; and Peterson, LA. Environ. Health Perspect., **99**, in press(1992).
25. Carmella, SG; Kagan, SS; Kagan, M; Foiles, PG; Palladino, G; Quart, AM; Quart, E; and Hecht, SS. Cancer Res., **50**, 5438(1990).
26. Hecht, SS; Carmella, SG; and Murphy, SE. Methods in Enzymology, in press(1992).
27. Foiles, PG; Akerkar, SA; Carmella, SG; Kagan, M; Stoner, GD; Resau, JH; and Hecht, SS. Chem. Res. Toxicol., **4**, 364(1991).
28. Castonguay, A; Tjälve, H; Trushin, N; d'Argy, R; and Sperber, G. Carcinogenesis, **6**, 1543(1985).
29. Adams, JD; LaVoie, EJ; O'Mara-Adams, KJ; Hoffmann, D; Carey, KD; and Marshall, MV. Cancer Lett., **28**, 195(1985).
30. Morse, MA; Eklind, KI; Toussaint, M; Amin, SG; and Chung, F-L. Carcinogenesis, **11**, 1819(1990).
31. Bos, JL. Cancer Res., **49**, 4682(1989).
32. Hollstein, M; Sidransky, D; Vogelstein, B; and Harris, CC. Science, **253**, 49(1991).

RECEIVED October 25, 1993

Chapter 18

Metabolism of N'-Nitrosonornicotine by Rat Liver, Oral Tissue, and Esophageal Microsomes

Sharon E. Murphy, Deborah A. Spina, and Rachel Heiblum

American Health Foundation, 1 Dana Road, Valhalla, NY 10595

N'-Nitrosonornicotine (NNN) is an esophageal, but not a liver, carcinogen in rats. This difference in susceptibility may in part be
explained by tissue-specific differences in metabolism. NNN is
metabolized by α-hydroxylation at the 2' or 5' position in the
pyrrolidine ring. 2'-Hydroxylation results in pyridyloxobutylation of
DNA. The rate of microsomal α-hydroxylation (pmol/mg/min) of 1
μM NNN was 1.8 for oral tissue, 3.8 ± 0.33 (n=3) for liver and
3.9 ± 1.9 (n=11) for the esophagus. The ratios of 2' to 5'
hydroxylation for the latter two tissues was 0.7:1.0 and 3.1:1.0,
respectively. The maximum rate of NNN α-hydroxylation by
esophageal microsomes was obtained at 6 μM NNN, suggesting the
presence of a high affinity enzyme in this tissue. Monoclonal
antibodies to P450 IIC11 inhibited NNN metabolism by liver
microsomes to 56% but had no effect on the metabolism by
esophageal microsomes. This is consistent with the presence of a
unique enzyme in the esophagus.

In the United States in 1991, 77% of the deaths due to esophageal cancer and
80% of the deaths due to head and neck cancer were attributable to cigarette
smoking (*1*). NNN is the most abundant esophageal carcinogen in tobacco and
tobacco smoke. NNN administered in the drinking water induced esophageal
tumors in 24/25 rats (*2*). When NNN was applied with 4-(methylnitrosamino)-1-
(3-pyridyl)-1-butanone (NNK) to the oral cavity of rats chronically for 2 yr, 8 of
30 animals developed oral cavity tumors (*3*). Application of NNK alone did not
induce oral cavity tumors (*4*). Therefore, either NNN alone or in combination
with NNK is responsible for the induction of oral cavity tumors in this model.
NNN is a likely causative agent for esophageal tumors and cancers of the oral
cavity in tobacco users (*5*).
    The esophagus is notoriously sensitive to tumor induction by nitrosamines.
In the rat, 61 of 170 tested nitrosamines induce a greater than 50% incidence of
esophageal tumors (*6,7*). Twenty two of the 61 also induce tumors in the oral
cavity. Only one of these 170 nitrosamines, ethanol-2-oxopropylamine, cause
tumors in the oral cavity but not in the esophagus. This suggests that these two
tissues are related in their susceptibility to tumor induction by nitrosamines.

0097–6156/94/0553–0223$08.00/0
© 1994 American Chemical Society

The esophagus is not sensitive to nitrosamines in general, but is sensitive to a particular subset of these carcinogens. This selectivity is striking if one looks at tumor incidence as a function of alkyl chain length for a series of methylalkylnitrosamines. No esophageal tumors were induced when the alkyl chain length was 1, 2, 7, 8, 9 or 10 carbons, while 95-100% incidence occurred when the alkyl chain was 3 to 6 carbons in length (*6-8*). This striking selectivity recently was reaffirmed in a study of over 4000 rats (*9-10*). N-Nitrosodimethylamine induced a high incidence of liver tumors but not a single esophageal tumor. Whereas, N-nitrosodiethylamine induced an approximately equal number of liver and esophageal tumors (*9*). In the same study, N-nitrosopyrrolidine did not induce any esophageal tumor, while N-nitrosopiperidine induced a high incidence of esophageal tumors (*10*). These two cyclic nitrosamines differ by only one carbon in the size of the nitrogen containing ring. NNN induces esophageal and nasal cavity tumors, but no liver tumors. NNK induces tumors of the lung, liver, pancreas and nasal cavity but not a single esophageal tumor (*5*).

Several investigators have suggested that this unique tissue selectivity with respect to tumor induction is due at least in part to an esophageal enzyme specific for the metabolism of nitrosamines (*8, 11, 12*). It is our hypothesis that oral cavity tissue is related to the esophagus in its ability to metabolically activate nitrosamines.

NNN Metabolism by Rat Esophagus and Oral Cavity Tissue

Tissue Culture Metabolism. NNN is metabolized primarily by α-hydroxylation at the 2' or 5' carbon of the pyrrolidine ring (Figure 1). In rat organ culture, the product of 5'-hydroxylation is hydroxy acid. The products of 2'-hydroxylation are keto alcohol, keto acid and diol. 2'-Hydroxylation results in the formation of 4-oxo-4-(3-pyridyl)butanediazohydroxide. This reacts with DNA to generate adducts thought to be important in tumor induction. Hecht and co-workers reported that cultured rat esophagus metabolized 5μM NNN to a 5 fold greater extent than cultured rat liver slices (*13*). The ratio of 2' to 5'-hydroxylation was 3.1 in the esophagus and 1.4 in the liver, suggesting that 2'-hydroxylation is the important pathway with regard to tumor induction.

We have compared the metabolism of NNN and NNK in cultured rat esophagus and oral tissue (*14*). In both tissues, the ratio of 2' to 5'-hydroxylation was between 3.1 and 3.6 over the range of concentrations studied. The total α-hydroxylation of NNN by rat esophageal tissue was 3 to 10 times that of NNK. The metabolism of NNN by α-hydroxylation is even greater in rat oral cavity tissue than the esophagus. The rate of NNN α-hydroxylation by cultured oral tissue compared to that of NNK was as much as 9 fold greater. This difference in metabolism persisted over a range of concentrations, from 1 to 100 μM, and is not related to the conversion of NNK to 4-(methylnitrosamino)-1-(3-pyridyl)-1-butanol.

Microsomal Metabolism. More recently, we isolated an active microsomal preparation from both rat esophageal and oral cavity tissue. Esophageal microsomes were prepared separately from both the epithelial layer and the supporting muscle. The microsomes and co-factors were incubated with 1μM [5-^{3}H]NNN and the products analyzed by radioflow HPLC (*14*). A representative chromatogram is presented in Figure 2. Only the epithelial layer contained a significant amount of NNN α-hydroxylation activity. One major radioactive peak was detected. This peak coeluted with both keto alcohol and the lactol of 4-hydroxy-4-(3-pyridyl)butanal. These are the primary products of 2' and 5'-

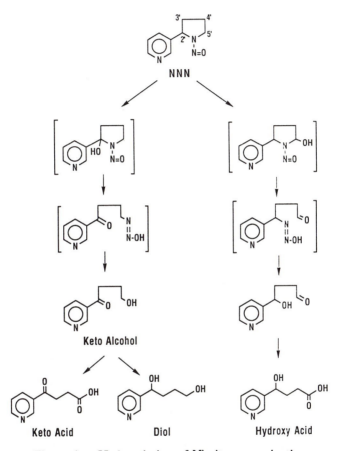

Figure 1. α-Hydroxylation of N'-nitrosonornicotine.

hydroxylation of NNN (2). The rate of formation of these products was linear with time (0 to 30 min) and protein concentration (0.05 to 0.4 mg/ml) and was dependent on the presence of NADPH. The presence of this enzyme activity in the epithelial tissue of rat esophagus is in agreement with the results of Tjälve and Castonguay (15). They reported the presence of bound radioactivity in this tissue following the administration of [³H]NNN.

The average rate of α-hydroxylation in 11 separate esophageal microsomal preparations was 3.9 +/- 1.9 pmol/mg/min. The average rate is the same as the average rate for liver microsomes (Table I). The actual rate of α-hydroxylation by esophageal microsomes is probably greater than that in the liver. The enzyme responsible for the metabolism of NNN in the esophagus was labile. The relatively large standard deviation reflects the wide range of activities, 1.2 to 8.7 pmol/mg/min, that were obtained in different preparations. This variability is in contrast to NNN α-hydroxylation by liver microsomes where the standard deviation of different preparations was quite small.

Table I. Microsomal Metabolism of 1 μM NNN

Tissue	α-Hydroxylation Products (pmol/mg protein/min)
Liver	3.8±0.3 (n=3)
Esophagus	
Total	0.3±0.07
Epithelial Tissue	3.9±1.9 (n-11)
Oral	1.8 (n=2)

Microsomes also were prepared from rat oral tissue. Tissue was obtained from the inside of both cheeks, the roof of the mouth and inside the lip. Although an effort was made to free the epithelial tissue from the muscle and connective tissue, the material used for microsomal preparations contained all tissue types. The homogenate (4.7 mg/ml), 9000 x g supernatant (2.4 mg/ml) and microsomes (0.5 mg/ml) from this tissue were incubated with 1 μM NNN for 40 min. The products formed were analyzed by radioflow HPLC. Chromatograms are presented in Figure 3. From organ culture studies, it is apparent that the oral tissue contains greater NNN α-hydroxylation activity than the esophagus (14). We were unable to obtain a microsomal preparation from the oral tissue which reflected this (Table I). This is in part because we are diluting the activity present in the epithelial layer by isolating microsomes from a mixture of tissue types. In addition, all the activity present in a homogenate of oral tissue was not recovered in the microsomal fraction. It will require some refinement of the techniques used here to obtain a microsomal preparation from oral tissue which is representative of the true activity present in this tissue.

Concentration Dependence of NNN Metabolism. The dependence of the rate of α-hydroxylation on the concentration of NNN was determined with both liver and esophageal microsomes. The rate with liver microsomes was not yet maximum at a concentration of 200 μM NNN. In contrast, the rate with esophageal microsomes reached a maximum at 6 μM (Table II). This suggests the presence of a high affinity (low K_M) enzyme in the esophagus that is responsible for α-hydroxylation of NNN.

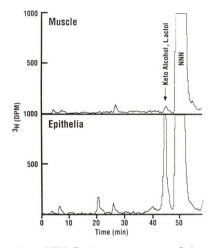

Figure 2. Radioflow HPLC chromatograms of the products of rat esophageal microsomal metabolism of NNN.

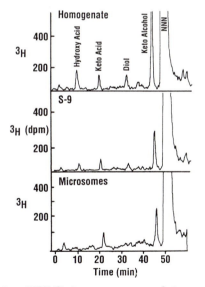

Figure 3. Radioflow HPLC chromatograms of the products of rat oral cavity metabolism of NNN.

Table II. Rate of NNN Metabolism with Varying Concentrations of NNN

Liver[a]		*Esophagus*[b]	
[NNN]	α-hydroxylation (pmol/mg/min)	[NNN]	α-hydroxylation (pmol/mg/min)
1 μM	3.5	1 μM	3.3
2 μM	5.3	3 μM	6.5
4 μM	10.4	6 μM	11
10 μM	26.4	10 μM	10
50 μM	100		
100 μM	170		
200 μM	285		

[a] 0.3 mg/ml protein, duplicate determinations.
[b] 0.1 mg/ml protein, duplicate determinations.

All rates reported in Tables I and II are for total α-hydroxylation of NNN. This includes both 2' and 5'-hydroxylation, that is the formation of keto alcohol and lactol. These two pathways are most likely catalyzed by two different enzymes (*16, 17*). The products of each pathway may be separated using normal phase HPLC.

The radioactive peak containing lactol and keto alcohol (Figure 2) was collected from the reverse phase column and the solvent removed. The residue was resuspended in 60% hexane 40% isopropanol: ethanol (2:1, solvent B) and injected on a silica column and eluted with a linear gradient from 100% hexane to 50% hexane:50% solvent B. Prior to reverse phase HPLC analysis, each sample was spiked with 18 μg lactol and 20 μg keto alcohol. After analysis by normal phase HPLC, the percent recovery of each metabolite was between 49 and 65%. The microsomal rates of 2'-hydroxylation and 5'-hydroxylation are reported for the esophagus and the liver in Table III. The ratio of 2' to 5'-hydroxylation was 3.1 for the esophagus and 0.71 for the liver. Therefore, the rate of keto alcohol formation by esophageal microsomes was 3 times greater than that by liver microsomes. This is the pathway considered to be important for tumor induction by this nitrosamine.

Table III. α-Hydroxylation of 1 μM NNN, 2' vs. 5'

	Esophagus (pmol/mg/min)	*Liver*
2'-Hydroxylation (Keto alcohol)	2.5	0.83
5'-Hydroxylation (Lactol)	0.8	1.17
Ratio 2'/5'	3.1	0.71

The Effect of Monoclonal Antibodies on Microsomal Metabolism of NNN. The involvement of particular cytochrome P450 enzymes (P450) in the metabolism of NNN was investigated by the use of antibodies to these enzymes. Monoclonal antibodies (Mab) which inhibit metabolism by P450 IA1/IA2, P450 IIB1/IIB2, P450 IIC11 and P450 IIE1 were obtained from Dr. Harry Gelboin of the National Cancer Institute. Antibodies to P450 IIC11 inhibited the α-hydroxylation of NNN by rat liver microsomes 56 % (Table IV). None of the other Mab had any effect on the metabolism of NNN by liver microsomes. The Mab to P450 IIC11 (1-68-11) is the same antibody that Mirvish and co-workers reported to inhibit α-hydroxylation of N-nitrosomethylamylamine (NMAA) by rat liver microsomes (18). NMAA is also an esophageal carcinogen. Therefore our results and their results indicate that this Mab is recognizing a hepatic P450 which metabolizes esophageal carcinogens. This antibody had no effect on the metabolism of NNN by esophageal microsomes (Data not shown). This result is consistent with the presence of a unique P450 enzyme in the esophagus which metabolizes these nitrosamines.

Table IV. Inhibition of Liver Microsomal Metabolism by Antibodies to Cytochrome P450[a,b]

Monoclonal antibody	α-Hydroxylation (percent of control)
P450 IA1/IA2 (1-7-1)[c]	96
P450 IIB1/IIB2 (4B-29-5)	100
P450 IIC11 (1-68-11)	44
P450 IIE1 (1-91-3)	100

[a] 0.3 mg/ml microsomal protein, (1.4 nmol P450/mg), 0.4 mg antibody.
[b] Average of 4 determinations.
[c] Numbers in parentheses are the designation assigned to a particular clone.

DNA Adduct Formation by NNN and NNK. Previously, we reported DNA binding studies with NNN and NNK in cultured rat esophageal and oral cavity tissue (14). Each tissue was incubated with either [5-^{3}H]NNN, [5-^{3}H]NNK or [C^{3}H$_3$]NNK. DNA pyridyloxobutylation was determined as the amount of keto alcohol released upon acid hydrolysis. DNA methylation was measured as 7-methylguanine. No methylation by NNN was detected in esophageal DNA (<0.4 pmol/μmol guanine). A small amount of pyridyloxobutylation by NNK was detected (0.17 pmol/μmol guanine) and a significant level of pyridyloxobutylation by NNN was detected (3.8 pmol/μmol guanine). This is consistent with NNN but not NNK being an esophageal carcinogen. The results obtained with oral cavity tissue were essentially the reverse, of those obtained with the esophagus. No pyridyloxobutylation by either NNN or NNK was detected (0.22 pmol/μmol guanine) but a significant level of 7-methylguanine was measured in tissue incubated with NNK (1.7 to 4.6 pmol/μmol guanine).
    Table V summarizes what is presently known about tumor induction by NNN and NNK and metabolic activation of these tobacco-specific nitrosamines in the oral cavity and the esophagus of rats.

Table V. Metabolic Activation and Tumor Induction by NNN and NNK

	Esophagus	*Tissue* Oral
NNN	Tumorigenic Low K_M P450 Pyridyloxobutylates DNA	Tumorigenic (?) Low K_M P450 (?) No DNA adducts identified
NNK	Non-tumorigenic Poorly metabolized relative to NNN Little DNA pyridyl- oxobutylation or methylation	Non-tumorigenic Poorly metabolized relative to NNN Methylates DNA (7 mG)

From this information the following conclusions can be drawn. Metabolism and DNA adduct formation data for NNN and NNK are consistent with the tumorigenicity of NNN and lack of tumorigenicity of NNK in the rat esophagus. Metabolism data for NNN and NNK in the oral cavity are similar to that in the esophagus suggesting that a common mechanism of tumorigenicity in these two tissues may apply. However, the mechanism of tumor induction in rat oral tissue by a mixture of NNN and NNK is presently unclear.

Acknowledgments

We wish to thank Dr. H. Gelboin and Dr. S. Park for providing us with monoclonal antibodies to cytochrome P450 enzymes. This work was supported by Grant CA-29580 from the National Cancer Institute.

Literature Cited

1.   Shopland, D.R.; Eyre, H.J.; Pechacek, T.F. *J. Natl. Cancer Inst.* **1991**, *83*, 1142-1148.
2.   Hoffmann, D.; Hecht, S.S. *Cancer Res.* **1985**, *45*, 935-944.
3.   Hecht, S.S.; Rivenson, A.; Braley, J.; DiBello, J.; Adams, J.D.; Hoffmann, D. *Cancer Res.* **1986**, *46*, 4162-4166.
4.   Prokopczyk, B.; Rivenson, A.; Hoffmann, D. *Cancer Lett.* **1991**, *60*, 153-157.
5.   Hecht, S.S.; Hoffmann, D. *Cancer Surveys* **1989**, *8*, 273-294.
6.   Preussmann, R.; Stewart, B.W. In Chemical Carcinogenesis; Searle, C.E., Ed.; *Am. Chem. Soc. Monograph 182*; Washington, DC, **1984**, Vol. 2, pp. 645-868.
7.   Lijinsky, W. *Cancer and Metastasis Reviews* **1987**, *6*, 301-356.
8.   Van Hofe, E.; Schmerold, I.; Lijinsky, W.; Jeltsch, W.; Kleihues, P. *Carcinogenesis* **1987**, *8*, 1337-1341.
9.   Peto, R.; Gray, R.; Brantom, P.; Grasso, P. *Cancer Res.* **1991**, *51*, 6415-6451.
10.   Gray, R.; Peto, R.; Brantom, P.; Grasso, P. *Cancer Res.* **1991**, *51*, 6470-6491.
11.   Labuc, G.E.; Archer, M.C. *Cancer Res.* **1982**, *42*, 3181-3186.
12.   Mirvish, S.S.; Wang, M.-Y.; Smith, J.W.; Deshpande, A.D.; Makary, M.H.; Issenberg, P. *Cancer Res.* **1985**, *45*, 577-583.

13. Hecht, S.S.; Reiss, B.; Lin, D.; Williams, G.M. *Carcinogenesis* **1982**, *3*, 453-456.
14. Murphy, S.E.; Heiblum, R.; Trushin, N. *Cancer Res.* **1990**, *50*, 4685-4691.
15. Tjälve, H. and Castonguay, A. *IARC Sci. Publ.* **1987**, *84*, 434-437.
16. Murphy, S.E.; Heiblum, R. *Carcinogenesis* **1990**, *11*, 1663-1666.
17. McCoy, G.D.; Chem, C.-H.B.; Hecht, S.S. *Drug Metb. Disp.* **1981**, *9*, 168-169.
18. Mirvish, S.S.; Huang, Q.; Chuon, J.; Wang, S.; Park, S.S.; Gelboin, H.V. *Cancer Res.* **1991**, *51*, 1059-1064.

RECEIVED October 11, 1993

Chapter 19

Mechanisms of Inhibition of Tobacco-Specific Nitrosamine-Induced Lung Tumorigenesis in A/J Mice

Fung-Lung Chung, Mark M. Morse[1], Karin I. Eklind[2], and Yong Xu[3]

Division of Chemical Carcinogenesis, American Health Foundation, 1 Dana Road, Valhalla, NY 10595

The mechanisms of inhibition by arylalkyl isothiocyanates and green tea polyphenol of lung tumorigenesis caused by the tobacco-specific nitrosamine 4-(methylnitrosamino)-1-(3-pyridyl)-1-butanone (NNK) were investigated. Our previous studies showed that pretreatment of A/J mice with phenethyl isothiocyanate (PEITC) inhibited the formation of lung tumors induced by NNK. However, PEITC's lower homologues, phenyl isothiocyanate (PITC) and benzyl isothiocyanate (BITC), were inactive. These results agreed with their effects on the formation of lung O^6-methylguanine (O^6-mG), a critical lesion in lung tumorigenesis resulting from the metabolic activation of NNK. We further tested a number of newly synthesized longer alkyl chain arylalkyl isothiocyanates for their inhibitory activities in A/J mice treated with NNK. The results showed that the potency of inhibition by arylalkyl isothiocyanates increased as the chain length increases. Treatment of mice with PEITC and BITC in the diet subsequent to NNK treatment had little effect on the formation of lung tumor induced by NNK. These results are consistent with the mechanism that inhibition of lung tumorigenesis by arylalkyl isothiocyanates is due largely to the inhibition of enzymes which activate NNK. In contrast to arylalkyl isothiocyanates, green tea and its major polyphenol, (-)-epigallocatechin gallate (EGCG), inhibited lung tumor formation in NNK-treated mice but exerted little effect on the formation of O^6-mG in lung DNA. These results suggest that one or more DNA lesions other than methylation are involved in NNK lung tumorigenesis. We

[1]Current address: Arthur James Cancer Hospital and Research Institute, Ohio State University, Columbus, OH 43210
[2]Current address: Astra AB, 15185 Södertälje, Sweden
[3]Current address: Institute of Nutrition and Food Hygiene, Chinese Academy of Preventive Medicine, 29 Nan-Wei Road, Beijing 10050, China

examined whether free radical-mediated oxidative DNA damage is induced in NNK-treated mice by measuring 8-hydroxydeoxyguanosine (8-OH-dG) formation. A dose-dependent increase of this lesion in lung was seen in NNK treated mice. These results provide the first direct evidence of oxidative DNA damage by a nitrosamine. Consistent with their ability to inhibit lung tumor formation by NNK, the increase of oxidation in DNA by NNK was suppressed by treatment with green tea extract or its EGCG. These results support the involvement of free radical damage in NNK lung tumorigenesis as well as the role of green tea and its polyphenol as antioxidants in inhibition of the NNK-induced lung tumorigenesis in A/J mice.

4-(methylnitrosamino)-1-(3-pyridyl)-1-butanone (NNK), a nicotine derived nitrosamine found in tobacco, is a potent and highly specific carcinogen for the induction of lung tumors in laboratory animals (*1,2*). It is believed that human exposure to NNK through smoking may be an important risk factor for the development of lung cancer (*2*). Since our goal is to use chemoprevention as a means to reduce the risk of lung cancer caused by smoking, we have focused on studies to identify both naturally occurring and synthetic agents capable of inhibiting the lung carcinogenesis by NNK and to understand the mechanisms of inhibition.

Results of these studies showed that the NNK-induced lung tumorigenesis in female A/J mice is effectively inhibited by pretreatment of arylalkyl isothiocyanates (*3*). Some of them, such as phenethyl isothiocyanate (PEITC), occur in cruciferous vegetables (*4,5*). The structure-activity relationship studies showed that the potency of arylalkyl isothiocyanates as inhibitors is influenced by the length of their alkyl chain (*6*). The inhibitory efficacy increases as the alkyl chain elongates up to 6 carbons. Thus, 6-phenylhexyl isothiocyanate PHITC, a synthetic compound, is considerably more active than PEITC. In addition to arylalkyl isothiocyanates, we also examined the potential inhibitory activity of green tea and its major polyphenol (-) epigallocatechin gallate (EGCG) against lung tumorigenesis induced by NNK in female A/J mice (*7*). The prevalence of green tea consumption in Japan is an intriguing dietary factor which may be involved in the reduced risk of lung cancer in male smokers in Japan as compared with their counterparts in the United States (*8*). We demonstrated that giving green tea as drinking water or EGCG in drinking water during NNK treatment inhibited lung tumor formation in A/J mice. These results provide an animal data base for the potential role of these agents as risk-lowering factors for lung cancer in humans.

Like most carcinogenic nitrosamines, NNK is metabolically activated *via* α-hydroxylation mediated primarily by cytochrome P-450s (*9*). This pathway generates alkylating agents such as methyldiazonium ion which methylates DNA. Numerous studies have shown that the formation of O^6-methylguanine (O^6-mGua) in lung DNA of A/J mice treated with NNK is a critical step for lung tumor induction (*10-12*). Although NNK treatment also results in pyridyloxobutylation, the importance of this DNA modification in NNK tumorigenesis has yet to be established (*13*). In this chapter, we will summarize results of our studies designed to better understand the mechanisms of inhibition of the NNK-induced lung tumorigenesis by arylalkyl isothiocyanates and green tea polyphenol. The structures of NNK, arylalkyl isothiocyanates and EGCG are shown in Figure 1.

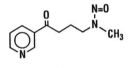

NNK

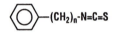

Arylalkyl Isothiocyanates

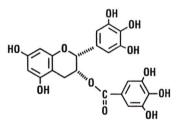

(-) Epigallocatechin-3-gallate (EGCG)

Figure 1   Structures of NNK, arylalkyl isothiocyanates, and EGCG.

Arylalkyl Isothiocyanates

Correlation of Inhibition of Lung Tumorigenicity and DNA Methylation Induced by NNK. In an earlier study, we tested the inhibitory activity of phenyl isothiocyanate (PITC), benzyl isothiocyanate (BITC), and PEITC against lung adenoma formation in female A/J mice-treated with NNK (3). Pretreatment with 4 consecutive daily doses of PEITC at 5 μmol or 25 μmol/dose by gavage inhibited by 76 or 97% the lung adenoma formation induced by a single dose of NNK (10 μmol, i.p.) given 2 h after the last gavage while PITC or BITC was inactive at 5 μmol and toxic at 25 μmol. Using identical treatments, PEITC reduced the levels of O^6-mGua in lung DNA while PITC and BITC had no effect on NNK-induced DNA methylation. Therefore, the inhibitory effects of the isothiocyanates on lung tumor formation correlated well with their effects on lung DNA methylation by NNK. Table I compares the tumor data with the levels of DNA methylation.

More recently, the alkyl chain length in arylalkyl isothiocyanates has been identified as a critical structural feature for inhibition (*14*). 6-PHITC is estimated to be at least 100-fold more efficacious than PEITC. We compared the effects of pretreatment with arylalkyl isothiocyanates of various alkyl chain length (up to 6 carbons) on lung tumorigenicity and DNA methylation in A/J mice treated with NNK. Table II shows that the potency of arylalkyl isothiocyanates in lung tumor inhibition is in general agreement with their effects on O^6-mGua formation. Therefore, pulmonary O^6-mGua appears to be a reasonable predictor of inhibitory potential in the NNK lung tumorigenesis. It should be noted that although PHITC completely blocked the formation of lung adenomas in NNK-treated mice, lung O^6-mGua formation was still evident. Therefore, the inhibition of O^6-mGua is quantitatively less than the corresponding inhibitory effect on lung tumor formation. It is possible that data obtained from a detailed time study may reveal a better correlation between DNA methylation and tumorigenicity in lung. It is also conceivable that additional lesion(s) other than DNA methylation are involved in NNK tumorigenesis. Unlike lung, DNA methylation in liver remains unchanged with isothiocyanate pretreatments. These results suggest that the lowered O^6-mGua levels in lung of the isothiocyanate-pretreated animals is not due to increased NNK activation in the liver which would result in decreased availability of NNK in the lung. The lack of inhibition of NNK activation in the liver by isothiocyanates is probably because of differences in the levels of expression of cytochrome P-450 isozymes involved in NNK metabolism in the liver as compared to the lung. Pretreatment of A/J mice with PEITC resulted in decreased lung microsomal activity for NNK metabolism (*3*). Smith *et al.* also showed that addition of arylalkyl isothiocyanates to mouse lung microsomes inhibited NNK metabolism in a dose-dependent manner (*9*). Analogous to the results of the tumor bioassay, the extent of inhibition seems to increase with increased alkyl chain length (*9*). These results are consistent with a mechanism that isothiocyanates exert their inhibitory activity mainly through inhibition of the enzyme responsible for the activation of NNK.

Lack of Tumor Inhibition by Isothiocyanates Upon Treatment Subsequent to NNK Administration. The mode of inhibition against lung tumorigenesis was further investigated to determine whether post-treatment of mice with isothiocyanates would affect NNK lung

Table I   Effects of PITC, BITC and PEITC on NNK-Induced Lung Tumor and O^6-mG Formation in A/J Mice

Pretreatment	Daily dose (μmol)	Tumors/mouse[a]	% of mice with tumors	O^6-mG[b](μmol/mol guanine)
1. None	–	10.7 ± 0.8^{1c}	100	30.9 ± 5.9^{1c}
2. PITC	5	9.5 ± 1.2^{1}	100	29.7 ± 4.4^{1}
3. BITC	5	7.6 ± 0.5^{1}	100	26.1 ± 6.7^{1}
4. PEITC	5	2.6 ± 0.4^{2}	89	3.9 ± 1.2^{2}
	25	0.3 ± 0.1^{3}	30[d]	N. D.[e]

[a]Groups of 20-30 A/J mice were administered corn oil or isothiocyanates by gavage daily for four consecutive days. Two h after the final gavage, a single dose of NNK (10 μmol/mouse) was administered i.p. Sixteen weeks after NNK administration, mice were sacrificed and pulmonary adenomas were quantitated.
[b]Groups of 5 mice were administered corn oil or isothiocyanates by gavage for four consecutive days. Two h after the final gavage, NNK was administered i.p. at a dose of 10 μmol/mouse. Mice were sacrificed 6 h after NNK administration.
[c]Mean ± SE. Means that bear different superscripts within a given column are statistically different ($p < 0.05$) from one another as determined by analysis of variance followed by Newman-Keuls' ranges test.
[d]Significantly ($P < 0.01$) less than that of group 1 as determined by the Chi-Square test.
[e]Not detected.

Table II Effects of the Longer Chain Arylalkyl Isothiocyanates on NNK-Induced Lung Adenomas and O^6-mG Formation in A/J Mice

Pretreatment	Daily Dose (μmol)	Tumors/Mouse[a]	% of Mice with Tumors	Adduct level[b] (μmol/mol guanine)		
				Lung	Liver	
				O^6-mG[c]	7-mG	O^6-mG
Corn oil	–	7.9 ± 0.4[1d]	100[1]	17.6 ± 0.9[1]	997±67[1]	177±11[1]
PEITC (n=2)	1	6.5 ± 1.4[1]	100[1]	17.2 ± 0.2[1]	890±39[1]	145±7[1,2]
PPITC (n=3)	1	1.2 ± 0.3[2]	75[1]	13.8 ± 0.4[2]	852±43[1]	150±9[1,2]
PBITC (n=4)	1	0.8 ± 0.3[2]	42[2]	13.4 ± 0.5[2]	960±66[1]	158±9[1,2]
PPeITC (n=5)	1	0.9 ± 0.3[2]	53[2]	11.1 ± 0.2[3]	786±40[1]	138±8[2]
PHITC (n=6)	1	0.0 ± 0.0[2]	0[3]	9.9 ± 0.1[3]	907±53[1]	155±6[1,2]

[a]Groups of 20 female A/J mice (corn oil/NNK controls: 60 mice) were administered corn oil vehicle or isothiocyanates (in 0.1 ml corn oil) by gavage for 4 consecutive days. At 2 h after the final pretreatment, mice were administered 10 μmol NNK (in 0.1 ml saline) i.p. Mice were killed at 16 weeks after NNK administration and pulmonary adenomas were quantitated.

[b]Groups of 15 A/J mice were administered corn oil vehicle or isothiocyanates (1 μmol in 0.1 ml corn oil) by gavage for 4 consecutive days. At 2 h after the final pretreatment, mice were administered 10 μmol NNK (in 0.1 ml saline) i.p. Mice were killed 6 h after NNK administration and the liver and lungs of each animal were excised. Following DNA isolation and purification O^6-mG and 7-mG were analyzed as described previously (3).

[c]Due to the relatively low levels of methylation in lung, 80% of the entire sample was used for the determination of O^6-mG. For this reason, 7-mG was not determined.

[d]Values within the same column that bear different superscripts are statistically different from one another ($P < 0.05$). Mean ± SE.

tumorigenesis (15). One week after a single dose of NNK (10 μmol/mouse, i.p.), mice were fed a diet containing 1 or 3 μmol/g diet of BITC or PEITC. The bioassay was terminated 16 weeks after NNK treatment. As expected, NNK-treated mice fed a control diet developed 100% tumor incidence, with 7.8 tumors/mouse. NNK treated mice fed PEITC at a concentration of 1 or 3 μmol/g diet developed 8.2 or 6.1 tumors/mouse. Post-feeding BITC at 1 μmol/g diet yielded 8 tumors/mouse while a 3 μmol/g diet gave 5.2 tumor/mouse, a small but statistically significant inhibition. However, at this dose a loss in weight gain was observed. The tumor incidences were not affected in all treated groups. This study clearly showed that post-treatment of arylalkyl isothiocyanates had little, if any, effect on NNK lung tumorigenesis. These results suggest that the basis of inhibition by PEITC and its homologues resides to a large extent on the initiation stage of NNK lung tumorigenesis.

Effect of Frequency of Isothiocyanate Administration on NNK Lung Tumorigenesis. In order to obtain more insight into the mechanism of inhibition by arylalkyl isothiocyanates, we examined whether there is a difference in the inhibitory activity by PEITC or PHITC between protocols using a single dose and 4 daily consecutive doses (16). In the original protocol, 4 consecutive daily doses of isothiocyanate were given and NNK was administered 2 h after the last dosing. Since multiple doses of isothiocyanate may inhibit or induce enzymes such as P-450s and glutathione transferase, it is not clear which effects would be dominant in tumor inhibition (17-19). Since isothiocyanates inhibited nitrosamine metabolism *in vitro*, a single dose shortly before NNK would most likely reflect an effect due to enzyme inhibition. Table III compares the effects of the single dose vs. the four-dose protocol on the tumor multiplicity and incidence. No significant difference was observed in lung tumor inhibition between these protocols. These results indicate that the final dose of isothiocyanate is a determinant of inhibition and the inhibition of NNK-induced lung tumorigenesis by PEITC and PHITC is due largely to inhibition of the enzymes for NNK activation.

Green Tea and Its Major Polyphenol EGCG

Lack of Correlation between Tumor Inhibition and DNA Methylation Induced by NNK. In a modified NNK lung tumor bioassay, A/J mice were treated with NNK at dose of 56 μmol/kg body weight or approximately 0.25 mg/mouse in corn oil by gavage 3 times weekly for 10 weeks. A 2% green tea solution (prepared by adding 1 g of dry green tea leaves in 50 ml of boiling water followed by filtration after standing at room temperature for 30 min) and its major polyphenol EGCG (560 ppm; same concentration as that found in the tea solution) administered as drinking water significantly inhibited the lung tumor multiplicity induced by NNK (7). Table IV shows that mice treated with NNK alone developed 22.5 tumors/mouse while NNK-treated mice that drank green tea or EGCG in water developed only 12.2 tumors or 16.1 tumors/mouse, respectively. Since the reduction of O^6-mGua in lung DNA is one indicator of reduced tumor formation, we examined whether the underlying mechanism of inhibition by green tea and EGCG is due to a reduction of O^6-mGua levels in the lung of NNK-treated mice. Mice were treated with an identical protocol used in the tumor bioassay, except NNK

Table III Effects of Frequency of Administration on Inhibition of Lung Tumorigenicity in NNK-Treated A/J Mice[a]

Pretreatment	Dose level (μmol)	% inhibition (multiplicity)			% inhibition (incidence)		
		1 dose	4 dose	Historic values[b] (4 doses)	1 dose	4 dose	Historic values (4 doses)
PEITC	5.0	62	79	48, 64, 76	5	25	7, 7, 11
PHITC	0.2	83	96	85	33	65	30

[a]Groups of 20 female A/J mice were administered corn oil or isothiocyanates by gavage either once, 2 h prior to NNK, or for 4 consecutive days with the final dose given 2 h before NNK administration (10 μmol/mouse by i.p. injection). Mice were sacrificed 16 weeks after NNK administration and lung adenomas were quantitated.
[b]See references 3 and 6.

Table IV   Effect of Tea, EGCG, and Caffeine on NNK-Induced Lung Adenomas in A/J Mice

Treatment group	No. of animals	Tumors/mouse[a]	% of mice with tumors
NNK	30	22.5 ± 4.7	100
Tea + NNK	25	12.2 ± 4.3^{b}	100
EGCG + NNK	25	16.1 ± 5.3^{b}	100
Tea	15	0.1 ± 0.2^{b}	7
EGCG	15	0.3 ± 0.6^{b}	20

[a]Mean ± SD
[b]Statistically different from NNK group, $P < 0.001$

was given only for 3 weeks. The formation of O^6-mGua in lung DNA 4 and 24 hours after the last NNK administration was quantified. Table V shows that levels of O^6-mGua at 4 and 24 h after NNK administration were not significantly altered by these treatments. These results show that neither green tea nor EGCG inhibited NNK activation or stimulated its repair, suggesting that additional mechanisms other than DNA methylation are involved in NNK lung tumorigenesis. Pyridyloxobutylation of DNA by NNK may be important. However, the exact role of this lesion is yet to be defined. Since EGCG is an antioxidant (*20, 21*), we investigated the possible roles of green tea and EGCG as antioxidants in inhibition of the lung tumorigenesis by NNK.

Oxidative DNA Damage Induced by NNK. Bartsch *et al.* reported that treatment of rats with tumorigenic doses of N-nitrosodimethylamine enhanced lipid peroxidation, suggesting the potential role of oxidative damage in nitrosamine carcinogenesis (*22*). However, direct evidence of oxidative DNA damage caused by nitrosamine has been lacking. We examined whether NNK treatment causes an increase of oxidative damage by analyzing 8-hydroxydeoxyguanosine (8-OH-dG), a common free radical-induced DNA lesion, as a marker using HPLC and electrochemical detection (*23*).

Our tumor bioassay showed that mice treated with NNK by gavage at a dose of 0.25 mg/mouse 3 times weekly for 10 weeks developed 22.5 tumors/mouse 6 weeks after the last NNK treatment. We used a similar protocol to study potential NNK-induced oxidative damage. A/J Mice were treated with NNK by gavage at doses of 0.25 or 0.50 mg/mouse 3 times weekly for 3 weeks or by a single dose of 4 mg/mouse. The single dose is comparable to the cumulative dose received at 0.50 mg/mouse. Animals were sacrificed 2, 4, and 24 hours after the last NNK dosing. Liver and lung DNA were isolated for the analysis of 8-OH-dG. Figure 2 shows that a dose-dependent increase in the pulmonary 8-OH-dG levels was observed in the NNK treated groups when compared with the control group. The single dose treatment also enhanced the levels of 8-OH-dG but the increase was not statistically significant. Four and 24 h after the final treatment, however, 8-OH-dG levels declined to the basal value observed in the control group, suggesting either efficient repair or rapid degradation of this oxidative lesion in lung DNA. In contrast to lung tissue, liver, a non-target tissue of NNK in A/J mice, showed an increase in the oxidative lesion only with multiple dosing of NNK at the higher dose. Lung appears to be more sensitive than liver to the oxidative damage induced by NNK. The better resistance of liver DNA to NNK-caused oxidative damage may be due to the greater activity in the liver than in the lung of some antioxidant enzymes such as superoxide dismutase or glutathione peroxidase in rodents (*24*). Similar results were observed in F344 rats treated with NNK (*23*). Table VI shows that levels of 8-OH-dG were significantly increased in the rat lung upon a single administration of NNK at a dose of 100 mg/kg body weight. However, the increase in liver was, again, not statistically significant and kidney, a non-target tissue, was inert to NNK treatment.

There are at least two plausible pathways by which free radicals can be generated via nitrosamine metabolism. These pathways include formation of the hydroperoxides by α-hydroxylation and release of nitric oxide by denitrosation (*25,26*). Regardless of the mechanism of free radical production, these results provide direct evidence of oxidative

Table V　Effects of Green Tea and EGCG on O^6-mG Levels in Lung DNA of A/J Mice 4 and 24 h after Treatment with NNK[a]

Treatment	O^6-mG (μmol/mol guanine)	
	4 h	24 h
NNK	16.8 ± 2.7[b]	17.8 ± 6.2
NNK + tea	23.3 ± 2.8	19.2 ± 5.0
NNK + EGCG	19.6 ± 1.0	15.0 ± 5.5

[a]Female A/J mice that drank water, green tea, or EGCG solution were given NNK in corn oil by gavage (56 μmol/kg body weight) 3 times weekly for 3 weeks. Mice were sacrificed 4 and 24 h after the last NNK treatment.
[b]Mean $\pm$ SD from 5 mice.

Table VI   Levels of 8-OH-dG in Lung, Liver and Kidney of Rats Treated with or without NNK[a]

	Hours after treatment	Adducts/10^5 dG	
		Control	Treated
Lung	2	3.0 ± 1.1[b]	5.12 ± 1.2[d]
	4	N. D.[c]	2.4 ± 0.6
	24	3.3 ± 1.0	2.3 ± 1.0
Liver	2	2.9 ± 0.4	4.3 ± 0.7
	4	N. D.	4.9 ± 2.4
	24	2.7 ± 0.6	2.6 ± 0.5
Kidney	2	4.8 ± 2.1	5.9 ± 1.6
	4	N. D.	5.4 ± 0.3
	24	4.3 ± 1.1	4.3 ± 0.6

[a] A single dose of NNK was administered i.p. at a dose of 100 mg/kg b.w.
[b] Mean ± SD from 3 or 4 animals
[c] Not determined
[d] Significantly different from the control, $P < 0.05$

damage in the target tissue DNA of rats and mice treated with NNK. While the actual role of this lesion in carcinogenesis requires further investigation, our study suggests that, in addition to alkylation, oxidative DNA damage may play a role in the NNK lung tumorigenesis.

Correlation of Inhibition of the Lung Tumorigenicity and Oxidative DNA Damage Induced by NNK. We examined the potential effects of green tea or EGCG on 8-OH-dG formation in lung and liver DNA of mice treated with NNK (7). This study was designed to mimic the tumor bioassay and, thus, a similar dosing regimen was used. However, in order to facilitate the detection of 8-OH-dG, the NNK dose was twice that used in the tumor bioassay. Mice were given 2% green tea or EGCG in water as drinking water for 5 weeks. Two weeks after these treatments began, NNK was administered by gavage at a dose of 112 μmol/kg body weight or approximately 0.5 mg/mouse 3 times weekly for 3 weeks. Mice were then sacrificed 2 h after the last NNK dosing and liver and lung DNA were isolated for the analysis of 8-OH-dG. Figure 3 shows that multiple doses of NNK caused a significant elevation of 8-OH-dG levels in the lung DNA from 1.7±1.2 to 3.2±1.7 adducts/10^5 deoxyguanosine, an approximately 2-fold increase from the background levels. Contrary to the lung, only a slight increase was seen in the liver and the increase was not statistically significant. Green tea and EGCG treatments suppressed the increased oxidative DNA lesion induced by NNK in the lung. The suppression of 8-OH-dG formation in the lung DNA of NNK-treated mice that drank green tea or EGCG is consistent with their ability to inhibit lung tumor formation. These results support the involvement of free radicals in lung tumorigenesis by NNK as well as the role of green tea and EGCG as antioxidants in the protection against NNK lung tumorigenesis.

Conclusion

Results of these studies show that different mechanisms are likely to be involved in the inhibition of NNK-induced lung tumorigenesis in A/J mice by arylalkyl isothiocyanates and by green tea EGCG. The former exert the inhibitory activity by their ability to reduce the cytochrome P-450 isozyme activity of NNK activation while the latter acts as an antioxidant by suppressing the oxidative damage induced by NNK. Arylalkyl isothiocyanates are considerably more potent than EGCG. A relatively small dose was required for inhibition. For example, with a dose of only one fiftieth of that of NNK, PHITC blocked approximately 80% of lung tumor formation (16). EGCG, on the other hand, needed more than 10 times the dose of NNK in order to achieve a 45% inhibition (7). It is conceivable, however, that on a molar basis, green tea polyphenol would be considerably less toxic than arylalkyl isothiocyanates.

These studies provide for the first evidence of the involvement of free radicals in nitrosamine tumorigenesis. The mechanism by which free radicals are generated by NNK treatment is not yet known. Metabolism by cytochrome P-450 is likely to be involved in the formation of free radicals. If this is true, it is expected that arylalkyl isothiocyanates treatment will decrease, in addition to DNA alkylation, the formation of 8-OH-dG induced by NNK. More studies are needed to examine the effects of isothiocyanates on oxidative damage. EGCG appears to have little effect on DNA methylation, indicating a lack of activity toward

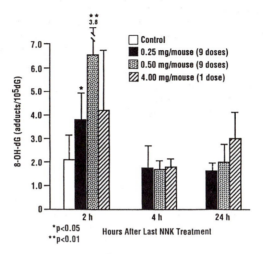

Figure 2   Levels of 8-OH-dG in lung DNA of mice treated with and without NNK. Five mice were used for each time point in each group.

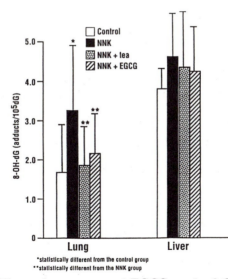

Figure 3   Effects of green tea and EGCG on the 8-OH-dG levels in lung and liver DNA of mice 2 h after NNK treatment. Data were obtained from 10 to 12 mice in each group.

cytochrome P-450 enzymes at the dose studied. The reduced levels of oxidative lesions in lung as a result of EGCG treatment may be related to its ability to reduce reactive oxygen species and/or to chelate iron ion resulting in a decreased production of hydroxyl radicals.

The different modes of action of arylalkyl isothiocyanates and green tea polyphenol in inhibition of the NNK lung tumorigenesis in A/J mice suggest the potential of combination treatment of these compounds as chemopreventive agents. Since the use of combination regimens has been shown to improve the inhibitory efficacy (*27,28*), a simultaneous treatment of isothiocyanates and green tea polyphenol may have an additive or synergistic inhibitory effect against the NNK-induced lung tumorigenesis. A bioassay designed to examine this possibility is in progress.

Acknowledgements

We thank Dr. Stephen S. Hecht for developing the single-dose NNK protocol in the A/J mouse lung adenoma bioassay, and Drs. Shantu G. Amin and Emerich S. Fiala for providing NNK and 8-OH-dG standard. This study is supported by grants CA-46535 and CA-51830 from the National Cancer Institute and a fellowship for Yong Xu from the Institute of Nutrition and Food Hygiene, Chinese Academy of Preventive Medicine, Beijing, China.

Literature Cited

1.   Hoffmann, D.; Rivenson, A.; Chung, F.-L.; Hecht, S. S. *CRC Crit. Rev. Toxicol.* **1991**, *21*, 305-311.
2.   International Agency for Research on Cancer *Tobacco Smoking*; IARC Monographs on the Evaluation of the Carcinogenic Risk of Chemicals to Humans ; IARC: Lyon, France, 1986, 38; .
3.   Morse, M. A.; Amin, S. G.; Hecht, S. S.; Chung, F.-L. *Cancer Res.* **1989**, *49*, 2894-2897.
4.   Tookey, H. L.; VanEtten, C. H.; Daxenbichler, M. E. In *Toxic Constituents of Plant Foodstuffs;* Liener, I. E., Ed.; Academic Press: New York, 1980 pp. 103-142.
5.   Sones, K.; Heaney, R. K.; Fenwick, G. R. *J. Sci. Food Agric.* **1984**, *35*, 720.
6.   Chung, F.-L.; Morse, M. A.; Eklind, K. I. *Cancer Res.* **1992**, *52*, 2719s-2722s.
7.   Xu, Y.; Ho, C.-T.; Amin, S. G.; Han, C.; Chung, F.-L. *Cancer Res.* **1992**, *52*, 3875-3879.
8.   Wynder, E. L.; Taioli, E.; Fujita, Y. *Jpn. J. Cancer Res.* **1992**, *83*, 418-423.
9.   Smith, T. J.; Guo, Z.; Thomas, P. E.; Chung, F.-L.; Morse, M. A.; Eklind, K.; Yang, C. S. *Cancer Res.* **1990**, *50*, 6817-6822.
10.   Belinsky, S. A.; Foley, J. F.; White, C. M.; Anderson, M. W.; Maronpot, R. R. *Cancer Res.* **1990**, *50*, 3772-3780.
11.   Peterson, L. A.; Hecht, S. S. *Cancer Res.* **1991**, *51*, 5557-5564.
12.   Devereux, T. R.; Anderson, M. W.; Belinsky, S. A. *Carcinogenesis* **1991**, *12*, 299-303.
13.   Hecht, S. S.; Spratt, T. E.; Trushin, N. *Carcinogenesis* **1988**, *9*, 161-165.

14. Morse, M. A.; Eklind, K. I.; Hecht, S. S.; Jordan, K. G.; Choi, C.-I.; Desai, D. H.; Amin, S. G.; Chung, F.-L. *Cancer Res.* **1991**, *51*, 1846-1850.
15. Morse, M. A.; Reinhardt, J. C.; Amin, S. G.; Hecht, S. S.; Stoner, G. D.; Chung, F.-L. *Cancer Lett.* **1990**, *49*, 225-230.
16. Morse, M. A.; Eklind, K. I.; Amin, S. G.; Chung, F.-L. *Cancer Lett.* **1992**, *62*, 77-81.
17. Ishizaki, H.; Brady, J. F.; Ning, S. M.; Yang, C. S. *Xenobiotica* **1990**, *20*, 255-264.
18. Benson, A. M.; Barretto, P. B. *Cancer Res.* **1985**, *45*, 4219-4223.
19. Vos, R. M. E.; Snoek, M. C.; van Berkel, W. J. H.; Muller, F.; van Bladeren, P. J. *Biochem. Pharmacol.* **1988**, *37*, 1077-1082.
20. Rush, R. J.; Cheng, S. J.; Klaunig, J. E. *Carcinogenesis* **1989**, *10*, 1003-1008.
21. Osawa, T.; Namiki, M.; Kawakishi, S. *Basic Life Sci.* **1990**, *52*, 139-153.
22. Ahotupa, M.; Bereziat, J.-C.; Bussacchini-Griot, V.; Camus, A. M.; Bartsch, H. *Free Radical Res. Commun.* **1987**, *3*, 285-291.
23. Chung, F.-L.; Xu, Y. *Carcinogenesis* **1992**, *13*, 1269-1272.
24. Rickett, G. M.; Kelly, F. J. *Development* **1990**, *108*, 331-336.
25. Streeter, A. J.; Nim, R. W.; Sheffels, P. R.; Heur, Y. H.; Yang, C. S.; Mico, B. A.; Bombar, C. T.; Keefer, L. K. *Cancer Res.* **1990**, *50*, 1144-1150.
26. Potter, D. W.; Reed, D. J. *Arch. Biochem. Biophys.* **1982**, *216*, 158-169.
27. Rao, C. V.; Tokumo, K.; Rigotty, J.; Zang, E.; Kelloff, G.; Reddy, B. S. *Cancer Res.* **1991**, *51*, 4528-4534.
28. Moon, R. C., Rao, K. V. N, Detrisac, C. J., and Kelloff, G. J. In Cancer Chemoprevention; Wattenberg, L., Lipkin, M., Boone, C. W., Kelloff, G. J., Ed; CRC Press: Boca Raton, Florida, 1992 pp. 83-93.

RECEIVED June 29, 1993

TOXIC, MUTAGENIC, AND CARCINOGENIC
EFFECTS OF *N*-NITROSO COMPOUNDS

Chapter 20

Chemical Structure of Nitrosamines Related to Carcinogenesis

William Lijinsky[1]

Division of Biometry and Risk Assessment, National Institute
of Environmental Health Sciences, Research Triangle Park, NC 27709

N-Nitroso compounds are the most broadly tested group of
carcinogens and are effective in all species. Of the several hundred
compounds examined most require metabolic activation, presumably
by enzymes that are present in only some organs of some species.
Directly acting nitrosamides and alkylnitrosoureas have no such
limitation, yet are similarly organ- and species-specific in their
carcinogenic action. Compounds giving rise to the same proximate
alkylating moiety often produce widely different carcinogenic effects;
alkylation of DNA is often similar in organs in which tumors arise
and in organs that are unresponsive. The results suggest that a
match between particular chemical structures and organ-specific
receptor molecules in a particular species is responsible for the
induction of characteristic cancers, such as pancreas duct tumors in
Syrian hamsters, esophageal tumors in rats and tumors of the
nervous system in rats and mice.

The discovery of the induction of liver tumors in rats treated with
nitrosodimethylamine (NDMA) by Magee and Barnes (1) led shortly to the
investigation of the biological properties of a large number of N-nitroso
compounds (2). Investigation of the mechanisms of carcinogenesis by N-
nitroso compounds began with studies by Magee and his colleagues (including
Heath, Hultin and Farber) of the properties of NDMA as an alkylating agent
(Figure 1)(3). It was generally agreed that N-nitroso compounds (including
directly acting alkylnitrosamides that do not require metabolic activation)
formed products that alkylated DNA, causing mutations which eventually gave

[1]Current address: 5521 Woodlyn Road, Frederick, MD 21702

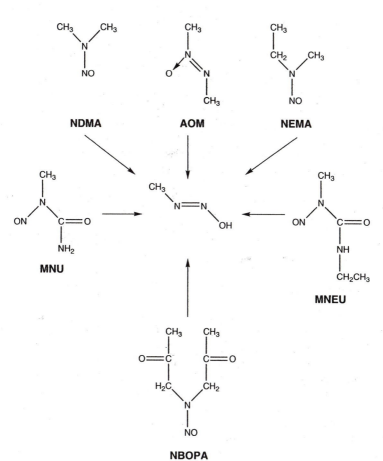

Figure 1. Methylating agents *in vivo*

rise to tumors and cancers. Those nitrosamines structurally restricted from forming an alkylating agent are not carcinogenic although there are some exceptions (*4*).

As work progressed it became apparent that treatment of animals with an N-nitroso compound led to alkylation of DNA and other macromolecules in many organs. For example, a number of compounds that are methylating agents through formation of a methyldiazonium intermediate (Figure 1) gave rise to a similar dose-related pattern of methylation in DNA in several organs of rats and hamsters. Methylation at the O^6-position of guanine is believed to be the important mutagenic lesion in DNA. However, treatment of the animals with those compounds gave rise to tumors in some, but not all, of the organs in which methylated DNA was found. In particular, methylnitrosourea, methylnitrosoethylurea and azoxymethane (isomeric with NDMA) did not induce liver tumors in rats, and nitrosobis-(2-oxopropyl)-amine did not induce liver tumors in male rats, although they methylated the liver DNA (Table I). It appears that alkylation of DNA is not sufficient to lead to formation of tumors, in the absence of other factors, which are presently unknown.

This is further borne out by observations with cyclic nitrosamines, which comprise more than 70 of the approximately 300 N-nitroso compounds that have been tested for carcinogenicity. The cyclic nitrosamines in Figure 2 show a similar variety of carcinogenic effects in different species to those of acyclic analogs, and are carcinogens of comparable potency. Yet those that have been examined thus far have produced very small (e.g. nitrosopyrrolidine) (*5*) or yet undetected amounts of DNA alkylation in tissues and organs of experimental animals. This suggests that actions of the cyclic nitrosamines other than alkylation of DNA are of prime importance in inducing tumors.

Metabolism of Nitrosamines

Many nitrosamines undergo complex metabolism in animals and several intermediate products are formed. These intermediates might be carcinogenic to different organs, perhaps explaining partially the variety of organ-specific effects among these structurally-related carcinogens. An example is the series of alkylmethylnitrosamines, which have different patterns of target organs in which they induce tumors in rats and in hamsters (Table II). The first member of the series is NDMA, which induces tumors of the liver, lung and kidney in rats, but only liver tumors in hamsters. The next member is ethylmethylnitrosamine, which induces tumors of the liver, lung and esophagus in rats, but again only liver tumors in hamsters. The C-3 to C-6 compounds are esophageal carcinogens in rats but they induce tumors of the liver, nasal mucosa and forestomach in hamsters. Those compounds with even-numbered carbon chains from n-octyl (C-8) to n-tetradecyl (C-14) induce bladder tumors in rats and, among those tested, in hamsters. Okada suggested that this periodicity (*6*) was due to metabolism of the long alkyl chains by successive beta oxidations, in the manner of the common fatty acids (Knoop). Finally, as shown for nitrosomethyl-n-octylamine in Figure 3, the common metabolite

Table I. Methylation of DNA by alkylating carcinogens 6 h after treatment

Compound	Dose (μmol)	Species	Organ	Methylation of DNA (pmol/mg DNA)		Tumors
				N7-Methylguanine	O^6-Methylguanine	
Nitrosodimethylamine	18	Rat	Liver	380	23	+
			Kidney	51	2.4	+
		Hamster	Liver	420	29	+
Azoxymethane	27	Rat	Liver	740	48	-
			Kidney	21	2	+
			Colon	19	1	+
	27	Hamster	Liver	1180	110	+
Nitrosoethylmethylamine	20	Rat	Liver	270	24	+
			Kidney	44	5.5	-
			Spleen	84	8	-
			Lung	22	5	+
	20	Hamster	Liver	1000	71	+
			Kidney	40	5	-
			Lung	30	5	-
Methylnitrosourea	20	Rat	Liver	103	11	-
			Kidney	111	16	-
			Lung	54	5	-
			Brain	23	2.4	+
			Spleen	180	14	+
	20	Hamster	Liver	130	16	-
			Kidney	108	11	-
			Lung	22	2.7	-
Methylnitrosoethylurea	20	Rat	Liver	54	3.4	-
			Lung	21	2.2	+
			Brain	17	1.8	+
Nitrosobis(2-oxopropyl) amine	17	Rat(♂)	Liver	43	6	-
	13	Rat(♀)	Liver	180	14	+
		Hamster	Liver	800	35	+

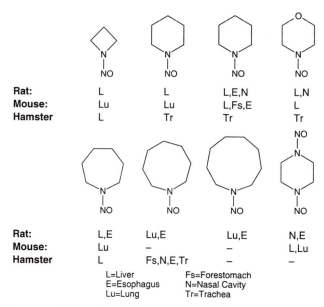

Rat:	L	L	L,E,N	L,N
Mouse:	Lu	Lu	L,Fs,E	L
Hamster	L	Tr	Tr	Tr

Rat:	L,E	Lu,E	Lu,E	N,E
Mouse:	Lu	–	–	L,Lu
Hamster	L	Fs,N,E,Tr	–	–

L=Liver   Fs=Forestomach
E=Esophagus   N=Nasal Cavity
Lu=Lung   Tr=Trachea

Figure 2. Tumors induced by cyclic nitrosamines: (l to r) nitrosoazetidine, nitrosopyrrolidine, nitrosopiperidine, nitrosomorpholine, nitrosohexamethyleneimine, nitrosoheptamethyleneimine, nitrosooctamethyleneimine, dinitrosopiperazine

Table II. Tumors in Rats and Hamsters by Alkylmethylnitrosamines**

N-Methyl-N-	Rats	Hamsters
Methyl	Liver, Lung, Kidney	Liver
Ethyl	Liver, Lung, Esophagus	Liver
n-Propyl	Esophagus	Liver, Nasal, Forestomach
n-Butyl	Esophagus	Liver, Nasal, Forestomach
n-Amyl	Esophagus	Lung, Nasal, Liver, Forestomach
n-Hexyl	Esophagus, Liver	Liver, Lung, Bladder
n-Heptyl	Liver, Lung	Liver, Lung, Nasal
n-Octyl	Liver, Lung, Bladder	Liver, Lung, Bladder
n-Nonyl	Liver, Lung	N.T.*
n-Decyl	Bladder, Lung	N.T.*
n-Undecyl	Liver, Lung	N.T.*
n-Dodecyl	Bladder, Lung	Bladder, Lung
n-Tetradecyl	Bladder	N.T.*

*N.T.=Not tested.
**Data derived from references 15 and 16.

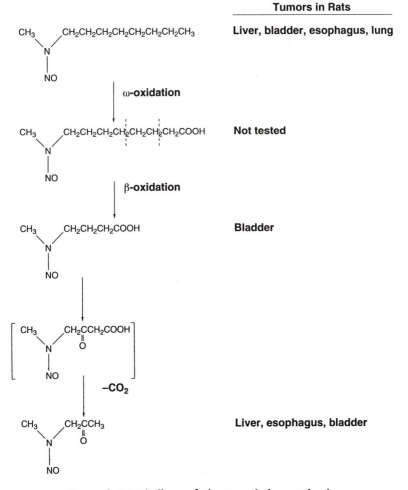

Figure 3. Metabolism of nitrosomethyl-n-octylamine

nitrosomethyl-3-carboxypropylamine is formed (and excreted in the urine) but is further beta oxidized and decarboxylated to produce nitrosomethyl-2-oxopropylamine (NMOPA), which induces bladder tumors in high incidence when given intravesically (*7*). The homologs with odd-numbered carbon chains do not induce bladder tumors but produce tumors of the liver and lung in rats, as does nitrosomethyl-n-octylamine. Liver and lung are also targets in hamsters of many compounds in this series, both even and odd numbered.

All of the compounds in the series do not give rise to the same tumors in a single species or to the same tumor in different species. Therefore, it is probable that the proximate carcinogens formed by metabolism of these compounds are not the same in every case. There are differences, qualitative and quantitative, in the products of metabolism among the compounds. Examination of methylation of DNA (O^6-MeG) by alkylmethylnitrosamines in a number of organs (*8*) showed that methylation parallels the effectiveness of tumor induction to some extent in particular organs. However, tumors often arose equally well in organs in which a compound produced low levels of DNA methylation (e.g. nitrosomethyl-n-octylamine in rat liver) as in those in which methylation was extensive. In other cases there was methylation of DNA but no tumors were induced in the organ (e.g. with nitrosoethylmethylamine and methylnitrosourea in the rat kidney). These results suggest that because of their chemical or physical properties nitrosamines and some of their metabolites may have an affinity for certain organs and receptor molecules which is responsible for the organ and species specificity. There have been few studies of the metabolism and pharmacokinetics of the more complex nitrosamines, which induce tumors in several organs (nitrosodi-n-butylamine, nitrosobis-(2-oxopropyl)amine, nitrosohexamethyleneimine) to elucidate the role of a number of metabolites. Attention usually has been focused on compounds with a single target organ, such as the induction by nitrosomethyl-n-amylamine of tumors in the rat esophagus (*9*).

Another example of important influences on carcinogenesis by N-nitroso compounds that suggest modulations of metabolism as factors in induction of particular tumors, is the effect of sex hormones. This is shown in Table III as the difference between male and female animals in the tumor response to certain N-nitroso compounds. Similar sex differences have been seen with other types of carcinogen, such as the response of the male but not the female hamster kidney to tumor induction by estrogens (*10*). The importance of the models in Table III lies in the rarity of such sex differences; most N-nitroso compounds induce tumors in their target organs equally well in males and females. There have been few studies comparing metabolism and activation of compounds like those in Table III, except for some examination of the effects of castration on alkylation of DNA in rat liver (*11*), which showed that the differences were small and insufficient to account for the absence of liver tumors from some of the rat groups.

Table III. Male/Female Differences in Carcinogenesis*

Compound	Species	Organ	Tumor Incidence (%)	
			Males	Females
N-Nitroso-				
Methylhydroxyethylamine	Rat	Liver	35	70
Methyl-2-oxopropylamine	Rat	Liver	10	75
Bis-(2-oxopropyl)amine	Rat	Liver	0	90
	Rat	Bladder	60	0
	Rat	Colon	40	7
	Rat	Thyroid	90	25
2-Methylpiperidine	Rat	Liver	0	50
Hydroxyethylurea	Rat	Lung	90	5
Dimethylamine	Rat	Lung	70	16
Butyl-4-Hydroxybutylamine	Mouse	Bladder	50	20

*Data derived from references 15 and 16.

Differences between Nitrosamines and Alkylnitrosoureas

Although it has been shown that alkylnitrosoureas produce similar levels of DNA alkylation in all organs examined (12,13) an expected result in light of their direct alkylating properties, they have very different patterns of organs in which they induce tumors compared with the analogous dialkylnitrosamines. This is apparent both in rats (Table IV) and in hamsters (Table V). The forestomach is the only organ which is the target of both nitrosamines and alkylnitrosoureas in both rats and hamsters. In rats, many tumors are induced in common by nitrosamines and by alkylnitrosoureas but tumors of the nervous system, glandular stomach, mammary gland, duodenum and intestines have not been reported in rats treated with nitrosamines, although they have been in rats treated with alkylnitrosoureas. Colon tumors are commonly induced in rats by alkylnitrosoureas but there have been only one or two instances of colon tumors as a result of nitrosamine treatments (14,15). The range of target organs in hamsters is much smaller than in rats (Table V).

The reason for the very narrow spectrum of hamster organs in which alkylnitrosoureas induce tumors contrasts sharply with the variety of rat organs susceptible to the carcinogenic action of alkylnitrosoureas (15,16). The fact that not every N-nitroso compound induces all types of tumor to which the species is susceptible implies that there are chemical structural reasons for the differences in organ specificity between compounds. This, in turn, suggests that there must be a 'fit' between the compound and some 'receptor' in particular cells. That one such receptor might be an enzyme is indicated by examination of some of the organs that do not respond to N-nitroso compounds of particular structures, some of which are shown in Table VI. For example, the rat esophagus is exquisitely sensitive to nitrosamines; more than half of the 200 nitrosamines tested are esophageal carcinogens in rats (16) and as little as a few milligrams of nitrosodiethylamine has induced a high incidence of esophageal tumors in rats (17). However, cyclic nitrosamines with 5 or fewer atoms in the ring and alkylmethyl-nitrosamines with more than 8 carbons in the alkyl chain do not induce esophageal tumors in rats. Also, no alkylnitrosourea, even when given by mouth and in drinking water, has induced tumors in the rat esophagus. Neither have they induced tumors in the rat nasal mucosa, another common site for nitrosamine-induced tumors in rats. These negative findings, together with the failure of any N-nitroso compound to induce tumors of the esophagus in hamsters, suggest that metabolism of the nitrosamine, following interaction with an enzyme (which might be a cytochrome P450), is a requirement for induction of tumors in the esophagus and in the nasal mucosa. The observation that there is DNA methylation by methylnitrosourea in the rat esophagus (18) shows that this is not the key event in induction of tumors of the esophagus in rats.

In the same vein, although the rat liver is susceptible to tumor induction by a large proportion of nitrosamines, few of the 26 nitrosopiperidines tested have induced liver tumors in rats and only one nitrosopiperazine derivative, dinitrosopiperazine, has induced liver tumors. Other than nitrosoazetidine and

Table IV. Differences in Target Organ Between Nitrosamines and Nitrosoureas* (RATS)

Organ	Nitrosamines	Alkylnitrosoureas
Liver	+	+
Esophagus	+	-
Nasal Mucosa	+	-
Lung	+	+
Forestomach	+	+
Glandular Stomach	-	+
Nervous System	-	+
Bladder	+	+
Trachea	+	-
Kidney	+	+
Thymus	+	+
Thyroid	+	+
Mammary Gland	-	+

*Data derived from references 15 and 16.

Table V. Differences in Target Organ Between Nitrosamines and Nitrosoureas* (HAMSTERS)

Organ	Nitrosamines	Alkylnitrosoureas
Liver	+	-
Forestomach	+	+
Nasal Mucosa	+	-
Lung	+	-
Pancreas	+	-
Spleen	-	+
Cervix	-	+
Bladder	+	-

*Data derived from references 15 and 16.

nitroso-oxazolidines, cyclic nitrosamines have not induced liver tumors in hamsters although many of them are liver carcinogens in rats. In contrast alkylmethylnitrosamines with 3 to 5 carbons or 10, 12 or 14 carbons in the alkyl chain have not induced liver tumors in rats although they induce tumors in the hamster liver.

The Importance of Chemical Structure for Carcinogenesis

Small changes in chemical structure frequently are accompanied by large changes in target organ specificity of carcinogenic N-nitroso compounds. The comparison between methylnitrosourea and ethylnitrosourea, which are directly acting compounds, reveals distinct differences in the types of tumor induced in several (but not all) species (Table VII). The differences between the two compounds suggest that in the rat, for example, the ethyl compound has a much broader set of affinities for different organs than does the methyl compound, since the former induces tumors in so many types of cell. This dichotomy between methyl- and ethyl-nitrosoureas is demonstrated further by examining the carcinogenic effects in rats of a number of directly acting dialkylnitrosoureas bearing a methyl or ethyl group, respectively, next to the nitroso function (Figure 4). There is a distinct and different pattern of tumors induced in rats of either sex by methylnitrosoureas and by ethylnitrosoureas, suggesting that there are organs for which the latter have affinity that do not respond to methylnitrosoureas (*19*). However, the ethylnitrosoureas induce tumors of the nervous system as do the methylnitrosoureas.

A major difference between rats and hamsters in their response to N-nitroso compounds is that alkylnitrosoureas invariably induce hemangiosarcomas of the spleen and forestomach tumors in hamsters and usually few or no other tumors. The nature of the alkyl group appears not to matter and tumors do not appear in the blood vessels elsewhere in the body of the hamster. Of the great variety of tumors induced in rats by alkylnitrosoureas, hemangiosarcomas of the spleen are not among them. On the other hand alkylnitrosocarbamates (Figure 5) such as methylnitroso-urethane, are equally effective alkylating agents and equally or more potent mutagens as the corresponding alkylnitrosoureas. Yet the former do not induce hemangiosarcomas of the spleen in hamsters and induce only forestomach tumors in rats by oral administration. Therefore, the second amino nitrogen in alkylnitrosoureas seems to be essential for their carcinogenic effect in the hamster spleen, suggesting that in those cells there is a structure with affinity for alkylnitrosoureas. Interaction with this "receptor" can be considered to begin the process of neoplastic transformation. The nature of this moiety is unknown but it is probably unavailable in the rat spleen and endothelial cells in other parts of the hamster body. This suggestion does not exclude the alkylating properties of alkylnitrosoureas as important in the induction of tumors.

As a further indication of the indispensible role of factors other than alkylation of DNA in carcinogenesis by N-nitroso compounds, the tumorigenic

Table VI. Nitroso Compounds Not Inducing Rat Tumors*

Esophagus	Liver	Nasal Mucosa
Cyclics with <6 atoms	Piperidines	Nitrosoureas
Alkyl methyl with >8 atoms	Piperazines	
Dialkyl with >3 carbons	Alkylmethyl with 3-5 carbons or 10, 12, 14 carbons	
Nitrosoureas		

*Data derived from references 15 and 16.

Table VII. Tumors Induced by Alkylnitrosoureas*

Compound	Species	Tumor Site
Methylnitrosourea	Rat	Brain, Stomach
	Mouse	Lung, Kidney
	Hamster	Spleen
	Guinea Pig	Stomach, Pancreas
	Rabbit	Brain, Intestine
	Monkey	Esophagus
	Pig	Stomach
Ethylnitrosourea	Rat	Mammary gland, Lung, Brain, Intestines
	Mouse	Mammary gland, Lung, Brain, Liver
	Hamster	Spleen
	Monkey	Ovary, Uterus, Osteosarcoma
	Gerbil	Melanoma

*Data derived from references 15 and 16.

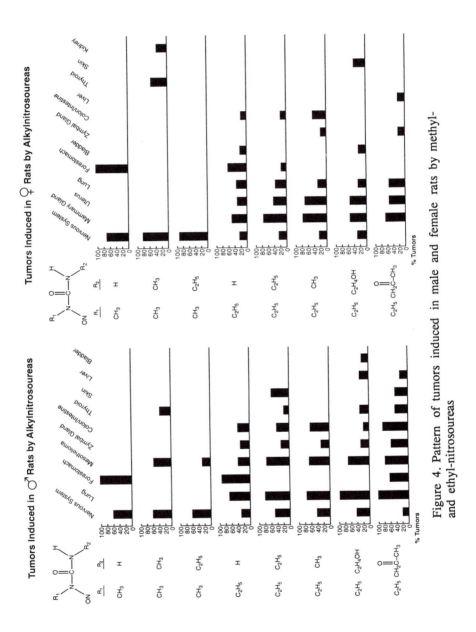

Figure 4. Pattern of tumors induced in male and female rats by methyl- and ethyl-nitrosoureas

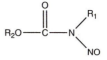

Nitrosoalkylurea   Nitrosoalkyl-
                          Carbamate Ester

Figure 5. Nitrosoalkylurea and nitrosoalkylcarbamate structures

effects in rats of some derivatives of hydroxyethylnitrosourea (HENU) can be considered (Figure 6). HENU itself is a multipotent carcinogen, inducing tumors in a large number of organs (*20*), but not including the liver or nervous system. Substitution of ethyl on the amino group results in a compound that also induces a wide range of tumors in rats (*21*) including those of liver and the nervous system. A chloroethyl group in place of ethyl changes the tumor spectrum dramatically. The only tumors induced (even though the rats lived longer than those treated with the other two compounds) were liver tumors and kidney tumors (*22*). The second substituent did not alter the potent mutagenicity of the hydroxyethylnitrosourea (*23*) although it strongly affected the organ specificity of the carcinogen, suggesting that a receptor with affinity for the entire molecule might be the limiting factor in each case. Furthermore, studies of the in vivo hydroxyethylation of DNA in the liver of rats by the three hydroxyethylnitrosoureas (*24*) showed that HENU formed ten times as much O^6-hydroxyethylguanine in DNA of liver (and several other organs) as the two dialkylnitrosoureas which induced liver tumors although HENU did not induce liver tumors.

Conclusions

The increasing number of non-mutagenic carcinogens identified (*25, 26*) make it unlikely that mutagenesis is the key event in carcinogenesis. The results discussed above suggest that organ-specific carcinogenic effects of N-nitroso compounds depend on the particular chemical structure of the molecule which, usually through metabolism, produces intermediates with particular affinities for certain cells or cell types. The combination of the chemical with some component of those cells begins or potentiates the process of neoplastic transformation. In conjunction with this - and perhaps superimposed on it - can be mutagenic events consequent on alkylation of DNA by the carcinogen. The mutations can change the character of the tumor. Some tentative structural "rules" governing organ- and species-specific carcinogenesis by N-nitroso compounds are shown in Figure 7.

Tumors Induced in Rats by Nitrosoalkylureas

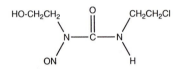

Liver
Kidney

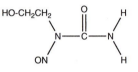

Lung
Forestomach
Colon
Zymbal's Gland
Ileum
Thyroid
Osteogenic Sarcoma

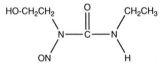

Liver
Lung
Brain
Thyroid
Mesothelioma
Zymbal's Gland
Colon

Figure 6. Tumors in rats by hydroxyethylnitrosoureas

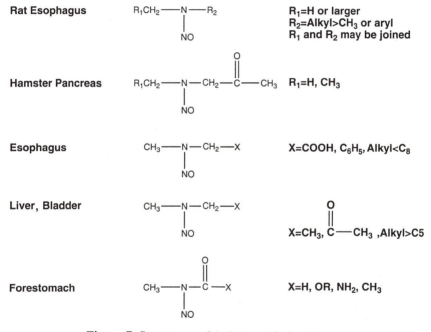

Figure 7. Structures related to particular tumors

Literature Cited

1. Magee,P.N., Barnes,J.M. Br.J.Cancer 1956, 10,114
2. Druckrey,H., Preussmann,R., Ivankovic,S.,Schmahl,D. Z.Krebsforsch. 1967, 69,103
3. Magee,P.N., Farber,E. Biochem.J. 1962, 83,114
4. Lijinsky,W. Cancer Metast.Rev. 1987, 6,301
5. Chung,F.-L.,Wang,M., Hecht,S.S. Cancer Res. 1989, 49,2034
6. Okada,M., Suzuki,E., Mochizuki,M. GANN 1976, 67,771
7. Thomas,B.J., Lijinsky,W.,Kovatch,R.M. Jpn.J.Cancer Res. 1988,79,309
8. Von Hofe,E., Schmerold,J., Lijinsky,W., Kleihues,P. Carcinogenesis 1987, 8,1337
9. Mirvish,S.S.,Ji,C.,Rosinsky,S. Cancer Res. 1988, 48,5663
10. Kirkman,H.,Bacon,R.L. J.Natl.Cancer Inst. 1952, 13,757
11. Lijinsky,W., Thomas,B.J., Kovatch,R.M. Chem.Biol.Interact. 1988, 66,111
12. Fong,L.Y.Y.,Jensen,D.E., Magee,P.N. Carcinogenesis 1990, 11, 411
13. Lijinsky,W. IARC Scientific Publications 1991, No. 105,305
14. Pour,P., Stepan,K. Cancer Lett. 1989, 45,49
15. Lijinsky,W. Chemistry and Biology of N-Nitroso Compounds 1992, Cambridge University Press, U.K.
16. Preussmann,R., Stewart,B.W. In: Searle,C.E., Ed. Chemical Carcinogens, American Chemical Society Monograph 182,1984,643
17. Lijinsky,W.,Reuber,M.D., Riggs,C.W. Cancer Res. 1981, 41,4997
18. Qin,X.,Nakatsuru,Y.,Kohyama,K.,Ishikawa,T. Carcinogenesis 1990, 11,235
19. Lijinsky,W. Environ.Mol.Mutagenesis 1989, 14(Suppl.16),78
20. Lijinsky,W.,Kovatch,R.M. Jpn.J.Cancer Res. 1988, 79,181
21. Lijinsky,W.,Singer,G.M.,Kovatch,R.M.Carcinogenesis 1985, 6,641
22. Lijinsky,W.,Kovatch,R.M.,Singer,S.SJ.Cancer Res.Clin.Oncol. 1986, 112,221
23. Lijinsky,W.,Elespuru,R.K.,Andrews,A.W.Mutation Res. 1987, 178,157
24. Ludeke,B.,Schubert,M.,Yamada,Y.,Lijinsky,W.,KleihuesP.Chem.-Biol. Interact. 1991,79,207
25. Tennant,R.W.,Margolin,B.H.,Shelby,M.D.,Zeiger,E., Haseman,J.K., Spalding,J.,Caspary,W.,Resnick,M.Ṣtasiewicz,S.Ạnderson,B.,Minor,R. Science 1987,236, 933
26. Lijinsky,W. Envir.Carcino.Revs. 1990,C8,45

RECEIVED December 8, 1993

Chapter 21

Formation of Tobacco-Specific Nitrosamines

Carcinogenicity and Role of Dietary Fat in Their Carcinogenicity

Dietrich Hoffmann, Abraham Rivenson, Ernst L. Wynder, and Stephen S. Hecht

American Health Foundation, 1 Dana Road, Valhalla, NY 10595

Seven tobacco-specific N-Nitrosamines (TSNA) have been identified in tobacco products. All of them were synthesized and bioassayed for carcinogenic activity in mice, rats and/or Syrian golden hamsters. In this group of compounds N′-nitrosonornicotine (NNN), 4-(methylnitrosamino)-1-(3-pyridyl)-1-butanone (NNK) and 4-(methylnitrosamino)-1-(3-pyridyl)-1-butanol (NNAL) were found to be organ-specific carcinogens which elicit benign and malignant tumors of the lung, upper aerodigestive tract and/or liver in the three species. NNK and NNAL are the only known tobacco components that also induce tumors of the pancreas. In mink, NNN induced in 100% of the animals carcinoma of the nasal cavity; most of these invaded the brain. No other tumors were observed though. In an assay in which rats received a low dose of NNK in the drinking water, the influence of the fat content of the diet was examined. The rats on a high-fat diet (23.5%) compared to rats on a low-fat diet (5.0%) had similar yields of tumors in the lung, nasal cavity, and liver; however, they had a significantly shorter survival span (77.8 vs. 85.5 weeks) and a significantly higher rate of pancreas tumors (47% and 31%).

Epidemiological studies have demonstrated that tobacco smoking causes cancer of the lung and upper aerodigestive tract and also cancer at distant sites, such as the pancreas, kidney and urinary bladder (*1*). A chewer of smokeless tobacco faces an increased risk for cancer of the oral cavity, pharynx and esophagus as well as for cancer of the pancreas and bladder (*2-4*). These observations not only suggest the presence in tobacco and tobacco smoke of "contact carcinogens", but also indicate a role of organ-specific carcinogens. The relationship of cigarette smoking with cancer of the bladder has been linked to aromatic amines and especially to the known bladder carcinogens 2-naphthylamine and 4-aminobiphenyl (*5-7*). However, tobacco products most likely contain additional organ-specific carcinogens, which, independent of the site of first contact, will primarily induce tumors at specific sites. On the basis of this working hypothesis, we began in 1974 our research program on the identification and formation of organ-specific carcinogens in tobacco and tobacco smoke and our studies on the carcinogenicity and biochemistry of these compounds (*8*).

In 1962, Druckrey and Preussmann (*9*) had suggested that tobacco smoke may contain nicotine-derived nitrosamines. In 1974, we first isolated NNN from cigarette smoke (*10*). In subsequent years, we documented the existence of a total of seven TSNA, all of which are formed by N-nitrosation of nicotine or of the minor tobacco alkaloids (Figure 1; *11*).

Bioassays of Tobacco-Specific N-Nitrosamines (TSNA)

Upon isolation and identification of the seven TSNA, we developed methods for the synthesis of each of them (*12-17*) and we bioassayed these compounds for carcinogenicity in mice, rats and Syrian golden hamsters. N´-Nitrosoanabasine (NAB) proved to be a moderately active carcinogen in the rat; it induces tumors in the esophagus and pharynx (*18,19*). N´-nitrosoanatabine (NAT) and 4-(methylnitrosamino)-4-(3-pyridyl)butyric acid (iso-NNAC) are inactive as carcinogens (*20,21*) and 4-(methylnitrosamino)-4-(3-pyridyl)-1-butanol (iso-NNAL) is currently being assayed.

Tables I and II summarize the strong organ-specific carcinogenic activity of NNN, NNK, and NNAL as evidenced by the induction of benign and malignant tumors of the lung, nasal cavity, esophagus, pancreas and/or liver in mice, rats, and hamsters (*8*). NNK induces lung tumors in all three animal species independent of mode and site of application. The lung tumors are primarily adenoma and adenocarcinoma; however, squamous cell carcinoma have also been observed in rats and hamsters (*20,22-27*). A single subcutaneous injection of 0.6 mg of NNK leads already to a significant number of lung adenocarcinoma. This type of tumor is histologically the same as the lung adenomas in cigarette smokers. During the past 20 years, the incidence of adenocarcinoma in the peripheral lung of smokers in the USA has risen much more drastically than that of squamous cell carcinoma in the bronchi, as its ratio has changed from 1 to 7 to 1 to 2-3 (*29*).

As shown primarily by S.S. Hecht and collaborators, in the lungs of laboratory animals and in human lung explants the metabolic activation of NNK to intermediates that bind to DNA occurs by α-hydroxylation of the methyl and methylene carbons (Figure 2; *30*). Methylene hydroxylation produces methyl diazohydroxide and methylated DNA bases, while methyl hydroxylation produces 4-(3-pyridyl)-4-oxobutyldiazohydroxide which is followed by pyridyloxobutylation of DNA. *In vivo* assays with NNK in rats have shown that 7-methylguanine, O^6-methylguanine and O^4-methylthymidine are formed in DNA of the lung, as well as of the liver and of the nasal cavity; the latter two are secondary target organs of NNK carcinogenicity. However, these types of DNA adducts were not found in the esophagus, spleen or kidney of rats, organs not susceptible to NNK carcinogenesis (*30-32*). Studies on O^6-methylguanine levels in lung DNA during chronic dosing with NNK have shown that this alkylated base accumulates and persists, in part due to inhibition of the repair enzyme, O^6-methylguanine-DNA methyltransferase by high doses of NNK (*32,33*).

These studies with the lung carcinogen NNK have enabled us to develop a mechanistic link between nicotine and NNK exposure and the formation of pro-mutagenic adducts, and finally, lung tumors. In the lung of mice, NNK forms O^6-methylguanine and subsequently causes miscoding due to the DNA adduct formation which then leads to lung adenoma; importantly, these lung adenoma contain activated K-*ras* proto-oncogene (*34*). Activated K-*ras* oncogene has also been found in specimen of lung adenocarcinoma from 13 of 45 cigarette smokers. Tobacco carcinogens appear to activate K-*ras* by point

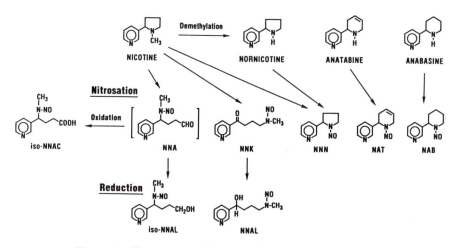

Figure 1.   Formation of Tobacco-Specific N-Nitrosamines

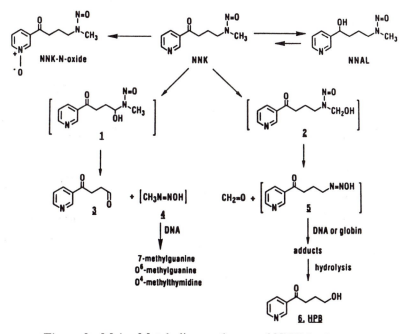

Figure 2.   Major Metabolism pathways of NNK in the rat.

Table I. Carcinogenicity of NNK in Laboratory Animals

TSNA	Animal (strain)	Route of Application	Target Organs	Total Dose mmol/animal	mmol/kg	Reference
NNK	Mouse(Sencar)	topical[a]	Skin*, lung	0.0056-0.028	0.24-1.22	LaVoie et al., 1987 (22)
	Mouse (A/J)	i.p.	Lung	0.01-0.1	0.06-0.6	Castonguay et al., 1983 (23)
	Mouse (A/J)	p.o.	Lung	0.01	0.06	Hong et al., 1992 (24)
	Rat (F-344)	s.c.	Lung*, nasal cavity liver	0.0025-3.6	0.009-7.0	Hoffmann et al.,1984 (20) Belinsky et al., 1990 (26)
	Rat (F-344)	p.o.	Lung*, pancreas, liver, nasal cavity	0.03-0.31	0.07-0.77	Rivenson et al., 1988 (25)
	Syrian Golden Hamster	s.c.	Lung*, trachea, nasal cavity	0.005-0.91	0.03-5.4	Hoffmann et al., 1984 (27)

* Major target organ.
[a] Tested as tumor initiator with TPA as tumor promoter.

Table II. Carcinogenicity of NNAL and NNN in Laboratory Animals

TSNA	Animal (strain)	Route of Application	Target Organs	Total Dose mmol/animal	mmol/kg	Reference
NNAL	Mouse (A/J)	i.p.	Lung	0.005-0.12	0.2-4.8	Castonguay et al., 1983 (23)
	Rat (F-344)	p.o.	Lung*, pancreas, liver	0.32	0.72	Rivenson et al., 1988 (25)
NNN	Mouse (Sencar)	topical[a]	None	0.028	1.22	LaVoie et al., 1987 (22)
	Mouse (A/J)	i.p.	Lung	0.1	4.0	Castonguay et al., 1983 (23)
	Rat (F-344)	s.c.	Nasal cavity*,	0.2-3.4	0.45-7.7	Hoffmann et al., 1984 (20)
	Rat (F-344)	p.o.	Esophagus*, nasal cavity	1.0-3.6	2.25-8.0	Hoffmann et al., 1975 (19)
	Rat (Sprague-Dawley)	p.o.	Nasal cavity*	8.8	22.0	Singer and Taylor, 1976 (28)
	Syrian Golden Hamster	s.c.	Trachea*, nasal cavity	0.91	5.4	Hoffmann et al., 1981 (27)

* Major target organ.
[a] Tested as tumor initiator with TPA as tumor promoter.

mutation on codon 12. In a recent study, K-*ras* point mutations have been observed at very high frequency in a subgroup of lung cancer patients with very poor prognosis *(35)*. Figure 3 depicts how nicotine is linked with lung tumor induction via the formation of NNK and its DNA adducts which cause the activation of the K-*ras* proto-oncogene.

This concept is supported by estimates of the doses of NNK which a heavy cigarette smoker inhales during his/her lifetime and by comparing such doses with the total dose required to induce lung tumors in rats. A 2 pack-a-day smoker of nonfilter cigarettes is exposed to 105 mg (6 μmol/kg) NNK over a 40-year period. The lowest dose of NNK that induces lung tumors in rats upon subcutaneous injection is 0.6 mg (9 μmol/kg) *(30)*.

NNK and its enzymatic reduction product, NNAL (Figure 1), are also the only tobacco smoke constituents known to induce pancreas cancer in animals. NNK, given to rats at a dose of 1 ppm in the drinking water (1 mg/L), induces tumors of the exocrine pancreas in addition to lung tumors. The dose of NNAL given to rats in the drinking water that induced pancreas tumors in addition to lung tumors was 5 ppm (5 mg/L; *25*). As will be discussed later, the multiplicity of the pancreas tumors in rats induced by NNK is greatly influenced by the amount of dietary fat given during the bioassay.

NNK is not only an organ-specific carcinogen but it is also a tumor initiator at the site of application. We have two examples of this. When applied in 10 subdoses (total 5.8 mg) to mouse skin, NNK does not induce skin tumors but elicits lung adenoma. However, when NNK-initiated mouse skin is subsequently treated three times weekly for 20 weeks with the tumor promoter TPA (12-O-tetradecanoylphorbol-13-acetate), a high skin tumor incidence is observed *(22)*. In another study, Syrian golden hamsters were pretreated with a single s.c. injection of a low dose of NNK and then exposed twice daily to air-diluted cigarette smoke for 69 weeks. The NNK pre-treated and smoke-exposed hamsters developed significantly more tumors in the respiratory tract than did the hamsters treated only with NNK and sham-smoked, or those exposed to smoke only. In this bioassay, NNK acted as a tumor initiator and cigarette smoke as tumor promoter *(37)*.

We also explored the carcinogenicity of TSNA in the oral cavity of rats by swabbing their oral surfaces twice daily for a total of 130 weeks with a highly diluted aqueous solution of a mixture of NNN (138 ppm) and NNK (28 ppm). Of 30 animals tested, 1 developed a lung adenoma, 4 had lung adenocarcinoma, and 8 rats developed tumors in the oral cavity; including tumors of the cheek, hard palate, and tongue. Since none of the 21 control rats developed oral tumors and only one animal had one lung adenoma, the data of this bioassay with NNN plus NNK were significant and support the concept that the TSNA in smokeless tobacco contribute to the carcinogenicity of this product in the oral cavity of rats *(38-40)* and that TSNA play a role in the increased risk of tobacco chewers and snuff dippers for cancer of the oral cavity and pharynx *(2,3)*.

N´-Nitrosonornicotine induces primarily papilloma and carcinoma of the nasal cavity in rats and in Syrian golden hamsters; nasal tumors are the second most frequent type of neoplasm, ranking after those in the trachea (Table I). In both animal species as well as in the mouse, NNK is a more powerful carcinogen than NNN. When NNN is given to F344 rats in the drinking water, it also causes carcinoma of the esophagus; this effect is not seen with NNK.

Bioassay of N´-Nitrosonornicotine in Mink

Although cigarette smokers and snuff dippers face an increased risk for cancer of the nasal cavity *(41)*, the major interest in nasal cancer relates to the occupational exposure to certain chemicals such as nickel compounds and isopropyl oil (43). It is known that nickel refinery workers in Canada, England, Norway and other countries have an excess rate of cancer of the nasal cavity (42). However, there are only a few animal models in which a chemical induces only tumors of the nasal cavity (44). Many N-nitrosamines, including N-nitrosodimethylamine, induce nasal tumors in laboratory animals; yet, all of these compounds also induce tumors at other sites such as trachea, lung and/or liver (44). In search of an animal model for nasal carcinogenesis, scientists from the University of Oslo, the Norwegian Radium Hospital, Oslo, and from our group decided after preliminary investigations that, because of its highly developed nasal cavity (Figure 4) and a sufficient susceptibility to carcinogens, the mink would be the animal of choice and NNN the most appropriate carcinogen.

The bioassay was set up with 20 mink. Beginning at the age of 3 weeks, the mink received twice weekly subcutaneous injections of NNN for 38 weeks for a total dose of 11.9 mmol (females, 9.5 mmol/kg; males 5.7 mmol/kg). After 3½ years, about one third of the life expectancy of the mink, all 19 animals at risk had developed malignant tumors of both the respiratory and the olfactory region of the nose. In most animals the malignant tumors were primarily esthesioneuroepithelioma which invaded the brain. Remarkably, NNN induced no other tumors in the animal model (52).

This bioassay demonstrates that NNN is a strong carcinogen in the mink. The absence of tumors at sites other than the nasal cavity makes this animal a unique model for the study of early diagnosis, prevention and therapy of nasal cancer. The larger area of the nasal cavity of the mink as compared to that of rodents makes this model also especially suitable for studies on the metabolism, binding, and molecular biology of nasal carcinogens.

The Role of Dietary Fat in the Carcinogenicity of NNK

During recent years a number of studies have explored the effect of natural and synthetic chemopreventive agents on the induction of tumors with NNK in mice and rats (24; 45-49). Certain chemopreventive agents are in fact remarkably effective in these model studies.

Our interest in the reduction of the carcinogenic effects of tobacco products per se is now focused on studying the influence of macronutrients, such as a high fat intake, on tobacco carcinogenesis, since population studies have revealed that the fat content of the diet appears to influence the risk of cancer among cigarette smokers (50,51).

For a bioassay, we placed male rats on isocaloric diets, groups I and III on a 23.5 % fat-containing diet and groups II and IV on a 5.0 % low-fat diet (Table III). The rats in experimental groups I and II received 2 ppm NNK (2 μg/L) in the drinking water, while the rats in the control groups III and IV were given tap water. We first observed (not expectedly) that the rats on the high-fat diet (groups I and III) gained significantly more weight and remained heavier throughout the lifetime bioassay than the rats on the low-fat diet (groups II and IV; Figure 5). Less expected was our observation that rats on the high-fat diet had a significantly shorter life span than the rats on the low-fat diet, independent of the presence or absence of NNK in the drinking water (Table III).

The average doses of NNK in drinking water ingested by rats in groups I and II were approximately equal, namely 18.06 ± 4.75 mg (0.17 mmol/kg) and 18.20 ± 4.10 mg (0.19 mmol/kg). This was so because the rats on the high-fat diet (group I) consumed more water during the 78.7 weeks (their average survival span) than the rats in group II did in the first 78.7 weeks of the

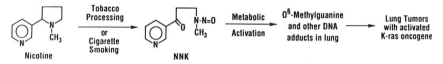

Figure 3. Scheme linking nicotine via NNK to formation of DNA adducts, to activation of K-ras oncogene to lung tumors.

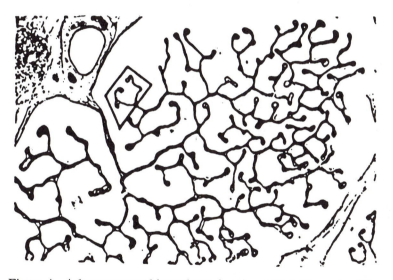

Figure 4. Arborescent turbinate bone forming a labyrith covered by respiratory mucosa. The corresponding surface of a rat turbinate would be about the size shown in the framed area.

Table III. Uptake of NNK , survival time, water consumption, and body weights of male F344 rats*

Parameters	Group I NNK, 2ppm High-fat (23.5%) 60 Rats	Group II NNK, 2ppm Low-fat (5.0%) 60 Rats	Group III - High-fat (23.5%) 20 Rats	Group IV - Low-fat (5.0%) 20 Rats
Body wt. 21 weeks	441.9 ± 23.5	380.0 ± 19.9	440.5 ± 20.1	409.8 ± 18.1
Body wt. 63 weeks	575.1 ± 32.4	522.1 ± 20.2	574.2 ± 40.3	517.9 ± 37.7
Body wt. 98 weeks	501.0 ± 28.5	457.4 ± 37.5	-	446.7 ± 68.5
Water consumption	9.65 ± 2.38	9.58 ± 2.04	8.75 ± 2.02	9.72 ± 2.14
Dose, mg/rat	18.06 ± 4.75	18.20 ± 4.10	-	-
Dose, mmol/kg	0.17	0.19	-	-
Avg. survival(wk)	78.7 ± 15.1	85.5 ± 13.4	75.3 ± 12.6	89.6 ± 12.7
Termination(wk)	97	100	95	105

*Average starting weights: 90.9-91.5 g.
The values in the columns are given with S.D.

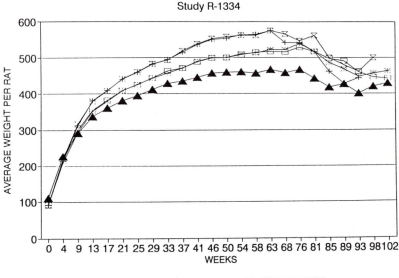

Figure 5. Body weights of male F344 rats treated with NNK in the drinking water (2 µg/L) and fed high fat diet (23.5%) and low fat diet (5.0%). Control rats were on tap water and high or low fat diet.

bioassay; the rats in group II consumed the difference in drinking water during the remaining 6.8 weeks of their average survival which was 85.5 weeks. The lung tumor yields in groups I (NNK; high-fat diet) and II (NNK; low-fat diet) were comparable, being 51 % and 48 % respectively. Tumors of the nasal cavity occurred in 8 % and 7 %, and tumors of the liver in 8 % and 4 % of the animals (Table IV). However, significant differences were observed in respect to tumors of the pancreas. Eleven animals had died with pancreas tumors (after 18 months) in group I as opposed to only 1 animal in group II (Table V). At the termination of the bioassay, 18 animals in group I had developed exocrine pancreas tumors (33%) and 10 had islet cell tumors (19%) compared to 14 rats (25%) and 5 rats (9%) in group II. Perhaps even greater significance lies in the multiplicity and the large average size of the exocrine pancreas tumors of the animals in group I. In regard to other types of tumors in rats in the NNK groups and in the control groups we did not observe statistically significant differences.

We conclude from this study that, in the rat model, the carcinogenic activity of the tobacco-specific nitrosamine NNK is not affected by the fat content of the diet with respect to incidence rates of cancer of the lung, nasal cavity and liver. However, the high-fat diet has a significant impact on the incidence rates of tumors of the pancreas. Rats given NNK and maintained on an isocaloric high-fat diet (23.5%) developed significantly higher rates of large-sized exocrine pancreas tumors and also higher rates of islet cell pancreas tumors than rats given the same dose of NNK but which were on a low-fat diet (5.0%). Studies will now be directed toward determining the nature of those fatty acids that have the greatest influence on NNK-related tumor development in the pancreas.

Table IV. Tumor Yields in Rats Treated with NNK and High-Fat vs Low-Fat Diet

Site and Type of Tumor	I 2 ppm NNK 23.5% Fat 60 Rats	II 2 ppm NNK 5.0% Fat 60 Rats	III - 23.5% Fat 20 Rats	IV - 5.0% Fat 20 Rats
Lung				
Adenoma	9	7	1	0
Adenocarcinoma	20	19	0	1
Adenosquamous Carcinoma	1	1	0	0
Nasal Cavity				
Papilloma	5	3	1	1
Carcinoma (squamous)	0	1	3	1
Liver				
Adenoma	3	2	0	0
Adenocarcinoma	2	0	0	0

Table V. Tumors of the Pancreas Observed Upon Treatment of F-344 Rats with NNK and High-Fat vs Low-Fat Diet

	I 2 ppm NNK 25.5% Fat	II 2 ppm NNK 5.0% Fat	III - 23.5% Fat	IV - 5.0% Fat
No. of Rats at Risk	54/60	56/60	17/20	19/20
Rats with Pancreas Tumors				
after 18 months	11[a]	1	2	1
after 24 months	28[b]	19	7	7
Exocrine Pancreas Tumors				
Acinar Adenoma	11	11	3	2
Acinar Adenocarcinoma	6	3	3	2
Ductal Adenocarcinoma	1	0	0	0
Rats with				
Multiple Tumor				
with 1 tumor	11	8	2	3
with 2 tumors	4	6	4	1
with 3+ tumors	3	0	0	0
Total No. of Exocrine				
Pancreas Tumors	18	14	6	4
Average Size of Pancreas				
Tumors (mm^2)	15.8 ± 11.0[a]	6.5 ± 6.1	8.2 ± 4.2	5.7 ± 3.2
Islet Cell Tumors	10	5	1	3

Group I compared with Group II: [a] $p<0.005$; [b] $p<0.05$.

Acknowledgements

Our research program in tobacco carcinogenesis is supported by the grant No. CA-29580 from the U.S. National Institutes of Health. The editorial assistance by Ilse Hoffmann and Laura DiSciorio is greatly appreciated.

Literature Cited

1. U.S. Surgeon General. Reducing the Health Consequences of Smoking. 25 Years of Progress. D.H.H.S. Publ. No. (CDC): Rockville, M.D., 1989; 89-8411.
2. International Agency for Research on Cancer. Tobacco Habits Other than Smoking: Betel-Quid and Areca-Nut Chewing and Some Related Nitrosamines; IARC Monogr. 37: Lyon, France, 1985.
3. U.S. Surgeon General. The Health Consequences of Using Smokeless Tobacco. Bethesda, M.D., 1986; NIH Publ: No. 86-2874.
4. Kabat, G.C.; Dieck, G.S.; Wynder, E.L. *Cancer* **1986**, *57*, 362-367.

5.   Doll, R. *Cancer Related to Smoking.* In The Second World Conference on Smoking and Health. R.G. Richardson, ed., Proc. Conf. Health Educ. Council, Imperial College, Pitman Medical London, **1971**; pp. 10-23

6.   Patrianakos, C.; Hoffmann, D. *J. Anal. Toxicol.* **1979**, *3*, 150-154.

7.   Vineis, P.; Caporaso, N.; Tannenbaum, S.R. *Cancer Res.* **1990**, *50*, 3002-3004.

8.   Hecht, S.S.; Hoffmann, D. *Cancer Survey* **1989**, *8*, 273-294.

9.   Druckrey, H.; Preussmann, R. *Naturwissenschaften* **1962**, *49*, 498-499.

10. Hoffmann, D.; Rathkamp, G; Liu, Y.Y. IARC Monogr. **1974**, 46, 885-889.

11. Brunnemann, K.D.; Hoffmann, D. *Recent Advan. Tobacco Res.* **1991**, *17*, 71-112.

12. Hu, M.W.; Bondinell, W.E.; Hoffmann, D. *J. Labelled Comp.* **1974**, *10*, 79-88.

13. Hecht, S.S.; Ornaf, R.M.; Hoffmann, D. *J. Natl. Cancer Inst.* **1975**, *54*, 1237-1244.

14. Hecht, S.S.; Chen, C.B.; Dong, M.; Ornaf, R.M.; Hoffmann, D.; Tso, T.C. *Beitr. Tabakforsch.* **1977**, *9*, 1-6.

15. Hoffmann, D.; Adams, J.D.; Brunnemann, K.D.; Hecht, S.S. *Cancer Res.* **1979**, *39*, 2505-2509.

16. Hecht, S.S.; Young, R.; Chen, C.B. *Cancer Res.* **1980**, *40*, 4144-4150.

17. Djordjevic, M.V.; Brunnemann, K.D.; Hoffmann, D. *Carcinogenesis* **1989**, *10*, 1725-1731.

18. Boyland, E.; Roe, F.J.; Gorrod, J.W.; Mitchley, B.C.V. *Brit. J. Cancer* **1964**, *18*, 265-270.

19. Hoffmann, D.; Raineri, R.; Hecht, S.S.; Maronpot, R.; Wynder, E.L. *J. Natl. Cancer Inst.* **1975**, *55*, 977-981.

20. Hoffmann, D.; Rivenson, A.; Amin, S.; Hecht, S.S. *J. Cancer Res. Clin. Oncol.* **1984**, *108*, 81-86.

21. Rivenson, A.; Djordjevic, M.V.; Amin, S.; Hoffmann, D. *Cancer Letters* **1989**, *47*, 11-114.

22. La Voie, E.J.; Prokopczyk, G.; Rigotty, J.; Czech, A.; Rivenson, A.; Adams, J.D. *Cancer Letters,* **1987**, *37*, 277-283.

23. Castonguay, A.; Lin, D.; Stoner, G.D.; Radok, P.; Furuya, K.; Hecht, S.S.; Schut, H.A.J.; Klauning, J.E. *Cancer Res.* **1983**, *43*, 1223-1229.

24. Hong, J.-Y.; Wang, Z.Y.; Smith, T.J.; Zhou, S.; Shi, S.; Pan, J.; Yang, C.S. *Carcinogenesis* **1992**, *13*, 901-904.

25. Rivenson, A.; Hoffmann, D.; Prokopczyk, B.; Amin, S.; Hecht, S.S. *Cancer Res.* **1988**, *48*, 6912-6917.

26. Belinsky, S.A.; Foley, J.F.; White, C.M.; Anderson, M.W.; Maronpot, R.R. *Cancer Res.* **1990**, *50*, 3772-3780.

27. Hoffmann, D.; Castonguay, A.; Rivenson, A.; Hecht, S.S. *Cancer Res.* **1981**, *41*, 2386-2393.

28. Singer, G.M.; Taylor, H.W. *J. Natl. Cancer Inst.* **1976**, *57*, 1275-1276.

29. Wynder, E.L.; Covey, L.S. *Europ. J. Cancer Clin. Oncol.* **1987**, *23*, 1491-1496.

30. Hecht, S.S.; Hoffmann, D. In: Orgins of Human Cancer: A Comprehensive Review. Cold Spring Harbor Press, Plainview, NY, 1991; 745-755.

31. Castonguay, A.; Foiles, P.G.; Trushin, N.; Hecht, S.S. *Environ. Health Perspect.* **1986**, *62*, 197-202.

32. Belinsky, S.A.; White, C.M.; Devereux, T.R.; Swenberg, J.A.; Anderson, M.W. *Cancer Res.* **1987**, *47*, 1143-1148.

33. Belinsky, S.A.; White, C.M.; Boucheron, G.A.; Richardson, F.C.; Swenberg, J.A.; Anderson, M.W. *Cancer Res.* **1986**, *46*, 1280-1284.

34.   Belinsky, S.A.; Deverux, T.R.; Stoner, G.D.; Anderson, M.W. *Cancer Res.* **1988**, *49*, 5303-5311.
35.   Rodenhuis, S.; Slebos, R.J.C.; Boot, A.J.M.; Evers, S.G.; Mooi, W.J.; Wagenaar, S.S.C.; van Bodegom, P.C.H.; Bos, J.L. *Cancer Res.* **1988**, *48*, 5738-5741.
36.   Slebos, R.J.C.; Kibbelaar, R.E.; Daleiso, O.; Kooistra, A.; Stam, J.; Meijer, C.J.L.M.; Wagenaar, S.S.C.; Vanderschueren, R.G.J.R.; van Zandwijk, N.; Mooi, M.J.; Bos, J.L.; Rodenhuis, S. *New Engl. J. Med.* **1990**, *323*, 561-565.
37.   Hecht, S.S.; Adams, J.D.; Numoto, S.; Hoffmann, D. *Carcinogenesis* **1983**, *4*, 1287-1290.
38.   Hecht, S.S.; Rivenson, A.; Braley, J.; DiBello, J.; Adams, J.D.; Hoffmann, D. *Cancer Res.* **1986**, *46*, 4162-4166.
39.   Hirsch, J.-M.; Johansson, S.L. *J. Oral. Pathol.* **1983**, *12*, 187-198.
40.   Johansson, S.L.; Hirsch, J.M.; Larsson, P.-A.; Saidi, J.; Osterdahl, B.-G. *Cancer Res.* **1989**, *49*, 3063-3069.
41.   Brinton, L.A.; Blot, W.J., Becker, J.A.; Winn, D.M.; Browder, J.P.; Farmer, J.C. Jr.; Fraumeni, J.F. Jr. *Am. J. Epidemiol.* **1984**, *119*, pp. 896-906.
42.   International Agency for Research on Cancer. *IARC Monogr. Suppl.* **1987**, *7*, 440 p.
43.   Hecht, S.S.; Castonguay, A.; Hoffmann, D. In: Nasal Tumors in Animals and Man. G. Reznik and S.F. Stinson, eds. C.R.C. Press, Boca Raton, FL, **1983**, *3*; pp. 201-232.
44.   Preussmann, R.; Stewart, B.W. *N*-Nitroso Carcinogens *In*: "Chemical Carcinogens" 2nd Edition, C.E. Searle, ed., Am. Chem. Soc. Monogr. American Chemical Society, Washington, DC, 1984, 182; 643-828.
45.   Morse, M.A.; Wang, C.-X.; Amin, S.G.; Hecht, S.S.; Chung, F.-L.; *Carcinogenesis* **1989**, *9*, 1891-1895.
46.   Morse, M.A.; Wang, C.-X.; Stoner, G.D.; Mandal, S.; Conran, P.B.; Amin, S.G.; Hecht, S.S.; Chung, F.-L. *Cancer Res.* **1989**, *49*, 549-553.
47.   Morse, M.A.; Eklind, K.I.; Amin, S.G.; Hecht, S.S.; Chung, F.-L. *Carcinogenesis* **1989**, *9*,1757-1759.
48.   Pepin, P.; Rossignol, G.; Castonguay, A. *Cancer J.* **1990**, *3*, 266-273.
49.   Wattenberg, L.W.; Coccia, J.B. *Carcinogenesis* **1991**, *12*, 115-117.
50.   Hebert, J.R.; Kabat, G.C. *European J. Clin. Nutr.* **1990**, *44*, 185-193.
51.   Hebert, J.R.; Kabat, G.C. *J. Natl. Cancer Inst.* **1991**, *83*, 872-874.
52.   Koppang, N.; Rivenson, A.; Reith, A; Dahle, H.K.; Evensen, O.; Hoffmann, D. *Carcinogenesis* **1992**, *13*, 1957-1960.

RECEIVED October 6, 1993

Chapter 22

Possible Mechanisms
of *N*-Nitrosodimethylamine Hepatotoxicity

Michael C. Archer, Wei Chin, and Valentia M. Lee

Department of Medical Biophysics, Ontario Cancer Institute, University
of Toronto, 500 Sherbourne Street, Toronto, Ontario M4X 1K9, Canada

Lethal cellular injury is a process which, via compensatory cell
proliferation, plays a critical role in carcinogenesis. Many studies have
addressed the mechanism by which NDMA induces hepatotoxicity, but
thus far its action remains an enigma. We have shown that the
covalent binding of the reactive intermediate formed by hepatic
metabolism of NDMA (the methyl diazonium ion) to protein and DNA
appears not to account for its hepatotoxicity. Furthermore, the
production of oxidative stress and a variety of other mechanisms cannot
yet account for lethal hepatocyte injury by NDMA. Until the
mechanism by which NDMA induces hepatotoxicity is understood, we
will not fully understand how NDMA and other nitrosamines produce
liver tumors.

The initiation, promotion and progression phases of cancer development all require
cell proliferation. Lethal cellular injury in target tissues frequently occurs following
administration of a chemical carcinogen to an experimental animal. Proliferation
is usually enhanced to compensate for such cell loss. Knowledge of the
mechanisms of cell death and the concomitant compensatory cell proliferation is,
therefore, crucial for our understanding of carcinogenesis.

N-Nitrosodimethylamine (NDMA) has been known to be hepatotoxic since
its adverse effects in animals and humans were first investigated (*1,2*). As with
many other hepatotoxicants, metabolism by the cytochrome P450 mixed function
oxidase system is required for NDMA-induced hepatotoxicity (*3-7*). There also is
evidence that tissue-specific metabolism is required for nitrosamine-induced
esophageal toxicity (*8,9*). Nitrosamine metabolism has been reviewed elsewhere
(*10*). We will focus here on the mechanism by which NDMA kills hepatocytes,
reviewing briefly our own data and those of others.

Despite numerous studies of liver cell injury by NDMA over almost 40 years,
no clear mechanism has emerged. In our studies we have utilized the properties of
two other N-nitroso compounds, N-nitrosomethylbenzylamine (NMBzA) and
methylnitrosourea (MNU), to elucidate the mechanism of action of NDMA.
NMBzA is an esophageal carcinogen in the rat (*11,12*), but, in rat liver, is
principally metabolized to the methyl diazonium ion (*8*), the same reactive
intermediate that is produced by the metabolic activation of NDMA (*10*).

Furthermore, NMBzA methylates hepatic DNA in a manner qualitatively indistinguishable from NDMA (13). Despite these similarities, studies to date have shown that NMBzA produces no liver tumors (11) and appears not to kill liver cells (14). Unlike NDMA and NMBzA, MNU decomposes spontaneously at neutral pH to the methyl diazonium ion (10). MNU methylates hepatic DNA (15), but is not hepatotoxic (16). It produces liver tumors in rats only when the animals are partially hepatectomized following treatment (17). Consequently, comparison of the biochemical and toxicological properties of NDMA, NMBzA and NMU in rat liver should provide useful information regarding the mechanism of action of NDMA as a hepatotoxicant.

Characteristics of NDMA-Induced Hepatotoxicity

It has been known for some time that severe centrilobular necrosis induced by NDMA in rodents, does not fully develop until 24-48 h following exposure (7,18-21). In experiments with rats, we showed that the hepatotoxicity of NDMA does not peak until about 48 h. A similar toxicokinetic behaviour has been observed for NDEA in rat liver (22). Since the half-life of NDMA in the rat is about 1 h (3, 23, 24), clearly the time lag between the production of toxic metabolites and the expression of irreversible damage is an important feature of the hepatotoxicity of this compound. We have observed a similar lag in the lethal injury of primary monolayer cultures of rat hepatocytes by NDMA (25).

NDMA produced a sigmoid dose-response curve when leakage of transaminases from dead liver cells into the serum of rats was used to measure hepatotoxicity (Figure 1). Doses less than 135 µmol/kg (10 mg/kg) caused no detectable biochemical or histological liver damage. At higher doses, the characteristic necrosis developed in the perivenous region of the liver lobule where these are high levels of cytochrome P450 IIE1 (NDMA-demethylase) (26,27). In contrast to NDMA, NMBzA produced no liver damage assessed either by serum enzyme measurement (Figure 1) or by histology, at doses as high as 667 µmol/kg (100 mg/kg, ~9xLD50). When rats were treated with 333 µmol/kg (50 mg/kg) NMBzA together with different doses of NDMA, however, hepatotoxicity was significantly elevated compared to that produced by NDMA alone (Figure 1). Treatment with the two nitrosamines produced hepatotoxicity at doses of NDMA as low as 67 µmol/kg (5 mg/kg).

Covalent Binding

A major hypothesis to explain the link between metabolic activation and cell death is that the injurious effects of xenobiotics are related to their ability to bind irreversibly to nucleophilic sites in macromolecules such as proteins and DNA.

Protein. Magee and Hultin (28) first observed methylation of proteins in rat liver by NDMA and concluded it might be related to acute hepatic injury. They isolated and identified 1- and 3-methylhistidine. Although the extent of methylation of the total mixed proteins was very small, Magee and Hultin pointed out that the inactivation of one essential enzyme might be sufficient to cause cell death, or alternatively, that sufficient abnormal protein might accumulate to induce an immune reaction. More recently, Diaz Gomez et al (7) found that several agents which decreased the hepatotoxicity of NDMA, also inhibited covalent binding of metabolites to the 9000 g supernatant proteins of rat liver. However, because the inhibitory compounds also inhibited NDMA metabolism, one can draw no conclusion regarding the role of protein methylation in hepatotoxicity from these experiments.

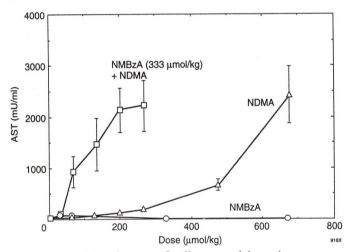

Figure 1. Dose dependency of liver toxicity (serum aspartate aminotransferase, AST) produced by NDMA, NMBzA or a combination of both compounds 48 h following treatment. Values are means ± SEM of 5 rats.

Using [^{14}C]-NDMA and [^{14}C-methyl]-NMBzA, we have compared the extent of covalent binding of the two nitrosamines to proteins in rat liver in vivo. The incorporation of radioactivity via the normal pathways of amino acid and protein biosynthesis utilizing the one carbon intermediates generated by metabolism of the nitrosamines, has already been shown to be small compared to the radioactivity derived from direct methylation of preexisting proteins (28). We first showed that the time course of binding for the two nitrosamines was similar, with a peak at around 4-6 h. The dose-response (Figure 2) measured 6 h following administration of the nitrosamines, showed a slight, but not significant, difference between NDMA and NMBzA. Although we cannot exclude the possibility that NDMA methylates specific proteins not methylated by NMBzA, such methylation seems unlikely since NDMA and NMBzA produce the same reactive intermediate (13). We conclude, therefore, that covalent binding of NDMA metabolites is unlikely to be the cause of its hepatotoxicity.

DNA. There recently has been a considerable amount of work to determine whether formation of O^6-methylguanine in DNA is related to the toxic effects of methylating agents in mammalian cells in culture (reviewed in *29, 30*). Observations from these experiments are generally consistent with the notion that O^6-methylguanine is a major cytotoxic lesion induced by direct acting S_N1-type methylating agents. Thus, compared to repair proficient cells, cells that have impaired ability to repair O^6-methylguanine show an increased sensitivity towards the cytotoxicity of methylating agents. Conversely, the cytotoxicity of methylating agents is decreased after transfection of the cloned O^6-methylguanine-DNA methyltransferase gene into mammalian cells lacking methyltransferase activity (reviewed in *31*). However, all of these studies utilized rapidly dividing cells in culture, while our interest is the liver, an organ in which few cells divide at any given time.

Therefore, in an effort to understand the mechanism of hepatocyte death produced by NDMA, we began by examining the methylation of hepatic DNA. The dose-response for DNA methylation produced in whole liver by NDMA 6 h after dosing appears in Figure 3, and is in general agreement with levels obtained by Pegg (32). It is clear from Figure 3 that the levels of O^6-methylguanine produced by NMBzA in hepatic DNA are significantly lower than those produced by NDMA. Since NMBzA is not hepatotoxic at any of the dose levels shown in Figure 3, if O^6-methyl-guanine is the toxic lesion, the results suggest that there is a threshold level of this base in hepatocytes below which toxicity is not produced. From Figures 2 and 3, it appears that this threshold would be about 500 µmol/mol guanine, the quantity produced by 135 µmol/kg NDMA.

Also shown in Figure 3 are the levels of O^6-methylguanine produced when NMBzA was given together with increasing doses of NMDA. Clearly the combined treatment produced additive amounts of O^6-methylguanine, an observation that agrees with our previous results using partially hepatectomized animals (33). We know from our immunohistochemical studies that both NDMA at doses ≤135 µmol/kg and NMBzA at doses up to 667 µmol/kg produce O^6-methylguanine in the same population of perivenous hepatocytes (34). Indeed, when NDMA and NMBzA were given together, we showed that the population of hepatocytes containing O^6-methylguanine was almost the same as when either of the nitrosamines was given alone (34). From Figure 3 it is clear that a combined dose of 68 µmol/kg NDMA and 333 µmol/kg NMBzA produces a level of O^6-methylguanine that is somewhat less than our predicted threshold for toxicity of about 500 µmol/mol guanine. However the combined dose gave almost a 20-fold increase in hepatotoxicity compared to NDMA alone. Thus, the data in Figure 3 suggest that DNA methylation by NDMA is unrelated to the production of toxicity. Of course it is possible that O^6-methylguanine is a toxic lesion and that the

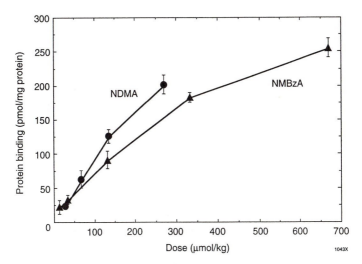

Figure 2. Dose dependency of protein methylation in rats treated with NDMA or NMBzA. Different doses of [^{14}C]-NDMA and [^{14}C-methyl]-NMBzA were injected ip, and the livers were taken 6 h later. Trichloroacetic acid-insoluble protein was prepared by standard methodology (29). Values are means ± SEM of 3 rats.

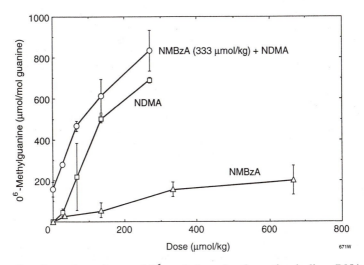

Figure 3. Dose dependency of O^6-methylguanine formation in liver DNA 6 h after treatment with NDMA, NMBzA or a combination of both compounds. Values are means ± SEM of 3 rats.

potentiation of the hepatotoxicity of NDMA by NMBzA occurs by an unrelated mechanism. Our work with MNU, however, suggests this is not the case. MNU given together with NDMA produced O^6-methylguanine levels that were approximately additive. In this case, however, there was no change in hepatoxicity from that observed from NDMA alone (Table I). The data in Figure 3

Table I. Hepatotoxicity (serum aspartate aminotransferase, AST) measured at 48h, and O^6-methylguanine formation measured at 6h in rat liver after treatment with NDMA, MNU or a combination of both compounds

Treatment (μmol/kg)	AST (mU/ml)	O^6-Methylguanine (μmol/mol G)
None	60 ± 33	-
NDMA (270)	197 ± 35	672 ± 158
MNU (971)	62 ± 3	253 ± 21
NDMA (270) + MNU (971)	202 ± 35	847 ± 111

Values are means ± SEM of 5 rats for AST levels and 3 rats for O^6-methylguanine levels.

and Table I reveal that an increase in toxicity would have been observed from the combination of agents if O^6-methylguanine was involved. Although it is unknown whether NDMA and MNU methylate the same or different populations of hepatocytes within the liver lobule, at the dose of NDMA used in this experiment (270 μmol/kg, 20 mg/kg), cells within the whole lobule will contain O^6-methylguanine. Therefore, if O^6-methylguanine is the toxic lesion, we might expect to see an effect of MNU on NDMA toxicity regardless of how the DNA methylation produced by MNU is localized within the liver lobule, assuming, of course, that all hepatocytes would be equally sensitive to the effects of this lesion. The inability of MNU to potentiate the toxicity of NDMA suggests that the potentiating effect of NMBzA is related to the metabolism of the nitrosamines or possibly to membrane effects caused by the lipophilicity of NMBzA.

Because the three compounds reacted identically when methylation of restriction fragments of the plasmid pBR322 were examined at nucleotide resolution by a sequencing assay (*13*), the differences in the hepatoxicity of NMDA, NMBzA and MNU are probably not related to methylation "hot spots" within the DNA.

Oxidative Stress

There is much evidence that oxidative stress imposed by the formation of partially reduced oxygen species (eg O_2^-, H_2O_2) or their equivalent such as organic hydroperoxides (*35*), causes the biological activity of many classical hepatotoxins. Toxic agents may oxidatively injure cells in 3 ways. They may be metabolized to free radicals that can reduce molecular oxygen, they may produce activated oxygen species by the action of cytochrome P450 as an oxidase, or they may impair the protective mechanisms of the cell so that the normal endogenous flux of activated oxygen becomes toxic.

Free Radicals. Evidence for free radical production during nitrosamine metabolism has been provided by the use of spin traps. Saprin and Piette (*36*) used phenyl-t-butylnitrone to demonstrate that peroxidizing rat liver microsomes incubated with NDMA produced the reactive dimethylamine free radical and nitric oxide. Floyd et al (*37*), using the spin trap 5,5-dimethylpyrroline-1-oxide and a number of nitrosamines incubated with microsomes or nuclei from rat liver, demonstrated formation of OH· and a free radical of unknown structure from the carcinogen itself. Dusan and Faljoni (*38*) provided evidence for the formation of singlet oxygen during the peroxidase-catalyzed degradation of nitrosodiisopropylamine. These studies suggest that during nitrosamine metabolism free radicals may form and possibly lead to the production of activated oxygen species.

Futile Metabolism. Cytochrome P450 can act as an oxidase in the presence of certain substrates and produce partially reduced forms of oxygen (*39*). When cytochrome P450 functions in this way, its normal metabolic cycle is uncoupled and bound oxygen is released without being incorporated into the substrate. While there is no direct evidence of this occurring during nitrosamine metabolism, Malaveille et al (*40*) have shown that the binding of a monoclonal antibody directed against NDMA-demethylase decreased the functional coupling between the reductase and the P450 complex. Their experiments suggest that antibody binding leads to an increased electron flow from the reductase towards molecular oxygen to form reduced oxygen species such as OH· at the expense of the monooxygenase function. This futile metabolism may possibly occur in the absence of antibody.

Glutathione Depletion. Impairing the protective mechanisms of the cell so that the normal flux of activated oxygen becomes toxic may produce oxidative stress. In particular, glutathione (GSH) depletion can sensitize cells to the partially reduced oxygen species present as a result of normal cellular processes. NDMA hepatotoxicity, however, is not preceded by appreciable GSH depletion in vivo in mouse liver (*41,42*), or in vitro in rat hepatocytes (*43*). NDMA toxicity also is not preceded by depletion of protein sulfhydryls (*42*). The alkylating intermediate produced by metabolism of NDMA is the methyl diazonium ion (*13*), a hard electrophile resulting in the preferential alkylation of hard nucleophilic sites as opposed to GSH, a soft site (*44*). Formation, via NDM metabolism, of sufficient reactive oxygen to deplete GSH also can be ruled out. Furthermore, use of agents such as buthionine sulfoximine to deplete GSH in hepatocytes has no effect on protein (*45*) or DNA (*46,47*) binding by NDMA. Our work also suggests that GSH depletion in vivo in the rat by buthionine sulfoximine has no effect on NDMA-induced hepatotoxicity (Lee and Archer, unpublished observations). One should note that Frei et al (*48*), using diethyl maleate to deplete hepatic GSH, reported increased methylation of hepatic DNA by NDMA. This finding, however, could be due to an effect on the activity of the drug metabolizing enzymes by diethyl maleate (*49*).

Lipid Peroxidation. Lipid peroxidation is perhaps the clearest manifestation of oxidative cell injury. There are several reports of nitrosamine-induced lipid peroxidation. Reynolds (*50*) first reported the induction of lipid peroxidation in vivo two hours after rats were given 26 mg/kg NDMA. In 1973, Jose and Slater (*51*) observed in a reaction that was dependent on nitrosamine metabolism, lipid peroxidation in hepatocytes cultured in the presence of NDMA. More recently, Ahotupa et al (*52*) observed lipid peroxidation in rats that received a single dose of NDMA. Lipid peroxidation in the liver, measured by 4 methods, increased rapidly, peaking 20 min - 1h after an NDMA dose of 133 µmol/kg (10 mg/kg). Two of the measurements of lipid peroxidation (diene conjugation and lucigenin-dependent chemiluminescence) fell to control levels within 24h, while the other two measurements (fluorescent products and thiobarbituric acid-reactive products) were still above control levels at this time. Interestingly, no hepatic lipid peroxidation was observed in rats given 133 μmol/kg (20 mg/kg) NMBzA. Although it seems clear that NDMA can produce hepatic lipid peroxidation, no causal relationship with hepatotoxicity has been established. If lipid peroxidation is responsible for NDMA-induced liver cell injury, it is unclear why this process peaks at 20 min - 1h while cell death does not peak until 24-48 h following nitrosamine administration. Furthermore, as mentioned previously, GSH depletion does not potentiate NDMA-induced hepatotoxicity, and NDMA does not deplete hepatic GSH. These results are not consistent with lipid peroxidation induced by oxidative cell injury. Some support for peroxidation of membrane lipids as a cause of cell death is provided by the report of Dashman and Kamm (53) that relatively high doses of vitamin E (55 mg/kg/day) given to rats for 3 days resulted in a significant decrease in the acute hepatotoxicity of NDMA. However, as they also observed a decrease of NDMA-demethylase activity in the vitamin E treated rats, one cannot draw a firm conclusion regarding mechanism.

  Calcium Transport. NDMA-mediated hepatotoxicity is not preceded by an inhibition of calcium transport mechanisms (42). This was unexpected in light of proposals that disruption of calcium homeostasis may represent a common mechanism through which a number of hepatotoxicants, acting through reactive metabolites, produce toxicity (54,55). However, since alkylation or oxidation of protein thiol groups critical to calcium transport is believed to be the mechanism which disrupts calcium homeostasis (54,56,57), that NDMA does not produce this effect is perhaps not surprising.

Other Mechanisms

In addition to covalent binding and oxidative stress, there are a number of other possible mechanisms by which NDMA may cause hepatotoxicity.

Protein Synthesis Inhibition. Inhibition of hepatic protein synthesis is one of the most characteristic biochemical effects of NDMA (58). A maximum inhibition of about 70% occurs at around 5 h. It seems likely that methylation of mRNA results in its less efficient utilization on polyribosomes (59). While inhibition of protein synthesis is a significant biochemical event, it seems unlikely that it induces necrosis. For example, cycloheximide and puromycin are potent inhibitors of hepatic protein synthesis but neither induces necrosis (60,61). Furthermore, although cycloheximide treatment protects against the hepatotoxic effects of NDEA (62), we were unable to observe any protection against NDMA hepatoxicity by a similar treatment (Giles and Archer, unpublished observations).

Other Metabolites. In addition to reactive metabolic intermediates such as the methyldiazonium ion and free radical species, four other metabolites possibly could be responsible for the hepatotoxicity of NDMA: formaldehyde, methanol, methylamine or nitric oxide.

  Formaldehyde and Methanol. The direct fragmentation of α-hydroxy-NDMA during the hepatic metabolism of NDMA produces formaldehyde in high yield while methanol is formed by reaction of the methyl diazonium ion with water (10). However, the metabolism and toxicity of formaldehyde in isolated rat hepatocytes are highly dependent on intracellular glutathione concentrations (63). In hepatocytes depleted of GSH by treatment with diethyl maleate, the rate of disappearance of formaldehyde significantly decreased. Formaldehyde toxicity was potentiated in cells pretreated with diethylmaleate and formaldehyde itself decreased the GSH concentration, probably by formation of the adduct hydroxymethylglutathione. One can rule out formaldehyde as a significant factor in NDMA hepatotoxicity due to the lack of sensitivity of NDMA toxicity to GSH depletion and to the lack of significant GSH depletion by formaldehyde formed during metabolism of NDMA. Methanol is rapidly metabolized to carbon dioxide in rat liver via formation of formaldehyde and formate (64) and it is unlikely to be involved in the development of hepatotoxicity. Furthermore, NMBzA is metabolized in the liver to formaldehyde and methanol, albeit at a significantly lower rate than NDMA (8).

Methylamine and Nitric oxide. Recent evidence suggests that methylamine and NO are formed during the normal oxidative metabolism of NDMA (65). Enzymatic denitrosation, normally thought of as a detoxification pathway, may account for as much as 20-30% of an administered dose of NDMA (66,67). Methylamine occurs naturally in man and animals and is rapidly metabolized (68,69). There is no evidence that it is hepatotoxic. Formation of NO, however, may lead to cellular injury. NO is cytotoxic due to its ability to bind to heme and non-heme iron (70). Further, NO is readily converted into nitrogen dioxide and nitrite by intracellular oxidizing agents. Nitrogen dioxide, a free radical, is toxic and causes peroxidative cleavage of membrane lipids (71). Two findings, however, suggest that NO is not involved in NDMA toxicity. Lorr et al (72) report that the rate of nitrite production is 6-7 times faster when NMBzA rather than NDMA is incubated with hepatic microsomes. This result, however, must be confirmed in vivo. Second, if NO or its oxidation products are the toxic species, depletion of GSH might be expected significantly to potentiate toxicity.

Vascular Damage. As recently pointed out by Reitman et al (42), the target of NDMA in the liver might not be the hepatocyte but, rather, the vasculature. Indeed, the hemorrhagic component of the necrosis induced by NDMA is unusual and is not observed for most other hepatotoxicants. Vascular damage would lead to nutrient and oxygen deprivation in hepatocytes. In 1956, Stoner (73) addressed this issue by measuring blood flow following NDMA adminstration. He concluded, however, that necrosis was not caused by damage to the vasculature though this conclusion possibly should be more thoroughly investigated.

Conclusion

In spite of almost 40 years of study, the mechanism by which NDMA induces hepatotoxicity remains an enigma. Its action is unlike any of the other classical hepatotoxicants. Covalent binding, oxidative stress, and a variety of other mechanisms cannot account for the action of NDMA in the liver. Since hepatotoxicity with its concomitant compensatory cell proliferation is an integral component of hepatocarcinogenesis, it is important that we understand the mechanism. The solution to this problem will provide new insights into toxicology and carcinogenesis.

Acknowledgements

These investigations were supported by Grant MT-7025 from the Medical Research Council of Canada and by the Ontario Cancer Treatment and Research Foundation.

Literature Cited

1.   Freund, H.A. *Ann. Intern. Med.* **1937**, *10*, 4-1155.

2.   Barnes, J.M., Magee, P.N. *Brit. J. Industr. Med.* **1954**, *11*, 167-174.
3.   Heath, D.F. *Biochem. J.* **1962**, *85*, 72-91.
4.   McLean, A.E.M.; Verschuuren, H.G. *Br. J. Exp. Pathol.* **1969**, *50*, 22-25.
5.   Stripp, B.; Sipes, I.G.; Maling, M.; Gillette, J.R. *Drug Metab. Dispos.* **1974**, *2*, 464-468.
6.   Godoy, H.M.; Diaz Gomez, M.I.; Castro, O.A. *J. Natl. Cancer Inst.* **1980**, *64*, 533-538.
7.   Diaz Gomez, M.I.; Godoy, H.M.; Castro, J.A. *Arch. Toxicol.* **1981**, *47*, 159-168.
8.   Labuc, G.E.; Archer, M.C. *Cancer Res.* **1982**, *42*, 3181-3186.
9.   Zucker, P.F.; Giles, A.; Chaulk, J.E.; Archer, M.C. *Carcinogenesis* **1991**, *12*, 405-408.
10. Archer, M.C.; Labuc, G.E. In *Bioactivation of Foreign Compounds*; Anders, M.W., Ed; Academic Press: Orlando, FL, **1985**; pp. 403-431.
11. Druckrey, H.; Preussmann, R.; Ivankovic, S.; Schmahl, D. *Z. Krebsforsch.* **1967**, *69*, 103-201.
12. Stinson, S.F.; Squire, R.A.; Sporn, M.B. *J. Natl. Cancer Inst.* **1978**, *61*, 1471-1475.
13. Milligan, J.P.; Catz-Biro, L.; Hirani-Hojatti, S.; Archer, M.C. *Chem.-Biol. Interactions* **1989**, *72*, 175-189.
14. Mehta, R.; Labuc, G.E.; Urbanski, S.J.; Archer, M.C. *Cancer Res.* **1984**, *44*, 4017-1022.
15. Swann, P.F.; Magee, P.N. *Biochem. J.* **1968**, *110*, 39-47.
16. Leaver, D.D.; Swann, P.F.; Magee, P.N. *Brit. J. Cancer*, **1969**, *23*, 177-187.
17. Craddock, V.M.; Frei, J.V. *Brit. J. Cancer* **1974**, *30*, 503-511.
18. Rees, K.R.; Shotlander, V.L.; Sinha, K.P. *J. Pathol. Bacteriol.* **1962**, *83*, 483-490.
19. Magee, P.N.; Barnes, J.M. *Adv. Cancer Res.* **1967**, *10*, 163-246.
20. D'Acosta, N.; Castro, J.A.; De Castro, C.R.; Diaz-Gomez, M.; De Ferreya, E.C.; De Fenos, O.M. *Toxicol. Appl. Pharmacol.* **1975**, *32*, 474-481.
21. Cardesa, A.; Mirvish, S.S.; Haven, G.T.; Shubik, P. *Proc. Soc. Exp. Biol. Med.* **1975**, *145*, 124-128.
22. Ying, T.S.; Sarma, D.S.R.; Farber, E. *Am. J. Pathol.* **1980**, *99*, 159-173.
23. Magee, P.N. *Biochem. J.* **1956**, *64*, 676-682.
24. Wishnok, J.S.; Rogers, A.E.; Sanchez, O.; Archer, M.C. *Toxicol. Appl. Pharmacol.* **1978**, *43*, 391-396.
25. Catz-Biro, L.; Chin, W.; Archer, M.C.; Pollanen, M.S.; Hayes, M.A. *Toxicol. Appl. Pharmacol.* **1990**, *102*, 191-194.
26. Ingelman-Sundberg, M.; Johansson, I.; Penttila, K.E.; Glaumann, H.; Lindros, K.O. *Biochem. Biophys. Res. Commun.* **1988**, *157*, 55-60.
27. Dicker, R.; McHugh, T.; Cederbaum, A.I. *Biochem. Biophys. Acta.* **1991**, *1073*, 316-323.
28. Magee, P.N.; Hultin, T. *Biochem. J.* **1962**, *83*, 106-114.
29. Yarosh, D.B. *Mutat. Res.* **1985**, *145*, 1-16.
30. Karran, P.; Lindahl, T. *Cancer Surveys* **1985**, *4*, 583-599.
31. Pegg, A.E. *Cancer Res.* **1990**, *50*, 6119-6129.
32. Pegg, A.E. *J. Natl. Cancer Inst.* **1977**, *58*, 681-687.
33. Silinskas, K.C.; Zucker, P.F.; Archer, M.C. *Carcinogenesis* **1985**, *6*, 773-775.
34. Dai, W-D.; Lee, V.; Chin, W.; Cooper, D.P.; Archer, M.C.; O'Connor, P.J. *Carcinogenesis* **1991**, *12*, 1325-1329.
35. Kyle, M.E.; Farber, J.L. *Handbook of Toxicologic Pathology;* Academic Press, Inc.: New York, NY, **1991**; pp 71-89.
36. Saprin, A.N.; Piette, L.H. *Arch. Biochem. Biophys.* **1977**, *180*, 480-493.

37. Floyd, R.A.; Soong, L.M.; Stuart, M.A.; Reigh, D.L. *Photochem. Photobiol.* **1978**, *28*, 857-862.
38. Duran, N.; Faljoni, A. *Biochem. Biophys. Res. Comm.* **1978**, *83*, 287-294.
39. White, R.E.; Coon, M.J. *Ann. Rev. Biochem.* **1980**, *49*, 315-356.
40. Malaveille, C.; Brun, G.; Park, S.S.; Gelboin, H.V.; and Bartsch, H. *Carcinogenesis* **1987**, *8*, 1775-1779.
41. Shertzer, H.G.; Tabor, M.W.; and Berger, M.L. *Expl. Molec. Path.* **1987**, *47*, 211-218.
42. Reitman, F.A.; Berger, M.L.; Minnema, D.J.; Shertzer, H.G. *J. Toxicl. Environ. Health* **1988**, *23*, 321-331.
43. Reitman, F.A.; Shertzer, H-G.; Berger, M.L. *Biochem. Pharmacol.* **1988**, *37*, 3183-3188.
44. Guengerich, F.P.; Liebler, D.C. *CRC Crit. Rev. Toxicol.* **1985**, *14*, 259-307.
45. Gorsky, L.D.; Hollenberg, P.F. *Chem. Res. Toxicol.* **1989**, *2*, 442-448.
46. Tacchi, A.M.; Jensen, D.E.; and Magee, P.N. *Mol. Toxicol.* **1987**, *36*, 881-885.
47. Prasanna, H.R.; Raj, H.G.; Lotlikar, P.D.; Magee, P.N. *Mol. Toxicol.* **1987**, *1*, 167-176.
48. Frei, E.; Bertram, B.; Wiessler, M. *Chem.-Biol. Interactions*, **1985**, *55*, 123-137.
49. Anders, M.W. *Biochem. Pharmacol.* **1978**, *27*, 1098-1101.
50. Reynolds, E.S. *Biochem. Pharmacol.* **1972**, *21*, 2555-2561.
51. Jose, P.J.; Slater, T.F. *Xenobiotica*, **1973**, *3*, 357-366.
52. Ahotupa, M.; Bussacchini-Griot, V.; Béréziat, J.-C.; Camuo, A.-M.; Bartsch, H. *Biochem. Biophys. Res. Comm.* **1987**, *146*, 1047-1054.
53. Dashman, T.; Kamm, J.J. *Biochem. Pharmacol.* **1979**, *28*, 1485-1490.
54. Bellomo, G.; Orrenius, S. *Hepatology* **1985**, *5*, 876-882.
55. Tsokos-Kuhn, J.O.; Todd, E.L.; McMillin-Wood, J.B.; Mitchell, J.R. *Mol. Pharmacol.* **1985**, *28*, 56-61.
56. Bellomo, G.; Mirabelli, F.; Richelmi, P.; Orrenius, S. *FEBS Letts.* **1983**, *163*, 136-139.
57. Thor, H.; Hartzell, P.; Svenson, S.; Orrenius, S.; Mirabelli, F.; Marimoni, V.; Bellomo, G. *Biochem. Pharmacol.* **1985**, *34*, 3717-3723.
58. Magee, P.N.; Vandekar, I. *Biochem. J.* **1958**, *70*, 600-605.
59. Nygard, O.; Hultin, T. *Biochem. J.* **1981**, *194*, 469-474.
60. Farber, E. *Ann. Rev. Pharmacol.* **1971**, *11*, 71-96.
61. Robinson, D.S.; Seakins, A. *Biochim. Biophys. Acta* **1962**, *62*, 163-165.
62. Flaks, B.; Nicoll, J.N. *Chem.-Biol. Interactions* **1974**, *8*, 135-150.
63. Ku, R.H.; Billings, R.E. *Chem.-Biol. Interaction* **1984**, *51*, 25-36.
64. Billings, R.E.; Tephly, J.R. *Biochem. Pharmacol.* **1979**, *28*, 2985-2991.
65. Keefer, L.K.; Anjo, T.; Wade, D.; Wang, T.; Yang, C.S. *Cancer Res.* **1987**, *47*, 447-452.
66. Streeter, A.J.; Nims, R.W.; Sheffels, P.R.; Heur, Y.-H.; Yang, C.S.; Mico, B.A.; Gombar, C.T.; Keefer, L.K. *Cancer Res.* **1990**, *50*, 1144-1150.
67. Burak, E.S.; Harrington, G.W.; Koseniauskas, R.; Gombar, C.T. *Cancer Letts.* **1991**, *58*, 1-6.
68. Asatoor, A.M.; Simenhoff, M.L. *Biochim. Biophys. Acta* **1965**, *111*, 384-392.
69. Zeisel, S.H.; Wishnok, J.S.; Blusztajn, J.K. *J. Pharmacol. Exp. Therapeut.* **1983**, *225*, 320-324.
70. Traylor, T.G.; Sharma, V.S. *Biochemistry* **1992**, *31*, 2847-2849.
71. Patel, J.M.; Block, E.R. *Am. Rev. Respir. Dis.* **1986**, *134*, 1196-1202.
72. Lorr, N.A.; Tu, Y.Y.; Yang, C.S. *Carcinogenesis* **1982**, *3*, 1039-1043.
73. Stoner, H.B. *Br. J. Exp. Pathol.* **1956**, *37*, 176-198.

RECEIVED June 29, 1993

Chapter 23

Genetic Tumor Epidemiology

Identifying Causative Carcinogenic Agents and Their Transforming Mutations

B. I. Ludeke, H. Ohgaki, and P. Kleihues

Department of Pathology, University Hospital, CH–8091, Zurich, Switzerland

Genotoxic carcinogens produce characteristic spectra of covalent DNA damage, which, upon DNA replication, induce distinctive mutations that can be identified in transformation-associated genes (oncogenes and tumor suppressor genes). Analyses of experimentally induced tumors have demonstrated that transforming mutations typically result from only one or several of the promutagenic adducts generated by a given carcinogen. Comparison of mutations observed in human tumors with mutations induced by suspected carcinogens in laboratory rodents has enabled causative carcinogens to be prediced in a variety of human malignancies. Genetic tumor epidemiology promises to be an increasingly useful tool in identifying pathobiologically significant carcinogens in complex mixtures typical for human exposure and in assessing the relative contributions of environmental carcinogens to the human tumor burden.

A multitude of genotoxic agents, including many known and suspected human carcinogens, have been characterized in great detail with respect to metabolism, reactions with DNA, occurrence and human exposure. Yet a correlation between specific types of DNA damage and tumor initiation has until recently relied on indirect evidence such as the initial abundance, persistence *in vivo* and miscoding properties *in vitro* of individual adducts. The identification of genetic changes in transformation-associated genes has now made it possible to investigate directly the contributions of specific DNA lesions to neoplastic processes. Many genotoxic agents produce characteristic spectra of physical DNA damage. These in turn generate typical mutation patterns which, provided they confer a growth advantage and are stably passed on to progeny cells, can be detected in tumors induced by these agents. This contribution reviews several examples which illustrate the potential of mutational profile anaysis in identifying chemical and physical carcinogens and discusses the caveats associated with the interpretation of mutational profiles.

0097–6156/94/0553–0290$08.00/0

ras Proto-Oncogenes and p53 and Rb Tumor Suppressor Genes as Diagnostic Markers for Carcinogen-Induced Mutations

The highly conserved *ras* proto-oncogenes code for membrane-bound proteins with intrinsic GTPase activity that mediate the transfer of signals from extracellular growth factors to intracellular effectors. In tumors, *ras* mutations are observed at codons 12 or 13, both of which code for Gly (GGX with X denoting any base), or at codon 61, which codes for Gln (CAG/A). The transforming *ras* mutations observed in tumors all produce constitutively activated *ras* proteins. Amino acid changes at residue 61 remove the catalytic Gln residue which is required for GTP hydrolysis and abrogation of *ras* function (reviewed in *1*). Replacing Gly12 with any amino acid other than Pro presumably induces conformational changes that prevent Gln61 from taking up the postulated catalytically acitve position (see *2*). The transforming efficiency of the mutant proteins, however, varies with substitution (*3*). Amino acid substitutions other than Ser at the adjacent Gly13 may similarly lead to structural distortions that interfere with GTP hydrolysis or GDP/GTP exchange (reviewed in *1*). The limited number of targets for mutational activation of *ras* genes enables rapid screening for *ras* mutations. These are restricted, however, to base substitutions within a small subset of sequence contexts which may not include preferred target sequences for some classes of carcinogens. Mutated *ras* genes have been detected in many types of human and rodent tumors (*4,5*).

Tumor suppressor genes play a central role in the control of cell proliferation and differentiation. There is evidence that the product of the p53 gene both activates the transcription of genes that suppress cell proliferation and regulates the assembly of DNA replication-initiation complexes (see *6*). It has been suggested that the acccumulation of p53 protein observed after many types of severe genetic damage either leads to apoptosis or serves to introduce a pause in the cell cycle during which DNA lesions are repaired (*7*). The retinoblastoma susceptibility gene (Rb) product appears to suppress cell proliferation at the level of expression of proliferation-associated genes by sequestering the required cellular transcription factors in inactive, reversible protein complexes (*8*). Oncogenic mutations in the p53 and Rb tumor suppressor genes generally result in a truncated or otherwise functionally aberrant or inactive protein. In contrast to mutational activation of *ras* genes, point mutational inactivation of tumor suppressor genes occurs at many sites and through a broad spectrum of physical changes including base substitutions, insertions and deletions (*9-11*). The p53 gene is one of the most frequently altered genes in human tumors (*9*).

Endogenous 5-Methylcytosine Induces a High Frequency of GC to AT Transitions at CpG Dinucleotides.

Retinoblastoma, a rare childhood eye malignancy, is caused by inactivation or loss of both alleles of the Rb gene (*12*). The disease occurs in a sporadic form, resulting from two independent somatic mutational events, and in a hereditary form, caused by the somatic inactivation of the wildtype allele in heterozygous individuals carrying a predisposing germline mutation. There are no known environmental risk factors for this malignancy but the incidence of hereditary retinoblastoma does show a weak correlation with paternal age (*13*). In retinoblastoma patients, CG to TA transitions at

Cpg dinucleotides are the predominant mutation among both somatic and germline point mutations (10,14). CpG pairs are also a common mutational hot spot in the p53 gene in a wide variety of neoplasms (9-11). The prevalence of transitions at CpG dinucleotides has been attributed to spontaneous hydrolytic (15) and nitric oxide-mediated (16) deamination of 5-methylcytosine to thymidine, and strongly suggests that 5-methylcytosine is the promutagenic lesion in a substantial fraction of "spontaneous" human cancers.

Physical and Chemical Carcinogens with Highly Characteristic Mutational Spectra

The mutational spectra of a large number of physical and chemical carcinogens have been characterized through the analysis of transformation-associated genes in experimentally induced animal tumors and in human tumors of defined etiology. Many genotoxic agents, including known and suspected human carcinogens, were found to induce highly characteristic mutational profiles in vivo which can be clearly differentiated from the pattern of "spontaneous" mutations.

UV Radiation. A high frequency of GC to AT transitions and GC to TA transversions at dipyrimidine sites, uncommon in internal cancers (4,9), and unique CpC to TpT double-base substitutions have been described in human squamous amd basal cell carcinomas on sun-exposed body sites (17,18). Among the DNA adducts produced by UV irradiation are highly stable dipyrimidine adducts created by covalent linkage of adjacent pyrimidines (see 19). Dipyrimidine photoadducts have been detected on the transcribed strand of ras genes at codon 12 in UV-irradiated cells (20), which correlate with mutations found in squamous and basal cell carcinomas (18). The predominance of base substitutions at dipyrimidine pairs in UV-associated tumors thus strongly suggests that the oncogenic mutations in UV-induced tumors are the result of pyrimidine photodimers. In E. coli, adenine is preferentially incorporated opposite dipyrimidine sites during replication (21), which leads to C to T transitions at modified cytosine residues but restores thymidine residues. The major repair pathway for photoadducts in eukaryotes involves the release of a short single-stranded fragment containing the lesion with the ensuing gap being filled by semi-conservative replication (22). UV-specific mutations may, however, be induced in eukaryotes by a bypass mechanism similar to that of prokaryotes since UV-induced mutagenesis is more pronounced in excision-deficient cells (23)

γ Irradiation. GC to AT transitions at the second position of codon 12 of the K-ras gene predominated in murine thymic lymphomas induced by γ irradiation (24). This mutation was also found in an irradiated human brain tumor (25). Since K-ras mutations have not been described in other human brain neoplasms (see 25), it is highly likely that this mutation was caused by therapeutic irradiation. In E. coli, γ irradiation also preferentially induces transitions (26). One of the major radiation-induced DNA adducts, 8-hydroxydeoxyguanosine (27), preferentially directs misincorporation of adenine (28), suggesting that other irradiation-induced lesions could have led to the observed point mutations. Possible mechanisms for the induction of GC to TA transitions include mispairing of irradiation-induced cytosine derivatives with thymidine and incorporation of adenine opposite radiation-induced apurinic sites (reviewed in *29*).

Simple Alkylating Compounds. As summarized in Table I, GC to AT transitions are the predominant or exclusive mutation in activated *ras* genes in diverse rodent tumors induced by simple alkylating *N*-nitroso compounds, including rat mammary tumors induced by methylnitrosourea (*30*), rat renal tumors induced by various methylating and ethylating agents (*31,32*), and mouse lung tumors induced by *N*-nitrosodimethylamine and 4-(methylnitrosamine)-1-(3-pyridyl)-1-butanone (NNK) (*33*). O^6-Alkyldeoxyguanosines, the major promutagenic DNA adducts of alkylating compounds (*34*), typically lead to G to A substitutions through miscoding with thymidine during replication (*35*). Transitions are also generated by 8-hydroxydeoxyguanosine (*36*) which is produced by oxygen free radicals generated during normal cellular oxidative processes and by various exogenous genotoxic agents (*37*), and by endogenous or nitric oxide-mediated deamination of cytosine (*38*) and 5-methylcytosine (*15,16*). It is not surprising, therefore, that GC to AT transitions are also frequently detected in "spontaneous" tumors (*9,10,32*). However, O^6-alkyldeoxyguanosines are generated preferentially at the second guanine residue of GpG dinucleotides (*39*) and are repaired less efficiently when preceded by a purine rather than by a pyrimidine (*40*). The second guanine residue in codon 12 of *ras* genes is a mutational hot spot for transitions in experimental tumors induced by simple alkylating compounds (Table I), implicating O^6-alkyldeoxyguanine adducts as a likely cause for the mutations since this sequence specificity is not observed for the formation of 8-hydroxyguanine or for deamination reactions (*15,36,38*).

Table I. Ras Mutations Induced by Simple Alkylating Compounds

Carcinogen	Substitution	Codon	Tumor	Species	Ref.
Methylnitrosourea	GGA→GAA	H-*ras* 12	mammary gland	rat	(*30*)
N-Nitrosodimethylamine	GGT→GAT	K-*ras* 12	kidney	rat	(*32*)
N-Nitrosodimethylamine	GGT→GAT	K-*ras* 12	lung	mouse	(*33*)
N-nitrosomethyl-(methoxymethyl)amine	GGT→GAT	K-*ras* 12	kidney	rat	(*31*)
NNK	GGT→GAT	K-*ras* 12	lung	mouse	(*33*)
Ethylnitrosourea	GGT→GAT	K-*ras* 12	kidney	rat	(*32*)
N-nitrosodiethylamine	CAA→AAA	H-*ras* 61	liver	mouse	(*41*)
	CAA→CTA	H-*ras* 61	liver	mouse	(*41*)
	CAA→CGA	H-*ras* 61	liver	mouse	(*41*)

AT to TA transversions and AT to GC transitions at the second base of codon 61 of H-*ras* were detected at high frequency in murine liver tumors induced by *N*-nitrosodiethylamine *(41)* (Table I), presumably through miscoding by O^4-ethyl-deoxythymidine *(42)*. O^4-Ethyldeoxythymidine is highly persistent in rat liver *(43)*, in contrast to O^6-ethyldeoxyguanosine which is rapidly repaired in this tissue *(44)*. O^4-Ethyldeoxythymidine is also very likely the promutagenic adduct leading to the activation of the *neu* gene by a TA to AT transversion in codon 664 which is consistently observed in schwannomas induced by a single transplacental application of nitrosoethylurea *(45)*.

Polycyclic Aromatic Hydrocarbons. A preponderance of GC to TA, AT to TA and GC to CG transversions is consistently observed in rodent tumors induced by benzo(a)pyrene, 3-methylcholanthrene and 7,12-dimethylbenz(a)anthracene *(46-50)*. Highly persistent bulky N^2-deoxyguanosine adducts are the major DNA lesions produced by both benzo(a)pyrene and 7,12-dimethylbenz(a)anthracene *(34)*. Although 3-methylcholanthrene DNA adducts have not yet been characterized, the similarity in mutational profiles of these three polycyclic aromatic hydrocarbons suggests that they produce similar promutagenic lesions. Bulky purine adducts are lost from DNA through spontaneous depurination or through the action of glycosylases, resulting in the formation of apurinic sites *(51)*. In *E. coli*, such non-informational sites preferentially direct the incorporation of adenine on the opposite strand during replication *(52)*. The high frequency of GC to AT transversions induced by benzo(a)pyrene, 3-methylcholanthrene and 7,12-dimethylbenz(a)anthracene suggests that eukaryotes have a similar, highly efficient bypass mechanism (see *51*). However, after transfection into COS7 cells of a shuttle vector containing a single apurinic site, misincorporation of guanine was twice as frequent as incorporation of adenine or cytosine, and five times as frequent as incorporation of thymidine *(53)*. Bulky DNA adducts may thus generate a wide array of single base substitutions in eukaryotes.

Use of Genetic Epidemiology for Identifying Causative Carcinogens in Human Tumors

Knowledge of the carcinogens involved in the etiology of human cancers is key to identifying sources of risk and for developing cancer prevention strategies. Genetic tumor epidemiology offers a new methodology for determining candidate carcinogens among the many environmental and lifestyle risk factors to which humans are continually exposed. As the following examples illustrate, mutational patterns support the causative role in some cancers of agents postulated earlier on the basis of human exposure and on the basis of metabolite and DNA adduct levels among populations at risk, and has implicated additional or alternate carcinogens in others.

Hepatocellular Carcinoma. In most human tumors, p53 mutations other than C to T transitions at CpG dinucleotides occur at many different sites dispersed throughout the evolutionarily conserved region spanning exons 5 through 8. A notable exception is hepatocellular carcinoma in the high incidence areas of China *(54)* and southern Africa *(55)*. These tumors show a striking cluster of GC to TA transversions at codon 249 not seen in liver cancers collected elsewhere *(56)*. Aflatoxin B$_1$, a potent liver carcinogen in laboratory animals (*57*), has been proposed to be the initiating carcinogen for these tumors since it is a major food contaminant in both high-risk regions (*58,59*). The mutational hot-spot at codon 249 is efficiently modified by aflatoxin B_1 *in vitro* (*60*). The major DNA lesion formed by aflatoxin B_1 is a bulky 7-alkylguanine adduct (*34*) which undergoes spontaneous depurination and enzymatic excision repair (*19*), and induces predominantly GC to TA transversions in rat liver tumors (*61*), presumably via incorporation of adenine opposite the ensuing abasic sites.

Tobacco-Related Lung Cancers. The major risk factor for human lung cancers is tobacco smoke (*62*). However, as summarized in Table II, different types of tobacco-related lung tumors exhibit different mutational profiles, suggesting that the carcinogenic components of tobacco smoke vary in the induction of lung tumors.

Non-Small Cell Lung Cancer. In non-small cell lung cancers, GC to TA transversions on the coding strand are the most common mutation (*63,64*). This suggests that the responsible promutagenic DNA lesions in tobacco-related small-cell lung cancers are predominantly bulky guanine adducts of polycyclic aromatic hydrocarbons found in tobacco smoke. A likely candidate is benzo(a)pyrene, which induced a high incidence of GC to TA transversions in murine pulmonary tumors (*46*). All of the GC to TA transversions which have been detected in the p53 gene in human lung tumors occur at target sites for DNA modification by benzo(a)pyrene (*60*), suggesting that they were induced by apurininc sites generated from bulky guanine adducts of benzo(a)pyrene or related polycyclic aromatic hydrocarbons.

Small Cell Lung Cancer. In contrast to non-small cell lung cancers, GC to AT transitions are frequent in small cell lung cancers (*9*), suggesting that different genotoxic agents are responsible for the two types of malignancies. Cigarettes contain nanogram amounts of simple alkylating nitrosamines, including *N*-nitrosodimethylamine and NNK (*65*). In rodents, the induction of lung tumors by NNK correlates with formation of O^6-methyldeoxyguanosine and persistence of this adduct (*66*). The

Table II. Predominant p53 Point Mutations in Human Lung Cancers

Risk factor	Tumor type	Substitution	
Cigarette smoking	Non-small cell lung cancer	G→T	(39%)[a]
		G→A	(17%)
		C→T	(17%)
	Small cell lung cancer	G→A	(46%)
Radon exposure	All lung cancers	C→A	(38%)

[a]Numbers in parentheses indicate percentage of mutations.
Data taken from Refs. 9,64 and 67.

exclusive occurrence of GC to AT transitions in murine pulmonary tumors after exposure to NNK (*33*) strongly suggests that O^6-methyldeoxyguanosine may also play a role in the induction of some tobacco-associated lung tumors in humans.

Lung Tumors after Exposure to Radon. In uranium miners exposed to high concentrations of radon, p53 point mutations on the coding strand included a large fraction of CG to AT transversions (Table II) but no GC to TA transversions, regardless of smoking habits (*67*). The unexpected mutational profiles in these tumors suggest that radon-induced genetic damage contributed significantly to oncogenesis.

Esophageal Cancer. Esophageal cancers display a wide spectrum of single base changes in the p53 gene (*9,68*). In addition to GC to TA transversions, these tumors also exhibit a high proportion of AT to TA transversions which are rare among p53 mutations in other tobacco related cancers (*63,64*). Epidemiological studies have revealed a strong synergism between smoking and alcohol consumption for esophageal malignancies in industrialized countries (*69*). The unusual pattern of point mutations could, therefore, reflect contributions from chemicals in alcoholic beverages. One common, albeit trace, impurity of alcoholic beverages known to selectively induce AT to TA transversions is ethyl carbamate (urethane) (*70,71*). Surprisingly, only two of five p53 mutations in esophageal tumors from Chinese patients were GC to AT transitions (*68*), even though O^6-methyldeoxyguanosine has been detected in human esophageal mucosa from high risk areas in China (*72*).

Caveats for the interpretation of mutational profiles.

One of the main problems associated with the interpretation of mutational profiles in tumors of uncertain etiology is posed by the overlapping mutational spectra of many genotoxic agents. Preferential induction of GC to TA transversions is observed with diverse compounds that form bulky guanine adducts, including a wide array of polycyclic aromatic hydrocarbons and aflatoxin B_1. Similarly, GC to AT transitions are induced by simple alkylating compounds, γ irradiation and endogenous oxidative processes. Thus even a seemingly striking mutational profile may not enable an unambiguous correlation with a suspected carcinogen. For example, mutational activation of K-*ras* through a GC to AT transition at the first guanine residue of codon 12 was observed in 40% of pulmonary tumors induced in rats by inhalation of $^{239}PuO_2$ aerosols (*73*). However, this mutation represented one of two K-*ras* mutations observed in spontaneous tumors in control animals (*73*). This suggests that the mutations at the first base of codon 12 may not have been irradiation-induced. Instead, ^{239}Pu-irradiation may have enhanced the biological selection for an existing *ras* mutation in preinitiated cells by providing a genetic or biochemical background, possibly by mutational activation or inactivation of other tumor-associated genes, in which this *ras* mutation causes a significant growth advantage. Fortunately, however, many chemical carcinogens exhibit disparate sequence selectivities for the formation of adducts. In addition, the local sequence also affects the repair and hence persistence of promutagenic adducts. Thus, candidate carcinogens may in many cases be deduced from the sequence context of oncogenic mutations.

Due to the tissue and species specific phenotypic selection of mutations in transformation-associated genes, the oncogenic mutations induced by a specific genotoxic agent or promutagenic DNA adduct can vary considerably between tissues and species. In rats, mutational activation of *ras* genes is common in tumors induced by methylating *N*-nitroso compounds in kidney (*31,32*), mammary gland (*30*), and esophagus (*74,75*), but is rare or absent in tumors induced by methylating agents in liver (*41,76*), lung (*77*) and stomach (H. Ohgaki, unpublished data). Oncogenic *ras* mutations have been implicated in a wide variety of human malignancies, most notably in colon carcinoma, pancreatic and pulmonary adenocarcinoma and myeloid disorders (reviewed in *4*) but are not or only infrequently involved in others, including breast (see *4*), esophageal (*78*) and small cell lung (*63*) cancers. There are similarly pronounced differences between species in the contributions of activated *ras* genes to neoplastic transformation. Although *ras* genes are frequently mutated in esophageal (*74,75*) and mammary (*30*) tumors in rats, they do not appear to be involved in the corresponding human neoplasms (*4,78*). Similarly, activated *ras* genes are observed in NNK-induced pulmonary tumors in mice but not in rats (*77*). In human hepatocellular carcinomas, a mutational hotspot consisting of a GC to TA transversion at codon 249 has been described in patients exposed to high levels of aflatoxin B_1 (*54,55*). However, in liver tumors induced by aflatoxin B_1 in non-human primates, only a low frequency of p53 mutations was observed, with none of the mutations occurring at codon 249 (*79*). This suggests that the GC to TA transversion, while characteristic for aflatoxin B_1 (see above), could also have been caused by other environmental agents. It is also possible, however, that mutations in p53 do not lead to a significant growth advantage in hepatic cells in non-human primates and thus they would not be expected to be selected for in liver tumors. The occurrence of this mutational hot spot in human hepatocellular carcinomas from areas with a high risk for exposure to aflatoxin B_1 but not in tumors from low exposure areas (*56*) strongly suggests that this mutation was indeed induced by aflatoxin B_1.

Neoplastic transformation appears to require the cooperation of a series of genetic events at multiple loci in most cell types (reveiwed in *80*). Although the same genes may be affected in many tumors, the order in which mutated forms appear during tumor progression can vary considerably. Thus, activation of *ras* genes or inactivation of p53 is an early event in some experimental and human cancers (*30,81,82*), whereas these events contribute to tumor progression in others (*76,83*). In some tumors, the accumulation rather than the temporal sequence of genetic changes appears to be critical for tumorigenesis (*83*). The underlying mutations may originate concurrently or at different time points during tumorigenesis. Potentially oncogenic changes may remain phenotypically "dormant" until an environment in which they lead to a selective growth advantage has been created by other genetic events (*67*). In addition, activated oncogenes may be lost with clonal expansion during tumor progression in some tumors (*84*), thereby potentially obscuring the initiating events. Since humans are intermittently exposed to complex mixtures of carcinogens throughout their entire lifespan, it may be difficult to correlate a particular mutational event in a given gene with exposure to a specific genotoxic agent.

Conclusions

Molecular epidemiology of experimental and human tumors has become an increasingly important tool for elucidating the molecular mechanisms of neoplastic transformation by genotoxic carcinogens despite the limitations outlined above. In addition to experimental studies, a number of potential practical applications appear within reach. As mutational patterns of additional carcinogens are identified and mutations in other transformation-associated genes are characterized, it may become feasible to develop strategies for identifying the disease-causing carcinogens and their promutagenic adducts in individual cancer patients. In particular, genetic epidemiology could become an important supplement to physicochemical determinations of exposure to suspected carcinogens and biochemical quantitations of DNA adducts in determining the causes of occupational cancers and of secondary tumors in cancer patients treated with cytostatic drugs or therapeutic radiation. Genetic epidemiology also promises to be of use in assessing the relative contributions to tumorigenesis of individual carcinogens in complex mixtures such as tobacco smoke, pyrolysis products in cooked foods and environmental pollutants, and should lead to a more detailed picture of the relative importance of environmental carcinogens to the overall human tumor burden.

Literature Cited

(1)   Grand, R. J. A.; Owen, D. *Biochem. J.* **1991**, *279*, 209-631.
(2)   Wittinghofer, A.; Pai, E. F. *Trends Biol. Sci.* **1991**, *16*, 382-387.
(3)   Seeburg, P. H.; Colby, W. W.; Capon, D. J.; Goeddel, D. V.; Levinson, A. D. *Nature* **1984**, *312*, 71-77.
(4)   Bos, J. L. *Cancer Res.* **1989**, *49*, 4682-4689.
(5)   Guerrero, I.; Pellicer, A. *Mutation Res.* **1987**, *185*, 293-308.
(6)   Montenarh, M. *Oncogene* **1992**, *7*, 1673-1680.
(7)   Lane, D. P. *Nature* **1992**, *358*, 15-16.
(8)   Chittenden, T.; Livingston, D. M.; Kaelin, W. G., Jr. *Cell* **1991**, *65*, 1073-1082.
(9)   Hollstein, M.; Sidransky, D.; Vogelstein, B.; Harris, C. C. *Science* **1991**, *253*, 49-53.
(10)   Ludeke, B. I.; Beauchamp, R. L.; Yandell, D. W. In *8th International Congress of Human Genetics*; Washington, DC, **1991**; pp 449.
(11)   Toguchida, J.; Yamaguchi, T.; Dayton, S. H.; Beauchamp, R. L.; Herrera, G. E.; Ishizaki, K.; Yamamuro, T.; Meyers, P. A.; Little, J. B.; Sasaki, M. S.; Weichselbaum, R. R.; Yandell, D. W. *New Engl. J. Med.* **1992**, *326*, 1301-1308.
(12)   Knudsen, A. G. *Proc. Natl. Acad. Sci. USA* **1971**, *68*, 820-823.
(13)   Vogel, F. *Z. Menschl. Vererbungs-Konstitutionslehre* **1958**, *34*, 398-399.
(14)   Yandell, D. W.; Campbell, T. A.; Dayton, S. H.; Petersen, R.; Walton, D.; Little, J. B.; McConkie-Rosell, A.; Buckley, E. D.; Dryja, T. P. *N. Engl. J. Med.* **1989**, *321*, 1689-1695.

(15) Coulondre, C.; Miller, J. H.; Farabaugh, P. J.; Gilbert, W. *Nature* **1978**, *274*, 775-780.

(16) Nguyen, T.; Brunson, D.; Crespi, C. L.; Penman, B. W.; Wishnok, J. S.; Tannenbaum, S. R. *Proc. Natl. Acad. Sci. USA* **1992**, *89*, 3030-3034.

(17) Brash, D. E.; Rudolph, J. A.; Simon, J. A.; Lin, A.; McKenna, G. J.; Baden, H. P.; Halperin, A. J.; Ponten, J. *Proc. Natl. Acad. Sci. USA* **1991**, *88*, 10124-10128.

(18) Pierceall, W. E.; Goldberg, L. H.; Tainsky, M. A.; Mukhopadhyay, T.; Ananthaswamy, H. N. *Mol. Carcinogen.* **1991**, *4*, 196-202.

(19) Friedberg, E. C. *DNA Repair*; W.H. Freeman and Co.: New York, 1985, 614 pages.

(20) Tormanen, V. T.; Pfeifer, G. P. *Oncogene* **1992**, *7*, 1729-1736.

(21) Tessman, I.; Liu, S.-K.; Kennedy, M. A. *Proc. Natl. Acad. Sci. USA* **1992**, *89*, 1159-1163.

(22) Weinfeld, M.; Gentner, N. E.; Johnson, L. D.; Paterson, M. C. *Biochemistry* **1986**, *25*, 2656-2664.

(23) McGregor, J. M.; Levison, D. A.; MacDonald, D. M.; Yu, C. C. *Lancet* **1992**, *339*, 1351.

(24) Newcomb, E. W.; Diamond, L. E.; Sloan, S. R.; Corominas, M.; Guerrero, I.; Pellicer, A. *Environ. Health Persp.* **1989**, *81*, 33-37.

(25) Brustle, O.; Ohgaki, H.; Schmitt, H. P.; Walter, G. F.; Ostertag, H.; Kleihues, P. *Cancer* **1992**, *69*, 2385-2392.

(26) Tindall, H. R.; Stein, J.; Hutchinson, F. *Genetics* **1988**, *118*, 551-560

(27) Kasai, H.; Crain, P. F.; Kuchjino, Y.; Nishimura, S.; Ootsuyama, A.; Tanooka, H. *Carcinogenesis* **1986**, *7*, 1849-1851.

(28) Shibutani, S.; Takeshita, M.; Grollman, A. P. *Nature* **1991**, *349*, 431-434.

(29) Breimer, L. H. *Br. J. Cancer* **1988**, *57*, 6-18.

(30) Zarbl, H.; Sukumar, S.; Arthur, A. V.; Martin-Zanca, D.; Barbacid, M. *Nature* **1985**, *315*, 382-385.

(31) Sukumar, S.; Perantoni, A.; Reed, C.; Rice, J. M.; Wenk, M. L. *Mol. Cell. Biol.* **1986**, *6*, 2716-2729.

(32) Ohgaki, H.; Kleihues, P.; Hard, G. C. *Mol. Carcinogen.* **1991**, *4*, 455-459.

(33) Devereux, T. R.; Anderson, M. W.; Belinsky, S. A. *Carcinogenesis* **1991**, *12*, 299-303.

(34) Singer, B.; Grunberger, D. *Molecular Biology of Mutagens and Carcinogens*; Plenum Press: New York, 1983, 347 pages.

(35) Saffhill, R.; Margison, G. P.; O'Connor, P. J. *Biochim. Biophys. Acta* **1985**, *823*, 111-145.

(36) Kamiya, H.; Miura, K.; Ishikawa, H.; Inoue, H.; Nishimura, S.; Ohtsuka, E. *Cancer Res.* **1992**, *52*, 3483-3485.

(37) Kasai, H.; Nishimura, S. *Env. Health Persp.* **1986**, *67*, 111-116.

(38) Wink, D. A.; Kasprzak, K. S.; Maragos, C. M.; Elespuru, R. K.; Misra, M.; Dunams, T. M.; Cebula, T. A.; Koch, W. H.; Andrews, A. W.; Allen, J. S.; Keefer, L. K. *Science* **1991**, *254*, 1001-1003.

(39) Richardson, F. C.; Boucheron, J. A.; Skopek, T. R.; Swenberg, J. A. *J. Biol. Chem.* **1989**, *264*, 838-841.

(40) Dolan, . E.; Oplinger, M.; Pegg, A. E. *Carcinogenesis* **1988**, *9*, 2139-2143.

(41) Stowers, S. J.; Wiseman, R. W.; Ward, J. M.; Miller, E. C.; Miller, J. A.; Anderson, M. W.; Eva, A. *Carcinogenesis* **1988**, *5*, 665-669.

(42) Klein, J. C.; Bleeker, M. J.; Lutgerink, J. T.; van Kijk, W. J.; Brugghe, H. F.; van den Elst, H.; van der Marel, G. A.; van Boom, J. H.; Westra, J. G.; Berns, A. J. M.; Kriek, E. *Nucleic Acids Res.* **1990**, *18*, 4131.

(43) Swenberg, J. A.; Dryoff, M. C.; Bedell, M. A.; Pop, J. A.; Huh, N.; Kirstein, U.; Rajewsky, M. F. *Proc. Natl. Acad. Sci. USA* **1984**, *81*, 1692-1695.

(44) Den Engelse, L.; De Graaf, A.; De Brij, R.-J.; Menkveld, G. J. *Carcinogenesis* **1987**, *8*, 751-757.

(45) Perantoni, A. O.; Rice, J. M.; Reed, C. D.; Watatani, M.; Wenk, M. L. *Proc. Natl. Acad. Sci. USA* **1987**, *84*, 6317-6321.

(46) You, M.; Candrian, U.; Maronpot, R. R.; Stoner, G. D.; Anderson, M. W. *Proc. Natl. Acad. Sci. USA* **1989**, *86*, 3070-3074.

(47) Halevy, O.; Rodel, J.; Peled, A.; Oren, M. *Oncogene* **1991**, *6*, 1593-1600.

(48) Manam, S.; Storer, R. D.; Prahalada, S.; Leander, K. R.; Kraynak, A. R.; Hammermeister, C. L.; Joslyn, D. J.; Ledwith, B. J.; van Zwieten, M. J.; Bradley, M. O.; Nichols, W. W. *Mol. Carcinogen.* **1992**, *6*, 68-75

(49) Andrews, D. F.; Collins, S. J.; Reddy, A. L. *Cancer Lett.* **1990**, *54*, 139-145.

(50) Carbone, G.; Borrello, M. G.; Molla, A.; Rizzetti, M. G.; Pierotti, M. A.; Della Porta, G.; Parmiani, G. *Int. J. Cancer* **1991**, *47*, 619-625.

(51) Loeb, L. A.; Preston, B. D. *Ann. Rev. Genet.* **1986**, *20*, 201-230.

(52) Schaaper, R. M.; Kunkel, T. A.; Loeb, L. A. *Proc. Natl. Acad. Sci. USA* **1983**, *80*, 487-491.

(53) Klinedinst, D. K.; Drinkwater, N. R. *Mol. Carcinogen.* **1992**, *6*, 32-42.

(54) Hsu, I. C.; Metcalf, R. A.; Sun, T.; Welsh, J. A.; Wang, N. A.; Harris, C. C. *Nature* **1991**, *350*, 427-428.

(55) Bressac, B.; Kew, M.; Wands, J.; Ozturk, M. *Nature* **1991**, *350*, 429-431.

(56) Ozturk, M. and collaborators *Lancet* **1991**, *338*, 1356-1359.

(57) Newberne, P. M.; Wogan, G. N. *Cancer Res.* **1968**, *28*, 770-781.

(58) Yeh, F. S.; Yu, M. C.; Mo, C. C.; Luo, S.; Tong, M. J.; Henderson, B. E. *Cancer Res.* **1989**, *49*, 2506-2509.

(59) van Rensburg, S. J.; Cook-Mozaffari, P.; van Schalkwyk, D. J.; van der Watt, J. J.; Vincent, T. J.; Purchase, I. F. *Br. J. Cancer* **1985**, *51*, 713-726.

(60) Puisieux, A.; Lim, S.; Groopman, J.; Ozturk, M. *Cancer Res.* **1991**, *51*, 6185-6189.

(61) McMahon, G.; Davis, E. F.; Huber, L. J.; Kim, Y.; Wogan, G. N. *Proc. Natl. Acad. Sci. USA* **1990**, *87*, 1104-1108.

(62) Doll, R.; Peto, R. *J. Natl. Cancer Inst.* **1981**, *66*, 1191-1308.

(63) Mitsudomi, T.; Viallet, J.; Mulshine, J. L.; Linnoila, R. I.; Minna, J. D.; Gazdar, A. F. *Oncogene* **1991**, *6*, 1353-1362.

(64) Mitsudomi, T.; Steinberg, S. M.; Nau, M. M.; Carbone, D.; D'Amico, D.; Bodner, S.; Oie, H. K.; Linnoila, R. I.; Mulshhine, J. L.; Minna, J. D.; Gazdar, A. F. *Oncogene* **1992**, *7*, 171-180.

(65) *Tobacco Smoking*; IARC Monographs Vol. 38, IARC: Lyon, France, 1986; 421 pages.

(66) Belinsky, S. A.; Foley, J. F.; White, C. M.; Anderson, M. W.; Maronpot, R. R. *Cancer Res.* **1990**, *50*, 3772-37780.

(67)   Vahakangas, K. H.; Samet, J. M.; Metcalf, R. A.; Welsh, J. A.; Bennet, W. P.; Lane, D. P.; Harris, C. C. *Lancet* **1992**, *339*, 576-580.
(68)   Bennett, W. P.; Hollstein, M. C.; He, A.; Zhu, S. M.; Resau, J. H.; Trump, B. F.; Metcalf, R. A.; Welsh, J. A.; Midgley, C.; Lane, D. P.; Harris, C. C. *Oncogene* **1991**, *6*, 1779-1784.
(69)   Wynder, E.; Mushinski, M. H.; Spivak, J. C. *Cancer* **1977**, *10*, 1872-1878.
(70)   Baumann, U.; Zimmerli, B. *Mitt. Gebiete Lebensm. Hyg.* **1986**, *77*, 327-332.
(71)   Ohmori, H.; Abe, T.; Hirano, H.; Murakami, T.; Katoh, T.; Gotoh, S.; Kido, M.; Kuroiwa, A.; Nomura, M. T.; Higashi, K. *Carcinogenesis* **1992**, *13*, 851-855.
(72)   Umbenhauer, D.; Wild, C. P.; Montesano, R.; Saffhill, R.; Boyle, J. M.; Huh, N.; Kirstein, U.; Thomale, J.; Rajewsky, M. F.; Lu, S. H. *Int. J. Cancer* **1985**, *36*, 661-665.
(73)   Stegelmeier, B. L.; Gillett, N. A.; Rebar, A. H.; Kelly, G. *Mol. Carcinogen.* **1991**, *4*, 43-51.
(74)   Wang, Y.; You, M.; Reynolds, S. H.; Stoner, G. D.; Anderson, M. W. *Cancer Res.* **1990**, *50*, 1591-1595.
(75)   Barch, D. H.; Jacoby, R. F.; Brasitus, T. A.; Radosevich, J. A.; Carney, W. P.; Iannaccone, P. M. *Carcinogenesis* **1991**, *12*, 2373-2377.
(76)   Watatani, M.; Parantoni, A. O.; Reed, C. D.; Enomoto, T.; Wenk, M. L.; Rice, J. M. *Cancer Res.* **1989**, *49*, 1103-1109.
(77)   Belinsky, S. A.; Devereux, T. R.; Anderson, M. E. *Mutat. Res.* **1990**, *233*, 105-116.
(78)   Hollstein, M. C.; Smits, A. M.; Galiana, C.; Yamasaki, H.; Bos, J. L.; Mandard, A.; Partensky, C.; Montesano, R. *Cancer Res.* **1988**, *48*, 5119-5123.
(79)   Fujimoto, Y.; Hampton, L. L.; Luo, L.-D.; Wirth, P. J.; Thorgeirsson, S. S. *Cancer Res.* **1992**, *52*, 1044-1046.
(80)   Weinberg, R. A. *Cancer Res.* **1989**, *49*, 3713-3721.
(81)   Kumar, R.; Sukumar, S.; Barbacid, M. *Science* **1990**, *248*, 1101-1104.
(82)   Chiba, I.; Takahashi, T.; Nau, M. M.; D'Amico, D.; Curiel, D. T.; Mitsudomi, T.; Buchhagen, D. L.; Cabone, D.; Piantadosi, S.; Koga, H.; Teissman, P. T.; Slamon, D. J.; Holmes, E. C.; Minna, J. D. *Oncogene* **1990**, *5*, 1603-1610.
(83)   Fearon, E. R.; Vogelstein, B. *Cell* **1990**, *61*, 759-767.
(84)   Tuck, A. B.; Wilson, S. M.; Khokha, R.; Chambers, A. F. *J. Natl. Cancer Inst.* **1991**, *83*, 485-491.

RECEIVED June 29, 1993

REACTIVE INTERMEDIATES:
BRIEF DISCUSSIONS OF RESEARCH

Chapter 24

Electrophilic Addition to "Aminononoate" (R¹R²NN(O)NO⁻) Ions

Joseph E. Saavedra[1], Tambra M. Dunams[1], Judith L. Flippen-Anderson[2], and Larry K. Keefer[1]

[1]Laboratory of Comparative Carcinogenesis, Frederick Cancer Research and Development Center, National Cancer Institute, Frederick, MD 21702
[2]Naval Research Laboratory, Washington, DC 20375

Nitric oxide (·NO) has long been known to be an important air pollutant, a harmful constituent of cigarette smoke and a nitrosamine precursor. In spite of these deleterious effects, this simple diatomic molecule is produced by numerous tissues of the body and is involved in many key bioregulatory processes (1). Several complex ions of structures $R_2N[N_2O_2]^-$ **1** and $RNH[N_2O_2]^-$ **2** have been prepared by reacting nitric oxide with secondary and primary amines, respectively (2). These adducts are of current interest as pharmacological probes by virtue of their ability to regenerate nitric oxide under physiological conditions (3). The decomposition to ·NO is a spontaneous, first order reaction where protonation of the anion is a key step. We have been interested in blocking the anionic oxygen with an alkyl group in order to provide stable, covalently bonded adducts, $Et_2N[N_2O_2]R'$ **3**. Compounds of this type might serve as prodrugs which cannot release nitric oxide until they are metabolically reconverted to **1a** (**1** in which R=Et) by enzymes specific to the desired target cell type.

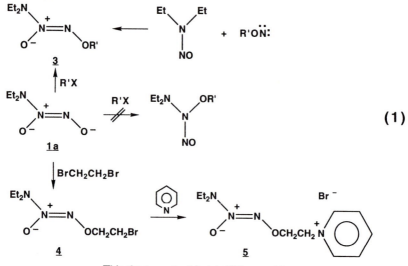

(1)

Anions of structure **1a** proved reactive toward a variety of electrophiles (*4-6*), such as alkyl halides, sulfate esters and epoxides, Table I. Alkylation had been reported (*4*) to occur at the interior oxygen of **1a** to form N-nitroso derivatives of structure $R_2NN(OR')NO$ as the major product. However, we discovered that alkylation of **1a** gives a product where the electrophile attaches to the exterior oxygen. This was established by comparing our product of ethylation to $Et_2NN(O)=NOEt$, a compound derived from the regiospecific trapping of ethoxynitrene by diethylnitrosamine, as described by Artsybasheva and Ioffe (*7*).

A species involving electrophilic attack on the interior oxygen was never observed in the alkylation of **1a**. The structure was unequivocally confirmed by single crystal X-ray analysis of **5**. This product was obtained from the reaction of **1a** and the bifunctional electrophile, 1,2-dibromoethane, to give **4** followed by addition to pyridine. Intermediate **4** proved to be a useful alkylating agent towards a variety of nucleophiles. The reaction of **4** with pyridine gave **5**, the crystalline product used for structure confirmation. The NONOate salt **1a** reacted with **4** to give the bis-adduct $Et_2N(N_2O_2)CH_2CH_2(O_2N_2)NEt_2$. Moreover, **4** was reacted with methylamine to give **3a** [**3** in which $R'=(CH_2)_2NHMe$], was hydrolyzed to the corresponding alcohol, and was dehydrobrominated to the vinyl derivative. These further reactions of **4** have been useful in providing us with structurally diverse prodrug candidates.

TABLE I. Alkylation products **3** by reacting **1a** with electrophiles

R'	electrophile	solvent	yield(%)
Et	Et_2SO_4	MeOH	33
n-Pr	*n*-PrI	DMF	33
$CH_2=CHCH_2$	allyl bromide	DMF	36
$MeCHOHCH_2$	propylene oxide	THF	6
$HOCH_2CH_2$	$BrCH_2CH_2OH$	THF	33
$MeOCH_2$	$MeOCH_2Cl$	THF	41

Conclusion.
A wide variety of potential prodrugs are now available by O-derivatization of the anionic functional group in secondary amine-nitric oxide complexes. The products of alkylation have the oxo-triazene structure, $R_2NN(O)=NOR'$. These agents are remarkably resistant to hydrolysis, and are thermally stable. Preliminary data with a few of these O-functionalized derivatives indicate that they can be used to lower blood pressure in rats and rabbits.

References.
1. Moncada, S.; Palmer, R. M. J.; Higgs, E. A. *Pharmacol. Rev.* **1991**, *43*, 109-142.
2. Drago, R. S.; Karstetter, B. R. *J. Am. Chem. Soc.* **1961**, *83*, 1819-1822.

3. Maragos, C. M.; Morley, D.; Wink, D. A.; Dunams, T. M.; Saavedra, J. E.; Hoffman, A.; Bove, A. A.; Isaac, L.; Hrabie, J. A.; Keefer L. K. *J. Med. Chem.* **1991**, *34*, 3242-3247.
4. Reilly, E. L. U.S. Patent No. 3,153,094, Oct. 13, 1964 (cf. *Chem. Abstr.* **1965**, *62*, 5192h).
5. Longhi, R.; Drago, R. S. *Inorg. Chem.* **1963**, *2*, 85-88.
6. Saavedra, J. E.; Dunams, T. M.; Flippen-Anderson, J. L.; Keefer L. K. *J. Org. Chem.* **1992**, *57*, 6134-6138.
7. Artsybasheva, Y. P.; Ioffe, B. V. *J. Org. Chem. USSR (Engl. Transl.)* **1987**, *23*, 1056-1060.

RECEIVED January 26, 1994

Chapter 25

Nitric Oxide–Nucleophile Complexes as Ligands

Structural Aspects of the Coordinated "NONOate" Functional Group in Novel Mixed-Ligand, Non-Nitrosyl Metal Complexes

Danae Christodoulou[1], David A. Wink[1], Clifford F. George[2], Joseph E. Saavedra[1], and Larry K. Keefer[1]

[1]Chemistry Section, Laboratory of Comparative Carcinogenesis, Frederick Cancer Research and Development Center, National Cancer Institute, Frederick, MD 21702
[2]Laboratory for the Structure of Matter, Naval Research Laboratory, Washington, DC 20375

"NONOates" are unusual N-nitroso compounds containing the anionic $N_2O_2^-$ functional group that have been recently shown to display versatile pharmacological activity *(1)*. Their biological effects are attributable to their ability to release the multifaceted bioregulatory agent *(2)*, nitric oxide (NO), when dissolved in physiological media *(3)*.

In comprehensively exploring the structural characteristics and reactivity of this functional group, we are investigating the coordination chemistry of $X-N_2O_2^-$ ions in which X is a secondary amine residue such as Et_2N. These $X-N_2O_2^-$ compounds should exhibit ligand properties if the oxygen atoms of the $N_2O_2^-$ group are in the Z-configuration. This type of structure would resemble the reported $[O_3S-N_2O_2]^{2-}$ *(4)* and $[O-N_2O_2]^{2-}$ *(5)* structures, which contain cis oxygens. One also would expect that coordination of the anions to metal centers would affect the stability, solubility and reactivity of the derivative molecules. In addition, the unique properties of metal ions, such as redox activity, ligand exchange properties or expansion of the coordination sphere, could be exploited to modulate the ligands' pharmacological activity. Thus metal derivatives of biologically active compounds could be useful in tissue targeting via potential interactions of vacant coordination sites on the metal center with ligand residues of biopolymers such as sulfhydryl, carboxyl or amino. In addition, metal complexes of various nuclearity and ligand content provide a discrete way of aggregating the drug (increasing the number of NO molecules per molecule of complex) in a single compound (equation 1).

Longhi and Drago *(6)* reported evidence for interaction of the $Et_2N-N_2O_2^-$ ion with certain metal ions, namely Co^{2+}, Fe^{2+} and Cu^{2+}. We have further investigated the interactions of this ion with different metal centers, in nonaqueous solutions, under conditions that prolong the half-lives of starting materials and products. Ligand substitution, or the use of an isolated complex as a synthon, generated a series of these complexes. This suggests lability of the products in solution. However, in some cases self assembly led to thermodynamically stable products suggesting self organization in the interaction of $X-N_2O_2^-$ ligands with metal ions. In the case of Cu^{2+} we have crystallized mixed-ligand complexes of various nuclearity, according to the general scheme of equation 1.

$$2Et_2N-N_2O_2^- \ + \ xCu^{2+} \ + \ yL \ \longrightarrow \ Cu_x(L)_y(Et_2N-N_2O_2)_2 \qquad (1)$$

$$L = OMe^-, \ OEt^-, \ MeOH, \ OAc^-$$

Among these copper complexes, we have isolated the mononuclear
$Cu(MeOH)(Et_2N-N_2O_2)_2$ (Fig. 1), previously prepared *(6)* via a different route, and
have determined its crystal structure by X-ray diffraction analysis. Structure
determinations on this and other metal complexes demonstrate the planar, bidentate
chelate nature of the $N_2O_2^-$ functional group. In the $N_2O_2^-$ group, the N-N linkage has
double bond character, in contrast to the $(Et_2)N$-N linkage which is a single bond
(structure in Fig. 1). The N-O distances are equivalent with single bond character; both
oxygens are functionalized and interact with the metal center. These compounds are
novel examples of metal complexes having nitric oxide as a ligand, oxygen bound rather
than in the form of a classical metal nitrosyl, L_nM-NO.

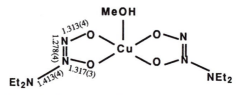

Figure 1. Mononuclear $Cu(MeOH)(Et_2N-N_2O_2)_2$.

Solution studies of the complexes in aqueous, nonaqueous and biological media
suggest potential applications in NONOate biology. Coordination of the anions to a
metal center in general prolonged the life of the derivative in nonaqueous medium,
whereas the free ligand was more stable in aqueous solution. In the latter solutions, the
behavior of the metal complexes was similar to that of the free ligand, suggesting that
nitric oxide release is controlled primarily by the hydrolysis of the ligand. The amount
of nitric oxide produced by these complexes is generally proportional to the number of
NONOate ligands present. Presumably this accounts for the increased vasorelaxant
effects observed for these compounds in isolated rabbit aorta.

In conclusion, aminoNONOates contain an $N_2O_2^-$ functional group best described
as a bidentate chelating ligand since the oxygens are found in the Z-configuration and
both interact with metal ions as demonstrated by crystallographic studies on select
copper complexes. These molecules illustrate how biologically active NONOate ligands
can coexist with other biologically relevant ligands in mixed-ligand complexes. Metal
complexes of this kind suggest possibilities of modulating the reactivity as well as
pharmacology of the $X-N_2O_2^-$ anions upon coordination.

References

(1) Keefer, L. K.; Wink, D. A.; Maragos, C. M.; Morley, D.; Diodati, J. G. In
The Biology of Nitric Oxide, Moncada, S.; Marletta, M. A.; Hibbs, J. B. Jr.; Higgs, E.
A. (eds), Portland Press, Colchester, UK, in press.
(2) Moncada, S.; Palmer, R. M. J.; Higgs, E. A. *Pharmacol. Rev.* **1991**, *43*,
109-142.
(3) Maragos, C. M.; Morley, D.; Wink, D. A.; Dunams, T. M.; Saavedra, J. E.;
Hoffman, A.; Bove, A. A.; Isaac, L.; Hrabie, J. A.; Keefer, L. K. *J. Med. Chem.*
1991, *34*, 3242-3247.
(4) (a) Cox, E. G.; Jeffrey, G. A.; Stadler, H. P. *Nature* **1948**, *162*, 770-771.
(b) Cox, E. G.; Jeffrey, G. A.; Stadler, H. P. *J. Chem. Soc.* **1949**, 1783-1793. (c)
Jeffrey, G. A.; Stadler, H. P. *J. Chem. Soc.* **1951**, 1467-1474.
(5) Hope, H.; Sequeira, M. R. *Inorg. Chem.* **1973**, *12*, 286-288.
(6) Longhi, R.; Drago, R. S. *Inorg. Chem.* **1963**, *2*, 85-88.

RECEIVED January 31, 1994

Chapter 26

Radical Cations in Nitrosation of *tert*-Dialkyl Aromatic Amines

S. Singh, R. Hastings, and Richard N. Loeppky

Department of Chemistry, University of Missouri, Columbia, MO 65211

N,N-Dialkylaromatic amines react surprisingly rapidly with nitrous acid to produce a nitrosamine and a nitro compound as shown in Scheme 1. At 25 ° C the $t_{\frac{1}{2}}$ of 0.19 mM N,N-dimethyl-4-chloroaniline (DMCA) in 1.9 mM HNO_2 (pH 3.7) is 8 min and that of the 4-carboethoxy analog (0.068 M) is 110 min. Most trialkyl amines undergo nitrosative dealkyation at very slow rates at 25 ° *(1,2)*. To determine if the more rapid reaction of N,N-dialkylaromatic amines could be due to their lower basicity, which permit higher concentrations of the key nitrosammonium ion intermediate, we nitrosated $N(CH_2CH_2OMe)_3$ and DMCA under conditions where the free amine concentrations were equal. It was found that DMCA nitrosated about 40 times faster. These data suggest that a new mechanism may be operative for aromatic dialkyl amines.

The concurrent formation of an aromatic nitro compound in these nitrosative dealkylation reactions suggests that a radical cation may be a reaction intermediate, because considerable data has recently been amassed to support radical cation involvement in nitration, and nitrite catalyzed nitration reactions of electron-rich aromatic systems *(3,4)*. Scheme 1 illustrates our mechanistic hypothesis which also allows for the formation of the nitrosamine by two routes, the classical mechanism *(1)* and another. Proof of a radical intermediate in the formation of the nitro compound was obtained from CIDNP ^{15}N NMR experiments. The nitro compound gave an enhanced emission signal *(3)* when the nitrosation of DMCA was done with Na[15]NO_2. The nitro compound could form from the radical cation either by direct reaction with [15]NO_2 produced from the decomposition of N_2O_3, or by its reaction with NO_2- followed by rapid oxidation. Reaction of the stable radical cation of N,N,N′,N′-tetramethyl-p-phenylenediamine (Wurster's blue) with aqueous $NaNO_2$ gave only a 2% yield of the nitro compound (Scheme 2) but a 72% yield of the nitrosamine. These data suggest that the nitro compound arises from NO_2 not NO_2- and also show that the nitrosamine can arise from a radical cation. The radical cation could be produced either from a homolytic dissociation of N-nitrosammonium ion or through an electron transfer reaction with an NO^+ donor *(3,4)*.

0097–6156/94/0553–0309$08.00/0

$$2HNO_2 \rightleftharpoons N_2O_3 + H_2O$$

$$N_2O_3 \longrightarrow {}_{\circ}NO_2 + NO$$

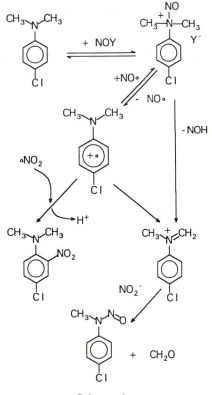

Scheme 1

CIDNP emission is not observed for the nitrosamine and is not expected because its immediate precursors are not radicals. The work of Dinocenzo (5) suggests that the nitrosamine arises from the radical cation by loss of a proton in the rate determining step to generate an α-amino nitrite ester which rapidly gives the nitrosamine (6). While this pathway is supported by out data and a deuterium isotope effect of 8.2 for the nitrosation of DMCA, we also observe the formation of N_2O which is formed in the classical mechanism from NOH eliminated by nitrosammonium ion. Other experimental data derived from the nitrosation of 4-carboethoxy-N,N-dimethylaniline show that the k_H/k_D values change with conditions suggesting that competing mechanisms are operative. Additional evidence in support of the competitive mechanism comes form the observation that the methyl vs. ethyl cleavage ethly ratios produced from the nitrosation of N-ethyl-N-methyl-4-carboethoxyaniline are time dependent.

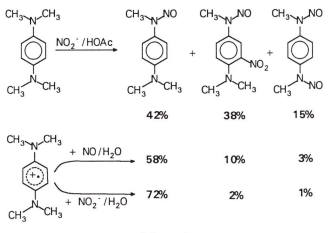

Scheme 2

Acknowledgement

The support of this research by grant R37 CA26914 from the National Cancer Institute is gratefully acknowledged.

References

(1) Smith, P. A. S.; Loeppky, R. N. *J. Am. Chem. Soc.* **1967**, *89*, 1147-1157.

(2) Gowenlock, B.; Hutcheson, R. J.; Little, J.; Pfab, J. *J. Chem. Soc. Perkin. Trans. II.* **1979**, 1110-1114.

(3) Clemens, A. H.; Helsby, P.; Ridd, J. H.; Omran, F. A. *J. Chem. Soc. Perkin. Trans. II.* **1985**, 1217.

(4) Kochi, J. K. *Acc. Chem. Res.* **1992**, *25*, 39.

(5) Dinocenzo, J. P.; and Banach, T. E. *J. Am. Chem. Soc.* **1989**, *111*, 8646.

(6) Keefer, L. K.; Roller, P. P. *Science.* **1973**, *181*, 1245.

RECEIVED November 15, 1993

N-NITROSO COMPOUND FORMATION AND INHIBITION: BRIEF DISCUSSIONS OF RESEARCH

Chapter 27

Nitrosamide Formation from Foodstuffs

P. Mende, R. Preussmann, and B. Spiegelhalder

Department of Environmental Carcinogens, FSP03, German Cancer Research Center, Im Neuenheimer Feld 280, D–6900 Heidelberg, Germany

During a study on *in vitro* nitrosation of foodstuffs and analysis of the nitrosation products by gas chromatography combined with NO-specific chemiluminescence detection we observed the formation of an unknown nitroso compound from cocoa and dried tomatoes. The instability of this nitrosation product both under alkaline and acidic conditions together with its strongly acid-catalyzed formation indicated the presence of a volatile nitrosamide. Its structure was determined by mass spectrometric analysis as N-nitrosopyrrolidin-(2)-one (NPyrO).

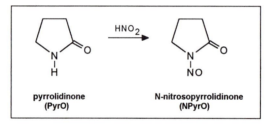

pyrrolidinone
(PyrO)   N-nitrosopyrrolidinone
                          (NPyrO)

When treated with 50 mM nitrite at pH 2, NPyrO forms up to 35 mg/kg from cocoa, coffee and coffee surrogates. Such high levels of nitrosamides have never been reported for nitrosated foodstuffs *(1)*. At nitrite concentrations closer to physiological conditions the NPyrO formation strongly depends on the nitrite/ sample ratio due to nitrite scavenging activity of the sample.

The presence of the NPyrO precursor pyrrolidin-(2)-one (PyrO) in food was shown using gas chromatography with nitrogen-specific chemiluminescence detection after conversion of PyrO into its trifluoroacetyl derivative. Only traces of PyrO were found in fresh food, but high amounts were detected in all dried samples (Table I). The increase in PyrO concentration does not correlate to the water loss during the drying process so it is evident that a reaction takes place which generates PyrO from a precursor compound. 4-Aminobutyric acid probably is involved in this reaction since it is known to form PyrO under dehydrating conditions.

0097–6156/94/0553–0314$08.00/0

Using a test system specific for detection of nitrosamides which release nonpolar diazoalkanes *(1)*, a nitrosamide-derived methylating activity was additionally detected in foodstuffs which contain caffeine (Table II). This activity has been reported previously in nitrosated Kashmir tea which is prepared by boiling tea leaves for several hours in the presence of sodium hydrogen carbonate *(2)*. The slightly alkaline treatment (pH 7.5-8) results in hydrolysis of caffeine and formation of the nitrosamide precursors caffeidine and *N,N'*-dimethylurea. However, the results from Table II indicate that a nitrosamide precursor is already present in the food sample and caffeine hydrolysis during boiling under alkaline conditions is not relevant for formation of the nitrosamide precursor(s). Furthermore, the concentrations of caffeidine (Table II) and *N,N'*-dimethylurea ($\leq$ 0.7 mg/l) are too low to explain the methylating activity found in nitrosated beverages. Other relevant nitrosamide precursors therefore remain to be identified.

Table I: *Occurrence of pyrrolidinone in foodstuffs*

food item	pyrrolidinone (mg/kg)		
	average	range	
fresh fruits / vegetable / meat / fish	0.1	0 -	0.4
beer	1.5	0.3 -	4
dried vegetables	22	5 -	48
coffee surrogates	36	30 -	41
coffee beans, instant coffee	40	29 -	50
cocoa powder	55	32 -	77

Table II: *Methylating activity due to unknown nitrosamides in nitrosated caffeine-containing beverages (prepared from 50 g dried food per liter). Nitrosation was done with 100 mM nitrite at pH 2 for one hour at 37°C. The methylating activity is compared to methylnitrosourea (MNU). Caffeidine was determined by gas chromatography/chemiluminescence detection (amine mode) after extraction with dichloromethane and derivatization with trifluoroacetic acid. The methylating activity of 1 mg caffeidine after nitrosation is equivalent to 0.002 mg MNU*

nitrosated sample	methylating activity (mg/l, MNU equivalents)	caffeidine (mg/l)
Kashmir tea (boiled with $NaHCO_3$)	3.0	0.1
black tea	2.4	0.3
black tea (boiled with $NaHCO_3$)	2.1	0.2
green tea	1.5	0.5
green tea (boiled with $NaHCO_3$)	1.7	0.4
Mate tea	4.4	0.2
coffee	2.1	0.3
cocoa	0.6	0.4

References:

(1) Mende, P., Spiegelhalder, B. and Preussmann, R. (1991) *Fd. Chem. Tox.* **27,** 475-478
(2) Kumar, R., Mende, P., Wacker, C.-D., Spiegelhalder, B., Preussmann, R. and Siddiqi, M. (1992) *Carcinogenesis* **13,** 2179-2182

RECEIVED January 26, 1994

Chapter 28

Tocopherol Inhibition of NO_2-Mediated Nitrosation

In Vitro and Biological Superiority of γ-Tocopherol

R. V. Cooney, A. A. Franke, L. J. Mordan, P. J. Harwood,
V. Hatch-Pigott, and L. J. Custer

Cancer Research Center of Hawaii, University of Hawaii at Manoa, 1236
Lauhala Street, Honolulu, HI 96813

Nitrogen oxides (NO_x) are potentially carcinogenic because of their ability to nitrosate amines. Exposure to NO_x can produce mutations directly through deamination of primary amino groups of DNA bases (1,2) or through the formation of stable N-nitrosamines which can act as DNA alkylating agents upon metabolic activation. Many antioxidants, including α-tocopherol, have been shown to prevent nitrosation; however, most assays look only at the immediate effects of an inhibitor on nitrosation of amines and do not take into account slower nitrosation reactions from other reaction products. Organic nitrous esters, formed from nitrosating agents, are known to react with amines to yield N-nitrosamines (3,4) and α-tocopherol has been reported to react with NO_2 to form a nitrous ester (5). This property is in contrast to the perception of α-tocopherol as a potent nitrosation inhibitor.

We report, in agreement with previous studies, that α-tocopherol is superior to both γ-tocopherol and δ-tocopherol in preventing nitrosation of morpholine by NO_2 when N-nitrosomorpholine is measured immediately after NO_2 exposure (99.6%, 88.4% and 84.2% inhibition of nitrosation respectively, for 10 mM tocopherol and 1 mM morpholine in dichloromethane exposed to 27.5 ppm NO_2/N_2 for 5 minutes). In contrast, measurement of N-nitrosomorpholine formation after extended incubation of the reaction mixture described above reveals α-tocopherol to be the least effective of the three (66%, 84.9% and 80.9% for α, γ and δ respectively, 168 hours after exposure). Figure 1 indicates that the difference in chemical reactivity between α-tocopherol and γ-tocopherol with NO_2 probably is due to the formation of an intermediate nitrosating agent from α-tocopherol which reacts with morpholine at 37°C to form an N-nitrosamine. The observed reaction is independent of light and does not require oxygen although nitrosation is slightly enhanced in the presence of O_2. γ-Tocopherol, which lacks a methyl group at C-5, does not form a nitrosating agent in its reaction with NO_2.

Exposure of γ-tocopherol to NO_2 resulted in the formation of a colored compound absorbing at 432 nm (in hexane) which was separated from the parent tocopherol by HPLC. As shown in Figure 2, nitric oxide (NO) formed during γ-tocopherol exposure to NO_2. γ-Tocopherol produced significantly more NO than α-tocopherol suggesting that greater efficiency of reduction of NO_2 to NO may, in part, account for the superiority of γ-tocopherol as a nitrosation inhibitor.

0097–6156/94/0553–0317$08.00/0

The C3H 10T1/2 mouse fibroblast transformation assay (7) was used to assess the relative effectiveness of α-tocopherol and γ-tocopherol to inhibit neoplastic transformation during the promotional phase of carcinogenesis. At a concentration of 30 μM, γ-tocopherol reduced the development of transformed foci by 90% while α-tocopherol was only 50% effective (p<.01 for the difference in means).

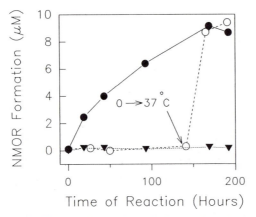

Figure 1. Nitrosating potential of tocopherols exposed to NO_2.
Five mls of tocopherol (1 mM) in hexane were exposed to 27.5 ppm NO_2 in N_2 by bubbling (60 ml/min) for 10 minutes. Morpholine and an internal standard were then added and the solution incubated at 37°C. Aliquots were removed at the indicated time and analyzed for N-nitrosomorpholine as previously described (6). ●————● , α-tocopherol; ▼········▼ ,γ-tocopherol; ○------○ , α-tocopherol at 0°C.

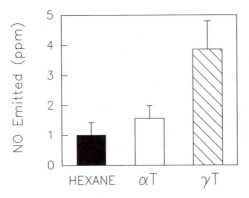

Figure 2. NO formation by α- and γ-tocopherols.
Reaction conditions are as described in Figure 1. The gas exiting the indicated solution was monitored for emitted NO by direct injection of aliquots into a thermal energy analyzer after 15 minutes exposure to NO_2 (6).

ACKNOWLEDGMENT

Supported by NIEHS Grant #ES04302.

REFERENCES

1. Wink, D.A., et al. (1991) *Science* **254**, 1001-1003.
2. Nguyen, T.T., et al.(1992) *Proc. Natl. Acad. Sci.* **89**, 3030-3034.
3. Loeppky, R.N, et al. (1984) *IARC Sci. Publ.* **57**, 353-363.
4. Mirvish, S. S., et al. (1986) *Cancer Letters* **31**, 97-104.
5. Selander, H., & Nilsson, J. L. G. (1972) *Acta Pharm. Suecica* **9**, 125-128.
6. Cooney, R.V., et al. (1992) *J. Env. Health Sci.* **A27**, 789-801.
7. Mordan, L.J. (1988) *Carcinogenesis* **9**, 1129-1134.

RECEIVED January 26, 1994

Chapter 29

pH Changes in Smokeless Tobaccos Undergoing Nitrosation

R. A. Andersen[1,2], P. D. Fleming[1,2], T. R. Hamilton-Kemp[3], and D. F. Hildebrand[2]

[1]Agricultural Research Service, U.S. Department of Agriculture, Department of Agronomy, University of Kentucky, Lexington, KY 40546–0091
[2]Department of Agronomy and [3]Department of Horticulture, University of Kentucky, Lexington, KY 40546

A previous study of moist and dry snuff indicated that nitrosated pyridine alkaloids accumulated at higher rates during storage at ca 50% moisture and 32°C as compared to less than 22% moisture and 24°C. Nitrite levels increased under these conditions but alkaloids such as nornicotine and nicotine decreased in concentrations *(1)*. Ghabrial *(2)* determined that air-cured burley tobacco incubated at 35°C and 30-40% moisture underwent a sharp increase in bacterial counts within 1 week. An increase in pH of the leaf followed the rise in bacterial counts. The purpose of our present investigation was to determine whether pH changes in smokeless tobaccos accompany nitrosation reactions during prolonged storage.

After chewing tobacco was stored for 48 weeks at each of four % moisture-temperature combinations (22.3%-24°C; 49.3%-24°C; 22.3%-32°C and 49.3%-32°C) pH levels in each combination decreased from 0-time values. The largest decrease was at 49.3%-32°C with increasingly smaller decreases observed in the following order: 22.3%-32°C, 49.3%-24°C and 22.3%-24°C. Chewing tobacco underwent little change in amounts of nitrite N and total nitrosated alkaloids in all four storage treatments. Storage of dry snuff at each of four similar % moisture-temperature combinations for 24 weeks resulted in some significant changes in pH, nitrite and N-nitroso compounds. pH levels and concentrations of nitrite and nitrosated alkaloids were higher in the stored high moisture dry snuffs but not in the low moisture counterparts at both temperatures. The pH's and concentrations of nitrite N and nitrosamine alkaloids in moist snuffs stored for 24 weeks at 55.5% moisture-24 or 32°C were markedly higher than in 21.9% moisture-24 or 32°C storage environments. Moisture-temperature interactions were observed in the high moisture-containing moist snuffs and dry snuffs.

Chewing tobacco pH as a function of storage time decreased with increasing duration for each moisture-temperature treatment; the greatest magnitude of change occurred in tobacco stored at 49.3% moisture-32°C and the least change occurred in tobacco at 22.3% moisture-24°C. Dry snuff pH's at the 51.4% moisture level

0097–6156/94/0553–0320$08.00/0

increased from 6.2 to an average final value of 8.0 at 48 weeks storage. Dry snuff stored at 12.3% moisture decreased from 6.2 to an average of 6.0 at 48 weeks. pH values of moist snuff stored at 55.5% moisture gradually increased from an initial value of 6.9 to an average final value of 7.2 at 48 weeks. Moist snuff at 21.9% moisture decreased from an initial pH of 6.7 to 6.4 at 48 weeks.

To test the hypothesis that increases of tobacco nitrosamines are mediated by bacterial growth at near neutral pH, we heated snuff tobaccos in a manner analagous to pasteurization prior to storage to determine whether chemical and pH changes associated with accumulations of nitrosamines would lessen after heating. There were no significant changes in pH and the chemical parameters in heat-treated dry snuff after 24 weeks storage at 51.4% moisture-24 or 32°C compared to 0-time controls. Significant increases in pH, total nitrosamines and nitrite were found for nonheat-treated dry snuff after 24 weeks storage in the same manner. As shown in Table I, significant increases in pH, total nitrosamines and nitrite were found for nonheat-treated moist snuff stored at 55.5% moisture at 24°C compared to 0-time controls. Heat pretreatment of moist snuff stored in the same manner did not differ from 0-time controls.

In summary: 1) pH's of high moisture content moist snuff (55.5%) and dry snuff (51.4%) increased 0.3 to 2.1 pH units during 48-week storage, 2) pH's of low moisture content moist snuff (21.9%) and dry snuff (12.3%) decreased (0.2 to 0.4 pH units) during 48-week storage, 3) pH's of chewing tobacco at each moisture-temperature treatment decreased during 48-week storage, 4) nitrosated alkaloids increased only in smokeless tobaccos that became more alkaline during storage, and 5) heat treatment of high moisture-containing snuffs at 100°C for 30 min just before storage prevented increases of pH, nitrosamines and nitrite that occurred in non-heat-treated controls after 24-week storage.

Table I. After Storage Effects of 30-Min Heat Treatment at 0-Time Storage on Moist Snuff (55.5% Moisture-24°C)

Heat treat? yes/no	°C	Weeks stored	pH	Nitrite-N	Total nitrosated alkaloids
				..............ug/g..................	
no	ambient	0	6.9 A*	12.8 A	38.0 A
yes	24	24	6.8 A	5.8 A	9.9 A
no	24	24	7.2 B	1203 B	532 B

*Means followed by different letter in column differ at p=05

REFERENCES

1.   Andersen RA, Fleming PD, Burton HR, Hamilton-Kemp TR, Sutton TG
     J Agric Food Chem, **1991**, *39*:1280-1287

2.   Ghabrial SA   *Tobacco Science*, **1976**, *20*:95-97

RECEIVED January 26, 1994

CHEMICAL AND BIOCHEMICAL MODELS
AND DNA ADDUCT FORMATION:
BRIEF DISCUSSIONS OF RESEARCH

Chapter 30

Oxidation of Alkylnitrosamines via the Fenton Reagent

Use of Nitrosamines To Probe Oxidative Intermediates in the Fenton Reaction

David A. Wink[1], Raymond W. Nims[1], Joseph E. Saavedra[1], Marc F. Desrosiers[2], and Peter C. Ford[3]

[1]Frederick Cancer Research and Development Center, National Cancer Institute, Frederick, MD 21702
[2]National Institute of Standards and Technology, Gaithersburg, MD 20899
[3]University of California, Santa Barbara, CA 98106

The activation of the carcinogen N-nitrosodimethylamine (NDMA) has been shown to be mediated by cytochrome P450. The biological oxidation of this nitrosamine can follow two distinct pathways, dealkylation and denitrosation. As previously postulated (*1*), an intermediate common to the two pathways is the alkylnitrosamino radical which is generated via hydrogen abstraction. This species can, via a rebound mechanism, form the α-hydroxynitrosamine which rapidly decomposes through the intermediacy of a diazonium salt to dealkylation products (formaldehyde, nitrogen gas, and methylating equivalents) or can rearrange to liberate NO and the Schiff's base, ultimately leading to denitrosation products (formaldehyde, nitric oxide, and monomethylamine). Heur et al. reported that the Fenton reagent was capable of oxidizing NDMA to give exclusively denitrosation products. This also was proposed to proceed via the alkylnitrosamino radical but in this case the absence of the hemoprotein pocket precludes dealkylation from occurring, resulting in denitrosation only (*2*).

The Fenton reaction (acidic mixture of ferrous ion and peroxide) has been stated to generate hydroxyl radical as the exclusive oxidizing intermediate (*3*). However, it recently has been suggested that metallo-oxo species capable of oxidation are also present (*4*). Using stopped-flow techniques, we demonstrated (Figure 1) that when peroxide, ferrous ion and NDMA were mixed, a transient (A) was formed at a rate 5 times faster than the rate of formation of ferric ion (*5*). This suggested the presence of an oxidizing intermediate that was intercepted by NDMA prior to the formation of hydroxyl radical, and this was substantiated by the oxidation of methylene blue when this redox dye was substituted for NDMA in the reaction mixture. The absorption spectrum of the transient suggested that it might be $((H_2O)_5Fe\text{-}NO)^{+2}$, implying that •NO might be generated in this reaction. As the concentration of NDMA was increased, the amount of transient A also increased,

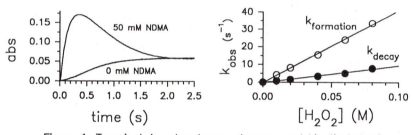

Figure 1. Transient A —absorbance changes and kinetic behavior

but there was little or no effect on the rate of appearance or disappearance of this species. These results can be explained by invoking the reactions of Scheme 1:

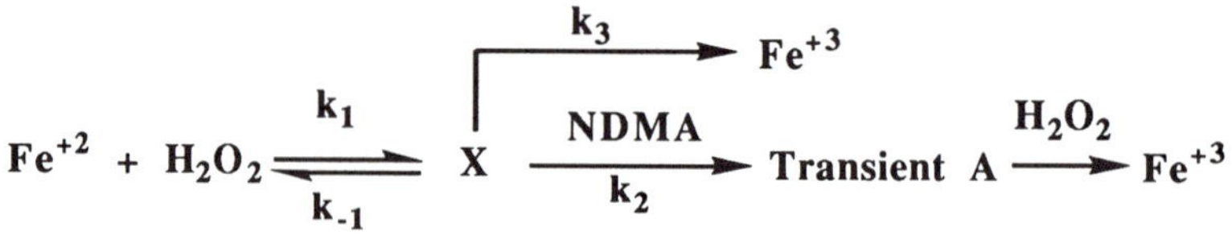

$$Fe^{+2} + H_2O_2 \underset{k_{-1}}{\overset{k_1}{\rightleftharpoons}} X \quad \overset{k_3}{\longrightarrow} Fe^{+3}$$
$$X \underset{k_2}{\overset{NDMA}{\longrightarrow}} Transient \; A \overset{H_2O_2}{\longrightarrow} Fe^{+3}$$

Scheme 1

A plot of $1/abs_{inf}$ vs $1/[NDMA]$ was linear, with $-1/X$-intercept $= 87$ mM (representing the concentration of NDMA required to generate half the maximal amount of transient A). Comparing this value with the predicted value for the hydroxyl radical ($-1/X$-intercept $<< 5$ mM) based on known rate constants (6) for the reaction of this radical species with the other components of the mixture (Fe^{+2}, H_2O_2, and NDMA), the intermediacy of hydroxyl radical was clearly ruled out. Replacing NDMA with fully deuteriated NDMA resulted in little or no kinetic effect on transient A, while causing a clear effect on the amount of transient formed. Comparing the two slopes from the plot of $1/abs$ vs $1/[nitrosamine]$ yielded an isotope effect of 4.5.

To ascertain the relationship between transient A formation and the formation of denitrosation products, a competitive kinetic study was done in the presence of MeOH. When [NDMA] was 100 mM, the addition of varying amounts of MeOH resulted in a decrease in the degradation of the nitrosamine. A plot of $1/[nitrate]$ vs [MeOH] was linear, with $-X$-intercept $= 20$ mM (a value representing the concentration of MeOH required to quench half the formation of denitrosation products). Under identical conditions, the formation of transient A was monitored by stopped-flow spectrophotometry, and the amount of the transient was also observed to decrease in the presence of increasing MeOH. A plot of $1/abs$ vs [MeOH] was linear, with $-X$-intercept $= 20$ mM, suggesting that under high concentrations of NDMA most of the nitrate was formed via transient A.

A plot of $1/[nitrate]$ vs $1/[NDMA]$, where [NDMA] < 10 mM, displayed a marked deviation from linearity, suggesting that another reactive intermediate capable of oxidizing NDMA was present. Examination of the quenching of product formation by methanol when [NDMA] $= 1$ mM yielded a linear plot of $1/[nitrate]$ vs [MeOH], with $-X$-intercept $= 4$ mM, such that $k_{MeOH}/k_{NDMA} = 0.25$. A value of 3 for k_{MeOH}/k_{NDMA} may be predicted, based upon the known rate constants for the reaction of hydroxyl radical with MeOH and NDMA. The 12-fold difference between this predicted value and the actual observed value suggested that this second oxidizing species was also not hydroxyl radical, but was instead some additional metallo-oxo species.

From the results presented above, it appears that two oxidizing intermediates (X and Y), neither of which is the hydroxyl radical, react with NDMA to cause denitrosation.   It appears that the two intermediates are related in that Y is apparently derived from X.

$$Fe^{II}(O_2^{-2}) \xrightarrow{\ H^+\ } Fe^{IV}O + OH^-$$

Intermediate X probably represents a peroxo-ferrous complex which, by hydrogen abstraction from NDMA, yields the alkylnitrosamino radical.   This radical decomposes to yield nitric oxide, which is bound to ferrous ion to form transient A. The transient species then can decompose in a peroxide-dependent fashion, yielding nitrate.   However, at lower NDMA concentrations, X is apparently converted to the second intermediate (Y), possibly $Fe^{IV}O$.   This intermediate is also capable of oxidizing NDMA.   The low deuterium isotope effect and the higher than expected concentration of MeOH required to quench NDMA oxidation by intermediate Y suggest that denitrosation in this case occurs via electron transfer instead of hydrogen abstraction, which was effected by intermediate X.   The similarities in isotope effects between the K_m values reported for the enzymatic oxidation of NDMA and those for the reaction of NDMA with species X suggest that the Fenton reaction is a surprisingly good model for the reactions taking place *in vivo*.

REFERENCES

(1) Wade, D.; Yang, C. S.; Metral, C. J.; Roman, J. M.; Hrabie, J. A.; Riggs, C. W.; Anjo, T.; Keefer, L. K.; Mico, B. A. (1987) *Cancer Res.* **47**, 3373-3377.
(2) Heur, Y.-H.; Streeter, A. J.; Nims, R. W.; Keefer, L. K. (1989) *Chem. Res. Toxicol.* **2**, 247-253.
(3) Walling, C. (1975) *Acc. Chem. Res.* **8**, 125-131.
(4) a) Sutton, H. C.; Vile, G. F.; Winterbourn, C. C. (1987) *Arch. Biochem. Biophys.* **256**, 462-471. b) Sutton, H. C.; Winterbourn, C. C. (1989) *Free Radic. Biol. Med.* **6**, 53-60.
(5) Wink, D. A.; Nims, R. W.; Desrosiers, M. F.; Ford, P. C.; Keefer, L. K. (1991) *Chem. Res. Toxicol.* **4**, 510-512.
(6) Buxton, G. V.; Greenstock, C. L.; Helman, W. P.; Ross, A. B. (1988) *J. Phys. Chem. Ref. Data* **17**, 513-886.

RECEIVED October 11, 1993

Chapter 31

Structures of Mutagens Formed from Fenton-Type Oxidation of N-Nitrosodialkylamines

M. Mochizuki, N. Tsutsumi, S. Hizatate, and E. Okochi

Kyoritsu College of Pharmacy, Shibakoen 1–5–30, Minato-ku, Tokyo 105, Japan

N-Nitrosodialkylamines require cytochrome P-450 mediated metabolic activation to elicit their carcinogenic and mutagenic activities. Since this metabolic system is complicated, many chemical model reactions have been investigated to elucidate the molecular mechanism. The Fenton reagent *(1)* is one of those chemical models which has been used with N-nitrosodialkylamines *(2)*. We reported on the conversions of N-nitrosodibutylamine by chemical models and comparison with rat liver S9 fraction *(3)*. During this study with the Fenton reagent, we found formation of a directly mutagenic product. In the present study, we report on the structure of the mutagen.

We used six N-nitrosodialkylamines: N-nitrosodimethylamine (NDMA), N-nitrosodiethylamine (NDEA), N-nitrosodipropylamine (NDPA), N-nitroso-N-methyl-propylamine (NMPA), N-nitrosodibutylamine (NDBA) and N-nitroso-N-methyl-butylamine (NMBA). In typical procedure the N-nitrosodialkylamine was mixed with ferrous sulfate, cupric acetate and hydrogen peroxide (Fe^{2+}-Cu^{2+}-H_2O_2) in pH 4.5 acetate buffer and the solution was incubated at 37°C for 2h. The reaction

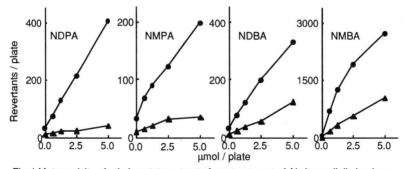

Fig. 1 Mutagenicity of ethyl acetate extracts from treatment of N-nitrosodialkylamines with Fe^{2+}-Cu^{2+}-H_2O_2 in *Salmonella typhimurium* TA1535(●) and *E.coli* WP2 hcr^-(▲);The X-axis corresponds the amount of original N-nitrosodialkylamines used in the reaction.

0097–6156/94/0553–0328$08.00/0

mixture was extracted with
ethyl acetate and was concentrated. The mutagenicity of
the extracts was tested in
Salmonella typhimurium
TA1535 and *E.coli* WP2 *hcr*⁻
in the absence of S9 mix
(Fig.1). No mutagens were
formed from NDMA and
NDEA. However, *N*-nitroso-
dialkylamines with propyl or
butyl groups gave mutagens.
The mutagenic activity was the
strongest in the extracts from
NMBA, followed by NDPA,

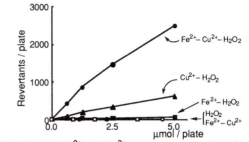

Fig.2 Effect of Fe^{2+} and Cu^{2+} on mutagenicity of products
from NMBA treated with Fe^{2+}-Cu^{2+}-H_2O_2 in *Salmonella*
typhimurium TA1535

NDBA and NMPA. In all cases, the mutagenicity was stronger in *Salmonella* than in
E.coli. The effect of metal ions on the mutagenicity also was investigated with the
products from NMBA, which showed the strongest mutagenic activity (Fig.2). Only
a very weak mutagenicity was observed by treatment with Fe^{2+}-H_2O_2 and some
mutagenicity was observed by treatment with Cu^{2+}-H_2O_2. Mutagenicity was greatly
enhanced by treatment with Fe^{2+}-Cu^{2+}-H_2O_2. These results showed that Fe^{2+} and Cu^{2+}
have synergistic effect on the formation of the mutagen.

The products from the reaction of NMBA with Fe^{2+}-Cu^{2+}-H_2O_2 were isolated
and identified. The ethyl acetate extracts was purified by silica gel column chromatography several times, and by silica gel preparative HPLC. A major product
isolated, which had no directly mutagenic activity, was identified by NMR spectra
as *N*-nitroso-*N*-methyl-3-oxobutylamine (yield 10%). The directly mutagenic
product "X" was isolated and crystallized as white plates (m.p. 46~47°C, yield
0.1%). "X" showed directly mutagenic activity in *Salmonella* and *E.coli* (Fig.3).
This activity can explain the entire
mutagenicity of the ethyl acetate
extracts of the reaction of NMBA.
Spectral data of "X" were: UV
(ethanol): λ_{max}=218nm(ε=11200);
IR (KBr): v_{max}=1570, 1526 cm⁻¹;
NMR (CDCl₃) ¹H-δ ppm: 2.13(3H),
2.58(1H), 3.03(1H), 4.29(2H):
¹³C-δ ppm: 20.9(methyl), 33.5
(methylene), 54.3(methylene),
114.1(quarternary carbon).
Elemental analysis, calculated for
$C_4H_7N_3O_3$: C 33.10, H 4.86, N
28.96, found: C 33.45, H 4.84, N
28.60. In ¹H-NMR spectrum, four

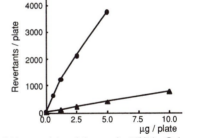
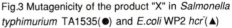

Fig.3 Mutagenicity of the product "X" in *Salmonella*
typhimurium TA1535(●) and *E.coli* WP2 *hcr*⁻(▲)

protons at 2.58, 3.03 and 4.29ppm had complicated couplings, which suggested the
existence of cyclic structure. In the IR spectrum, representative absorptions were
assigned to a nitrite ester. "X" was observable on TLC after development with
diphenylamine and UV (λ=254nm) irradiation. From the above results, the structure
of "X" was estimated as 5,6-dihydro-6-methyl-6-nitrito-4*H*-1,2,3-oxadiazine.

This structure was presumed to be synthesized from a diazotate intermediate
by cyclization in the presence of a nitrosating agent. A possible synthetic route to
"X" through the use of *N*-nitroso-*N*-(3-oxobutyl)-*p*-toluenesulfonamide has been
explored. Diketene was treated with ammonia gas to give acetoacetamide. After the
carbonyl group was protected with dithioacetal, the amide group was reduced to give
a primary amine. This primary amine was tosylated and deprotected to give *N*-(3-

oxobutyl)-*p*-toluenesulfonamide which was nitrosated to give *N*-nitroso-*N*-(3-oxo-butyl)-*p*-toluenesulfonamide. Treatment of the *N*-nitrosoamide with potassium ethoxide, however, did not give a diazotate. A different pathway for synthesizing "X" is now in progress.

ACKNOWLEDGMENTS

We deeply appreciate Ms.Kei Ito, Ms.Satoko Ishikawa, Ms.Shiho Ishimura and Ms.Tomoko Kawata for their valuable advice and technical assistance.

REFERENCES

1. Walling,C. *Acc.Chem.Res.* **1975**, *8*, 125.
2. Heur,Y.; Streeter,A.J.; Nims,R.W.; Keefer,L.K. *Chem.Res.Toxicol.* **1989**, *2*, 247.
3. Suzuki,E.; Mochizuki,M.; Shibuya,K.; Okada,M. *Gann*, **1983**, *74*, 41.

RECEIVED September 13, 1993

Chapter 32

Metabolism of Methylbutyl- and Methylamylnitrosamine by Rat and Human Esophagus and Other Tissues and Induction of Esophageal Adenocarcinoma in Rats

S. S. Mirvish[1], Q. Huang[1], S. C. Chen[1], S. S. Park[2], and H. V. Gelboin[2]

[1]Eppley Institute for the Research of Cancer, University of Nebraska Medical Center, Omaha, NE 68198
[2]National Cancer Institute, Bethesda, MD 20892

We examined whether microsomes of human and rat esophagus and liver can activate methyl-n-amylnitrosamine (MNAN, a rat esophageal carcinogen) and dimethylnitrosamine (DMN, a rat liver carcinogen) (1). Microsomes were prepared from human esophagi and samples with 0.6 mg protein were incubated for 20 min with MNAN and cytochrome P450 cofactors. Incubation with 12 mM MNAN gave mean values of 0.64 nmol formaldehyde, 0.21 nmol pentaldehyde and 0.56 nmol total hydroxy- (HO-) MNANs/min/mg protein. CO inhibited 81% of NMAA demethylation, indicating P450 involvement. Rat esophageal microsomes dealkylated MNAN similarly to human esophageal microsomes but with 2-6 x activity. Human and rat esophageal microsomes demethylated 6 mM NMAA 18-20 x as rapidly as they demethylated 5 mM DMN, whereas liver microsomes of these species demethylated 6 mM MNAN only 0.9-1.4 x as rapidly as they demethylated 5 mM DMN. However, liver microsomes of both species were more active than esophageal microsomes for MNAN depentylation, which leads to DNA guanine methylation. Hence, both human and rat esophagus may contain P450s that specifically dealkylate unsymmetric dialkylnitrosamines.

Like MNAN, methyl-n-butylnitrosamine (MBN) induces esophageal squamous cancer (ESC) in rats. Using tissue slice techniques (2), incubation of freshly removed rat esophagus with MBN produced mainly 2- and 3-HO-MBN. Liver slices produced mainly 3-HO-MBN and lung slices produced mainly 4-HO-MBN. These new metabolites of MBN were identified and determined by gas chromatography-thermal energy analysis and comparison with the independently synthesized compounds. We examined MBN metabolism by rat liver microsomes as before (3). The microsomes mainly catalyzed 3-hydroxylation and the previously studied demethylation and debutylation of MBN. Pretreatment with phenobarbital induced all 5 reactions (especially debutylation) of MBN. In similar tests, 3-methylcholanthrene induced 3-hydroxylation and debutylation, and isoniazid induced demethylation and debutylation of MBN. Monoclonal antibodies that inhibit P450 isozymes with specific epitopes were employed to identify isozymes involved in 50-100% of each reaction, as done before (4). Antibody 4-7-1 was used for inhibiting P450 2B1/2B2 because it appeared more specific than the previously used antibody 2-66-3.

For metabolism of both MBN and MNAN, rat esophagus effected hydroxylation chiefly at C-3 and also at C-2 (attributed to an esophagus-specific P450), rat lung slices carried out ω-hydroxylation (perhaps due to P450 4B1), and rat liver slices carried out ω - 1 hydroxylation. Both the induction and the antibody results indicated that, for both MBN and MNAN, P450 2C11 mainly catalyzes demethylation and ω - 1 hydroxylation, 1A1 or 1A2 mainly catalyzes 3-hydroxylation and debutylation of MBN or depentylation of MNAN, 2E1 catalyzes demethylation, and 2B1 or 2B2 mainly produces ω - 1 hydroxylation, demethylation and debutylation or depentylation. Note that both P450 1A1/1A2 and P450 2C11 catalyzed 3-hydroxylation of MBN, whereas 1A1/1A2 catalyzed 3-hydroxylation and 2C11 effected 4-hydroxylation of MNAN. Apparently, the lung enzyme and 2C11 determine hydroxylation site by its distance from the end of the alkyl chain, whereas the esophageal enzyme, 1A1/1A2 and 2E1 define hydroxylation site by its distance from the nitrosamine group.

Esophageal adenocarcinoma (EAC) is increasing in Western countries. EAC occurs in lower esophagus associated with Barrett's esophagus (replacement of squamous by glandular mucosa). Excess exposure to bile salts (which are irritant detergents) could play a role in etiology. To induce EAC in rats, Pera et al. (5) subjected rats to jejunoesophagostomy (joining jejunum to esophagus) with the lower esophageal sphincter tied off. 2,6-Dimethylnitrosomorpholine (DMNM) was then injected once weekly. ESC occurred in 40% of test rats receiving the higher of 2 doses of DMNM. Lower esophageal EAC occurred in 31% of test rats but not in control groups. In studies by us (6), male Sprague-Dawley rats were subjected to duodenoesophagostomy with the lower esophageal sphincter tied off, or to esophagogastroplasty (widening the lower esophageal sphincter) to increase gastric reflux into esophagus. Some groups were injected s.c. once weekly with DMNM or i.p. 3 x with MNAN (25 mg/kg). After 22 weeks, ESC incidence was about 30% in all groups that received a nitrosamine. Lower esophageal EAC occurred in 35% of rats that were subjected to duodenoesophagostomy and received a nitrosamine, 7% of rats with duodenoesophagostomy alone and 0% in other groups. Therefore, reflux of duodenal contents (including bile) to esophagus was more important than reflux of gastric contents for EAC induction in rats and perhaps also in humans. Grant support: R01-CA-35628 and core grant CA-36727 from NIH, grant 92-33 from Nebraska Health Dept., core grant SIG-16 from Am. Cancer Soc.

References

1. Huang, Q.; Stoner, G.; Resau, J.; Nickols, J.; Mirvish SS. *Cancer Res.* 1992, *52*, 3547-3551.
2. Mirvish, S.S.; Wang, M.Y.; Smith, J.W.; Deshpande, A.D.; Makary, M.; Issenberg, P. *Cancer Res.* 1985, *45*, 577-583.
3. Ji, C.; Mirvish, S.S.; Nickols, J.; Ishizaki, H.; Lee, M.J.; Yang, C.S. *Cancer Res.* 1989, *49*, 5299-5304.

4. Mirvish, S.S.; Huang, Q.; Ji, C.; Wang, S.; Park, S.S.; Gelboin, H.V. *Cancer Res.* 1991, *51*, 1059-1064.
5. Pera, M.; Cardesa, A.; Bombi, J.A.; Ernst, H.; Pera, C.; Mohr, U. *Cancer Res.* 1989, *49*, 6803-6808.
6. Attwood, S.E.A.; Smyrk, T.C.; DeMeester, T.R.; Mirvish, S.S.; Stein, H.J.; Hinder, R.A. *Surgery* 1992, *11*, 503-510.

RECEIVED January 26, 1994

Chapter 33

Chemistry of Putative Intermediates in Bioactivation of β-Oxidized Nitrosamines

Richard N. Loeppky, Eric Erb, Aloka Srinivasan, and Li Yu

Department of Chemistry, University of Missouri, Columbia, MO 65211

1,2,3-Oxadiazolinium cations **1** and **2** have been proposed as intermediates in the bioactivation of β-hydroxynitrosamines. Enzymatic sulfation of the OH of methylethanolnitrosamine **3a** followed by cyclization to **1** has been proposed as mechanism for the *in vivo* methylation of DNA by **3a** *(1)*. 3-Alkyl-5-hydroxy-1,2,3-oxadiazolinium cations **2** (R_1=H) form from α-nitrosamino aldehydes upon protonation *(2)*, and may be intermediates in the transnitrosation reactions of these aldehydes which are formed enzymatically from the oxidation of their corresponding alcohols *(3)*. A goal of this research is the elucidation of the poorly understood chemistry of 1,2,3-oxadiazolinium cations toward nucleophiles of biological significance.

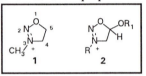

3-Alkyl-5-hydroxy-1,2,3-oxadiazolinium cations are too unstable for definitive studies with bases and nucleophiles. Their 5-methoxy analogs **2** (R_1=CH$_3$) were prepared by treatment of the methyl acetals of the respective α-nitrosamino aldehydes with trimethylsilyl triflate. The $t_{1/2}$ of **1** in water at 25° C is 5 days, (**3a** is the exclusive product) whereas that of **2** (R,R_1=CH$_3$) is less than 5 minutes. The reactions of these compounds were studied in aqueous as well as organic solvents.

In order to determine whether oxadiazolinium cations of structure type **2** could be intermediates in transnitrosation reactions of α-nitrosamino aldehydes, we examined the reactions of **2** (R = CH$_3$ or C$_6$H$_5$,R_1=CH$_3$) with morpholine. N-nitrosomorpholine (NMOR) forms from both substrates in CH$_2$Cl$_2$ (CH$_3$,18%; C$_6$H$_5$,54%). Reaction of **2** (R = CH$_3$) with morpholine in CH$_2$Cl$_2$ also generates a 7% yield of N-methylmorpholine. Attack of morpholine at the O-CH$_3$ of **2** would yield methylethanalnitrosamine which is known to transnitrosate morpholine. But no N-methylmorpholine is found among the reaction products of **2** (R=C$_6$H$_5$, R_1=CH$_3$) ruling out this pathway. These results suggest a possible role for 3-alkyl-5-hydroxy-1,2,3-oxadiazolinium cations in the interesting transnitrosation chemistry of α-nitrosamino aldehydes.

0097–6156/94/0553–0334$08.00/0

The reactions of **1** and **2** toward nucleophiles are complicated because the protons adjacent to the positively charged N are

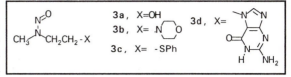

acidic enough to be easily abstracted at pH 9-10 (*4*). Basic nucleophiles attack as both nucleophiles and bases. Morpholine does not react with **1** in either H_2O or CH_2Cl_2 to produce NMOR. In CH_2Cl_2 it attacks C-5 to give a 25% yield of the open chained nitrosamine **3b** and attack at the methyl group to give N-methyl-morpholine is relatively minor (13%). We have also shown that thiophenol (in CH_2Cl_2, but not H_2O) attacks C-5 to give **3c** (33%). This is the only carbon attacked by water (giving **3a**), and Michejda has also shown that it is the primary position for attack by acetic acid and aniline (*5*). Guanine derivatives and guanine residues in DNA react with **1** to incorporate the entire nitrosamine fragment at N^7 (**3d**). This is the major reaction of **1** with DNA *in vitro*. Only small amounts of base methylation are observed. Despite numerous experiments designed to specifically detect the products of nucleophilic attack at C-4 in **1** and related compounds we have not observed any attack at this position.

The major product of the reaction of **1** with PhSH in H_2O is diphenyldisulfide, PhS-SPh (92%) the other isolated products were $PhSCH_3$ (5%) and **3c** (1%). In CH_2Cl_2 the yield of PhS-SPh is decreased (13%) and the products of nucleophilic substitution are increased (47%). Because of the difficulty of product isolation with **1** the reaction of the analogous 3-phenyl-1,2,3-oxadiazolinium triflate with PhSH was examined. The yield of PhS-SPh in aqueous buffer at pH 7.4. was 41%. Other key products included benzene (60%), azobenzene (8%), and glycolaldehyde (75%). These products and the observations of the Hünig group (*4*) suggest the involvement of phenyldiazene (Ph-N=NH) as a reaction intermediate. Generation of this unstable compound from potassium phenyldiazocarboxylate in the presence of PhSH also led to the disulfide and many of the other reaction products produced from the reaction of **2** ($R=C_6H_5$, $R_1=CH_3$) with PhSH. Further work showed that the

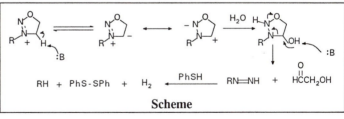

Scheme

reaction of **1** with morpholine in H_2O generated significant quantities of methane, which we believe is formed from methyldiazene. We believe that these compounds are arising through abstraction of the C-4 H of proton of oxadiazolinium cation as shown in the scheme. The oxidation rate of N-acetylcysteine by **1** in aqueous buffer occurs most rapidly at higher pH's. It is probable that base generates the diazene which oxidizes the thiol. If **1** were generated *in vivo* its reduction by abundant thiols is very likely. Collectively, these data cast doubt on the ability of **1** to methylate DNA *in vivo*.

Acknowledgment: The support of this research by grant RO1-ES03953 from the National Institute of Environmental Health Sciences is gratefully acknowledged.
References:
(1) Kroeger-Koepke, M. B.; Koepke, S. R.; Hernandez, L.; Michejda, C. J. *Cancer Res.* **1992**, *52*, 3300-3305 and references contained therein.
(2) Loeppky, R. N.; Fleischmann, E. D.; Adams, J. E.; Tomasik, W.; Schlemper, E. O.; Wong, T. C. *J. Am. Chem. Soc.* **1988**, *110*, 5946-5951.
(3) Loeppky, R. N.; Tomasik, W.; Kerrick, B., E. *Carcinogenesis (London)* **1987**, *8*, 941-946.
(4) Hünig, S. *Helv. Chim. Acta* **1971**, *54*, 1721-1747..
(5) Michejda, C. J.; Koepke, S. R.; Kupper, R. *IARC Sci. Publ.*,**1980**, *31* (N-Nitroso Compd.: Anal., Form. Occurrence), 155-67.

RECEIVED November 15, 1993

Chapter 34

Modulation of N-Nitrosomethylbenzylamine Metabolism in Rats by Concurrently Administered Ethanol and Diallyl Sulfide

B. I. Ludeke, Y. Yamada, F. Dominé, and P. Kleihues

Institute of Neuropathology, University Hospital, CH–8091 Zurich, Switzerland

The development of malignancies in humans is determined by exposure not only to complex mixtures of exogenous and endogenous carcinogenic agents, but also to modulators of carcinogen action. Both factors which enhance the carcinogenicity of known carcinogens as well as chemoprotective compounds have been identified in common foodstuffs. The elucidation of their modes of action and the determination of their efficacies should be useful in increasing our knowledge of the mechanisms of multistep carcinogenesis and in designing improved cancer prevention strategies. We have investigated the effects of ethanol and various alcoholic beverages, established risk factors for esophageal cancer in humans, and of diallyl sulfide (DAS), a major fragrance and flavor component of garlic oil with chemopreventive properties, on the bioactivation in rats of N-nitrosomethylbenzylamine (NMBzA), a highly potent and selective esophageal carcinogen in this species.

Esophageal cancer is uncommon in Western countries, except in the French provinces of Brittany and Normandy, where mortality rates for esophageal cancer in males are elevated three to four times above the European average (1). Consumption of alcoholic beverages and use of tobacco products are the leading causes of esophageal cancer in industrialized countries. Although the effects of ethanol and tobacco are multiplicative, the relative risk for esophageal cancer was shown to increase significantly with daily intake of alcohol even in non-smokers (reviewed in 1). The relative risk is furthermore dependent on the type of beverage consumed, and an exceptionally high risk has been attributed to apple-based distillates (calvados), which are traditionally popular in Brittany and Normandy (2). Ethanol has previously been shown to competitively inhibit hepatic nitrosamine metabolism in vivo (3). It has been suggested that consumption of alcoholic beverages increases extrahepatic carcinogen concentrations and may thus contribute to the ethanol-associated increase in tumors of the esophagus and other extrahepatic tissues in which neoplastic transformation in humans is otherwise infrequent (3). We have now determined the dose-activity relationship of concurrently administered ethanol on the interorgan shift in the bioactivation of NMBzA in rats (4)..

Male F344 rats received a single intragastric dose of NMBzA (2.5 mg/kg body wt; 7.4 ml/kg body wt) in tap water containing 0-20% (v/v) ethanol. After a survival time of 3 h, concentrations of the promutagenic base O^6–methyldeoxyguanosine

0097–6156/94/0553–0337$08.00/0

(O^6-MEdG) were quantitated by an immunoslot-blot assay. In controls, levels of O^6-MEdG were similar in the esophagus, lung and liver (11-14 µmol/mol dG). In the esophagus, simultaneously administered ethanol increased concentrations of O^6-MEdG from 15.2 µmol/mol (0.1% ethanol) to 46 µmol/mol (20%). This increase was dose-dependent for 1-20% ethanol, with low doses causing a larger effect per gram of ethanol than high doses. In lung, levels of O^6-MEdG increased from 11 µmol/mol (0.1%) to a plateau value of 24 µmol/mol ($\geq$ 5%). In liver, a slight but statistically nonsignificant trend toward decreased levels of DNA methylation was observed with increasing doses of ethanol.

In order to be able to detect both enhancing and inhibitory effects of alcoholic beverages on NMBzA metabolism, the beverages were adjusted to an ethanol content of 4% prior to administration. Increases in esophageal O^6-MEdG were similar (+50% to +116%) with pear brandy, sake, farm-made calvados, gin, Scotch whisky, white wine, Pilsner beer and aqueous ethanol. Significantly higher increases in DNA methylation were observed for commercially distilled calvados (+125%) and red burgundy (+162%). In contrast to its effects at 4% ethanol, farm-made calvados at 20% ethanol elicited a significantly higher increase (+200%) in esophageal DNA methylation than aqueous ethanol at 20% (+148%). The effects of the various beverages on the formation of O^6-MEdG in liver were statistically nonsignificant. Although diluted to 4% ethanol, the total amounts of beverage administered were equivalent to a person of 80 kg drinking 470 ml of beer, 180 ml of wine, 120 ml of sake, 50 ml of pear brandy, gin, whisky or commercially distilled calvados, or 30 ml of farm-made calvados, and were thus comparable to average-size single servings. Our results show that simultaneously ingested ethanol is an effective modulator of nitrosamine bioactivation at consumption levels typical of moderate social drinking, and that some alcoholic beverages contain congeners that amplify the effects of acute ethanol. In view of the close synergism between tobacco smoke and alcohol, it is therefore conceivable that short-term modulation of nitrosamine metabolism by ethanol plays a role in the etiology of some forms of human cancer.

DAS is an effective inhibitor of tumorigenesis by various metabolically activated carcinogens. In rats, preadministration of DAS resulted in the complete suppression of neoplastic transformation in the esophagus by NMBzA (5). In vitro studies demonstrating competitive inhibition by DAS of microsomal oxidation of NMBzA and other nitrosamines suggested that the anticarcinogenic properties of DAS may include the modulation of nitrosamine bioactivation in situ. In the chemoprevention assay, rats received five weekly treatments consisting of a single intragastric dose of 200 mg/kg of DAS followed 3 h later by a s.c. injection of NMBzA (3.5 mg/kg) (5). We have assessed the effects of doses of 10-200 mg/kg of DAS on DNA methylation by NMBzA in various rat tissues under conditions equivalent to a single treatment cycle of the chemoprevention experiment (6). After a survival time of 6 h, levels of O^6-MEdG and 7-MEdG were determined by immunoslot-blot assays. Pretreatment with 200 mg/kg of DAS led to decreases in concentrations of O^6-MEdG in esophagus (-26%), nasal mucosa (-51%), trachea (-68%) and lung (-78%). In liver, levels of 7-MEdG were reduced by 43%. Inhibition of DNA methylation was proportional to dose for >25 mg/kg in esophagus, liver and nasal mucosa, for 25-200 mg/kg in trachea and 10-50 mg/kg in liver. The dose-activity relationship for inhibition of NMBzA bioactivation suggests that short-term modulation of carcinogen bioactivation by DAS is only one of multiple mechanisms by which DAS suppresses nitrosamine carcinogenesis.

Literature Cited
1. Driver, H.E.; Swann, P.F. *Anticancer Res.* **1987**, *7*, 309-320.
2. Tuyns, A.J.; Péquinot, G.; Abbatucci, J.S. *Int. J. Cancer* **1979**, *23*, 443-447.
3. Swann, P.F.; Coe, A.M.; Mace, R. *Carcinogenesis* **1984**, *5*, 1337-1343.
4. Yamada, Y.; Weller, R.O.; Kleihues, P.; Ludeke, B.I. *Carcinogenesis* **1992**, *13*, 1171-1175.
5. Wargovich, M.J.; Woods, C.; Eng, V.W.S.; Stephens, L.C.; Gray, K. *Cancer Res.* **1988**, *48*, 6872-6875.
6. Ludeke, B.I.; Dominé, F.; Ohgaki, H.; Kleihues, P. *Carcinogenesis* **1992**, *13*, 2467-2470.

RECEIVED January 31, 1994

Chapter 35

DNA Adducts Induced by Pancreas-Specific Nitrosamines

Demetrius M. Kokkinakis[1] and Jeffrey R. Norgle[2]

[1]Southwestern Medical Center, University of Texas, 5323 Harry Hines Boulevard, Dallas, TX 75235–9036
[2]Northwestern University, 710 Fairbanks Court, Chicago, IL 60611

Pancreas specific nitrosamines (PSNs) are β-oxidized derivatives of N-nitrosodiisopropylamine inducing adenocarcinomas of ductal origin in the pancreas of Syrian hamsters *(1)*. PSNs methylate and hydroxypropylate DNA yielding 7- and O^6-methylguanines (7-mG and O^6-mG) and 7- and O^6-(2-hydroxypropyl)guanines (7-hpG and O^6-hpG) *(2,3)*. Levels of adducts and ratios of methylation *versus* hydroxypropylation depend on the tissue and oxidation state of the β-carbon of the nitrosamine *(3)*. One of the most potent PSN, N-nitrosbis(2-oxopropyl)amine (BOP), yields methylating agents while reduced PSNs such as N-nitroso(2-hydroxypropyl)(2-oxopropyl)amine (HPOP) and N-nitrosobis(2-hydroxypropyl)amine (BHP) mainly yield 2-hydroxypropylating agents *(3,4)*. Accordingly, high ratios of methyl to hydroxypropyl adducts observed in the DNA of liver and kidney of hamsters treated with BOP, indicate rapid activation of this compound. On the other hand, the low ratios observed in the pancreas indicate that this organ is exposed and activates not only the parent compound, but also its reduced metabolites HPOP and BHP.

RESULTS

Initiation of pancreatic cancer was studied using a 9-day continuous treatment of female hamsters with 210 mg/kg HPOP which induces a high incidence of pancreatic ductal tumors and a lower incidence of liver cholangiocarcinomas within 25 wks *(5)*. During exposure to HPOP, methyl and hydroxypropyl adducts gradually rose to reach maximum levels at the 9th day. Ratios of O^6-mG to O^6-hpG were 3.8 in liver and 2.0 in pancreas, while respective ratios of 7-mG to 7-hpG were 8.3 and 3.2. Subsequent removal of DNA adducts was slower than that observed after a single injection of PSN (Table I), indicating that continuous treatment with the carcinogen resulted in the reduction of the capacity

0097–6156/94/0553–0340$08.00/0
© 1994 American Chemical Society

TABLE I.
LEVELS OF DNA ADDUCTS IN HAMSTER TISSUES AND HALF LIVES OF THEIR REPAIR

	LIVER		KIDNEY		LUNG		PANCREAS	
	s.i.	*c.i.*	*s.i.*	*c.i.*	*s.i.*	*c.i.*	*s.i.*	*c.i.*
7-mG (nmol/mmol G)	690	1701	130	68	160	221	30	72
half-life (hrs)	44	46	28	55	32	72	32	96
7-hpG (nmol/mmol G	65	206	19	45	40	36	7	22
half-life (hrs)	110	216	70	132	72	140	72	140
O^6-mG (nmol/mmol G)	77	519	28	55	23	55	22	45
half-life (hrs)	100	240	66	>240	70	>240	62	>240
O^6-hpG (nmol/mmol G)	22	138	7	44	13	33	3	22
half-life (hrs)	100	NR	ND	NR	ND	NR	ND	NR

s.i.: single injection of 27 mg/kg HPOP; *c.i.*: continuous infusion of HPOP at a daily dose of 27 mg/kg;
NR: not repaired; ND: not determined.
(Reproduced with permission from reference 7. Copyright 1993 American Association for
Cancer Research, Inc.)

of certain tissues to repair DNA damage. Removal of methyl adducts was markedly faster than that of 2-hydroxypropyl adducts. O^6-mG was repaired in liver and to some extent in other tissues, while O^6-hpG was not repaired for at least 10 days after termination of carcinogen treatment. The above differences in repair resulted in a gradual decrease of the ratios of methyl *versus* hydroxypropyl adducts. The slow repair of O^6-hpG contrasts that of O^6-hydroxyethylguanine which is repaired faster than O^6-mG *(6)*.

TABLE II.
LEVELS OF DNA SYNTHESIS IN THE PANCREAS OF HPOP TREATED HAMSTERS

	DAYS AFTER INITIATION OF HPOP TREATMENT				
	7	10	14	20	25
[^{3}H] THYMIDINE INCORPORATION IN DNA:	160	225	240	237	175

HPOP was mitogenic in the pancreas of Syrian hamsters. Levels of DNA synthesis were greater in HPOP treated animals than in controls and reached a maximum level between 5 to 11 days after termination of carcinogen treatment (Table II). Although large numbers of acinar and ductular cells synthesized DNA during the administration of the carcinogen and were likely to mutate, such cells did not appear to subsequently subdivide (Figure 1). In fact, most of these cells were not present 12 days after termination of such treatment (Figure 1). Increased rates of cell division at the post-treatment stage (days 10 to 21) indicated that mutations leading to cancer may arise from division of cells with partially repaired DNA. Duct cells synthesizing DNA after termination of carcinogen treatment did not only survive, but subsequently subdivided (data not shown). The above results suggest that methyl adducts contribute to the toxicity of HPOP in the pancreas, while hydroxypropyl adducts may contribute significantly to its mutagenicity.

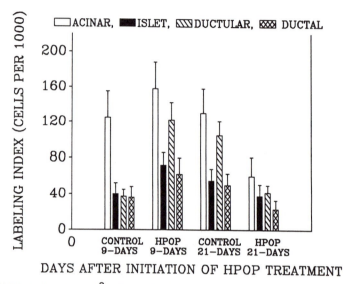

DAYS AFTER INITIATION OF HPOP TREATMENT

FIGURE 1. Levels of [³H]-Thymidine labeling in various cell populations of the pancreas of Syrian hamsters 9 and 21 days after treatment with 210 mg/kg HPOP administered continuously for 9 days. Controls were treated with saline. Single injections of [³H]-Thymidine were given at days 2, 5 and 8 during the carcinogen treatment.
(Reproduced with permission from reference 7. Copyright 1993 American Association for Cancer Research, Inc.)

REFERENCES

[1] Pour, P., Kruger, F.W., Althoff, J., Cardessa, A., and Mohr, U. (1975) J. Natl. Cancer Inst. *54*, 141-145.
[2] Kokkinakis, D.M. and Scarpelli, D.G. (1989) Cancer Res. *49*, 3184-3189.
[3] Kokkinakis, D.M. (1992) Carcinogenesis *13*, 759-765.
[4] Kokkinakis, D.M. (1991) Chem-Biol. Interactions *78*, 167-181.
[5] Kokkinakis, D.M. (1990) Carcinogenesis *11*, 1909-1913.
[6] Koepke, S.R., Kroeger-Koepke, M.B., Bosan, W., Thomas, B.J., Alvord, W.G., and Michejda, C.J. (1988) Cancer Res. *48*, 1537-1542.
[7] Kokkinakis, D.M. Subbarao, V. (1993) Cancer Res. *53*, 2790-2795.

RECEIVED January 26, 1994

Chapter 36

DNA Pyridyloxobutylation

4-(Acetoxymethylnitrosamino)-1-(3-pyridyl)-1-butanone Inhibits the Repair of O^6-Methylguanine

Lisa A. Peterson, Xiao-Keng Liu, and Stephen S. Hecht

Division of Chemical Carcinogenesis, American Health Foundation,
1 Dana Road, Valhalla, NY 10595

4-(Methylnitrosamino)-1-(3-pyridyl)-1-butanone (NNK) both methylates and pyridyloxo-butylates DNA (Figure 1). Mechanistic studies in the A/J mouse demonstrated that O^6-methylguanine (O^6-mG) persistence was correlated to NNK-induced lung tumor activity (1). While only weakly carcinogenic, the pyridyloxobutylating agent, 4-(acetoxymethyl-nitrosamino)-1-(3-pyridyl)-1-butanone (NNKOAc), enhanced the tumorigenicity of the methyl-ating agent, acetoxymethylmethylnitrosamine (AMMN). This increase in tumorigenic activity was associated with an increase in O^6-mG levels.

One mechanism by which pyridyloxobutylation may increase O^6-mG levels is through inhibition of the repair protein, O^6-alkylguanine-DNA alkyltransferase (AGT). We tested this hypothesis by determining whether NNKOAc could inactivate AGT. Transferase activity was determined by measuring the amount of [^{3}H]methyl groups transferred from [^{3}H]Me-DNA to the protein (2). Co-incubation of NNKOAc (0-5 mM) with semipurified rat liver AGT and [^{3}H]Me-DNA led to a concentration-dependent decrease in transferase activity (Figure 2). Pre-incubation of NNKOAc with AGT prior to the addition of the [^{3}H]Me-DNA substrate led to a modest increase in inhibitory activity (Figure 2), suggesting that NNKOAc may react directly with the protein.

Pre-incubation of AGT with NNKOAc-treated DNA also resulted in AGT inactivation. The extent of AGT inhibition was related to the concentration of NNKOAc (0-5 mM) used to alkylate the DNA (Figure 3). Little or no inhibition was observed when AGT was incubated with DNA which had been reacted with the hydrolysis products of NNKOAc, including form-aldehyde, 4-hydroxy-1-(3-pyridyl)-1-butanone (HPB), or 1-(3-pyridyl)but-2-en-1-one. Furthermore, DNA incubated with NNKOAc in the absence of esterase did not significantly affect transferase activity. The inhibitory activity of NNKOAc-treated DNA was lost when the DNA was subjected to neutral thermal hydrolysis. This treatment releases the unstable pyridyloxobutyl DNA adducts from DNA as HPB (3). These data indicate that pyridyloxo-butyl adducts are capable of inactivating rat liver AGT. When the ratio of pmol AGT inac-tivated to pmol of HPB-releasing adducts present was determined, 2-4% of the HPB-releasing adducts were responsible for the inhibition reaction. These inhibitory adducts were stable for at least 4 days in pH 7 buffer at 37°C, suggesting that they have sufficient lifetime in DNA under physiological conditions.

The ability of pyridyloxobutyl adducts to compete with O^6-mG was determined by co-incubating NNKOAc-treated DNA with [^{3}H]MedNA and AGT. The presence of pyridyloxo-butyl DNA adducts diminished the amount of [^{3}H]methyl groups transferred to AGT (Table I). Therefore these adducts compete well with O^6-mG for reaction with AGT.

0097–6156/94/0553–0343$08.00/0

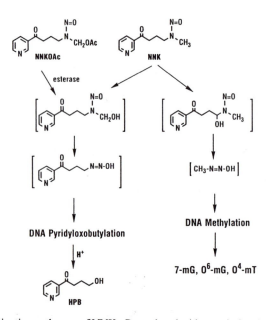

Figure 1. Bioactivation pathways of NNK. Reproduced with permission from Reference 4. Copyright 1993 American Association for Cancer Research.

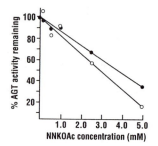

Figure 2. Inhibition of AGT by NNKOAc; ●, co-incubation; ○, pre-incubation.

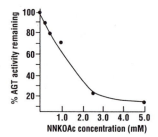

Figure 3. Inhibition of AGT by NNKOAc-Treated DNA. Reproduced with permission from Reference 4. Copyright 1993 American Association for Cancer Research.

Table I. Competition of NNKOAc-Treated DNA with [³H]MeDNA
for Reaction with AGT[a]

pmol HPB-releasing adducts in DNA[b]	pmol O⁶-mG in [³H]MeDNA	pmol [³H]methyl transferred to AGT	% inhibition of transfer
0	1.5	0.85 ± 0.07	0
5.5 (0.17)	1.5	0.79 ± 0.03	7
11.0 (0.33)	1.5	0.55 ± 0.01	35
16.5 (0.50)	1.5	0.45 ± 0.06	47
27.5 (0.83)	1.5	0.37 ± 0.01	56

[a]NNKOAc-treated DNA and [³H]MeDNA were incubated with AGT for 30 min at 37°C
[b]Number in parentheses is estimated level of inhibitory adducts (3% of total).

Our results demonstrate that the pyridyloxobutylation pathway can inactivate AGT either directly or via a DNA adduct. Furthermore these adducts can compete with O⁶-mG for reaction with AGT. If this competition occurs *in vivo*, more O⁶-mG adducts will persist to potentially initiate tumors when both DNA alkylation pathways are operative. Therefore we propose a co-carcinogenic role for pyridyloxobutylation in which a pyridyloxobutyl DNA adduct is capable of increasing the levels of O⁶-mG through competition for reaction with AGT. These studies are supported by CA-44377 from the National Cancer Institute.

References

1.   Peterson, L.A.; Hecht, S.S. *Cancer Res.* 1991, *51*, 5557-5564.
2.   Myrnes, B., Norstrand, K., Giercksky, K.-E., Sjunneskog, C., and Krokan, H. *Carcinogenesis* 1984, *5*, 1061-1064.
3.   Hecht, S.S., Spratt, T.E., and Trushin, N. *Carcinogenesis* 1988, *9*, 161-165.
4.   Peterson, L.A.; Liu, X.-K.; and Hecht, S.S. *Cancer Res.* 1993, *53*, 2780-2785.

RECEIVED January 26, 1994

Chapter 37

Sequence-Specific Methylation of Single- and Double-Stranded DNA by Methylnitrosourea

R. W. Wurdeman and B. Gold

Eppley Institute and Department of Pharmaceutical Sciences, University of Nebraska Medical Center, Omaha, NE 68198

DNA damage induced by chemical and physical agents and "spontaneous" processes can lead to base pair mismatches and bulge sites (1). These non-Watson-Crick base pairing motifs may be converted into somatic mutations during DNA repair or replication. The impact of these abnormal base pairing arrangements on DNA structure was probed using the sequence selective methylation of DNA at N7-G by N-methyl-N-nitrosourea (MNU) (2). The reactions of MNU with 5'-[^{32}P]P-labeled single-stranded (s-s) DNA $\underline{1}$ [$\underline{1}$, 5'-d(CACTG^{5}G^{6}G^{7}ACTG^{11}C)], complementary double-stranded (d-s) DNA $\underline{1+2}$ [$\underline{2}$, 3'-d(GTGACCCTGACG)], d-s DNA's with a G-G mismatch ($\underline{1+3}$, [$\underline{3}$, 3'-GTGAC$\underline{G}$CTGACG]), d-s DNA with a G-A mismatch ($\underline{1+4}$ [$\underline{4}$, 3'-GTGAC$\underline{A}$CTGACG]), d-s DNA with a G-T mismatch ($\underline{1+5}$ [$\underline{5}$, 3'-GTGAC$\underline{T}$CTGACG]), d-s DNA $\underline{1+6}$ ($\underline{6}$, 3'-GTGAC_CTGACG) with a single bulged G and d-s DNA $\underline{1+7}$ ($\underline{7}$, 3'-GTGA_C_TGACG) with two bulged G's are reported. In addition, the effect of temperature on the methylation of $\underline{1+2}$ is reported along with the T$_m$'s of all the duplexes and the CD spectra of $\underline{1}$, $\underline{2}$ and $\underline{1+2}$.

Experimental

Oligomers $\underline{1}$-$\underline{7}$ were prepared on a DNA synthesizer and purified by LC. $\underline{1}$ was 5'-[^{32}P]-end-labeled with T4 kinase and purified by electrophoresis on a 2% polyacrylamide gel. The 5'-[^{32}P]-end-labeled $\underline{1}$ (with or without $\underline{2}$-$\underline{7}$) was incubated with 500 μM MNU in 10 mM buffer (pH 7.8) containing 100 mM NaCl and 1 mM EDTA for 2 h at 20 °C. The DNA was treated with piperidine to selectively convert the N7-MeG lesions into strand breaks (3) and the DNA was run on 20% polyacrylamide denaturing gel. The gel then was exposed to X-ray film and the autoradiogram analyzed on a scanning densitometer (Fig. 1). The effect of temperature on the methylation of $\underline{1+2}$ was performed as described above except for the change in temperature (Fig. 2). The denaturation of $\underline{1+2}$-$\underline{7}$ as a function of temperature was followed by monitoring their UV absorbance at 260 nm: T$_m$ of $\underline{1+2}$ (56 °C), $\underline{1+3}$ (44 °C), $\underline{1+4}$ (51 °C), $\underline{1+5}$ (46 °C), $\underline{1+6}$ (43 °C) and $\underline{1+7}$ (31 °C). The CD spectra of 168 μM $\underline{1}$, $\underline{2}$ and $\underline{1+2}$ are shown in Fig. 3.

0097–6156/94/0553–0346$08.00/0

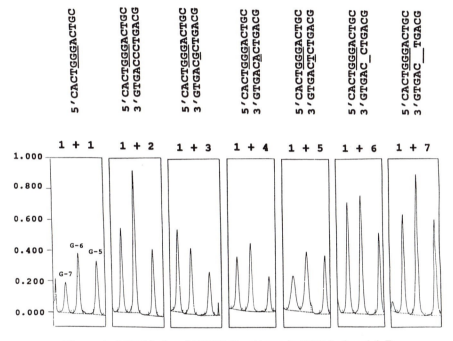

Figure 1. MNU-induced N7-MeG patterns in DNA's 1 and 1-7.

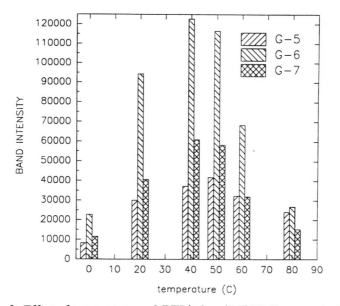

Figure 2. Effect of temperature on MNU-induced N7-MeG pattern in 1+2.

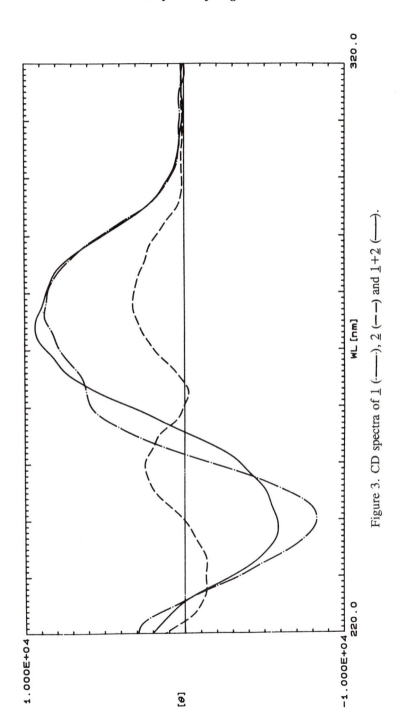

Figure 3. CD spectra of 1 (·——·), 2 (——) and 1+2 (——).

Results and Discussion

The results for N7-MeG at $G^5G^6G^7$ show that: (a) there is a distinct methylation pattern in $\underline{1}$ (Fig. 1); (b) the methylation in $\underline{1}$ is reduced by ~60% as compared to duplex target $\underline{1}+\underline{2}$; (c) as the temperature of the reaction exceeds 50 °C the cleavage pattern in $\underline{1}+\underline{2}$ (T_m of 50 °C) and the absolute intensities of the bands convert to those seen in s-s target (Fig. 2); and (d) the cleavage patterns in duplex $\underline{1}+\underline{3}$-$\underline{7}$ with mismatched or bulged G's are, to varying degrees, quantitatively and qualitatively different from normal duplex $\underline{1}+\underline{2}$ (Fig. 1). The methylation pattern seen with $\underline{1}$ indicates persistent s-s DNA structure and is confirmed by the CD which shows that $\underline{1}$ and $\underline{1}+\underline{2}$ have similar Cotton bands indicative of B-form DNA (Fig. 3). The methylation pattern in $\underline{1}$ may result from the presence of G residues in the syn-conformation. We also propose that the origin of the patterns in the DNA with the mismatches and bulges is related to perturbations in the stacking of the G's (1,4) and/or the presence of G's in the syn-conformation (5).

We have shown for the first time that the methylation pattern induced by MNU at G can distinguish between normal and abnormal DNA structures. This method may be useful in footprinting mismatches, bulges and other non-Watson-Crick base pairs in large DNA fragments as well as in monitoring the repair of different abnormal DNA base pairing motifs.

REFERENCES

1. (a) Karran, P.; Lindahl, T. *Biochemistry* **1980**, *19*, 6005-6011. (b) Loeb, L. A.; Preston, B. D. *Annu. Rev. Genet.* **1986**, *20*, 201-230. (c) Kunkel, T. A. *Biochemistry* **1990**, *29*, 8003-8011.

2. (a) Wurdeman, R. L.; Gold, B. *Chem. Res. Toxicol.* **1988**, *1*, 146-147. (b) Wurdeman, R. L.; Church, K. M.; Gold, B. *J. Am. Chem. Soc.* **1989**, *111*, 6408-6412.

3. Maxam, A. M.; Gilbert, W. *Methods Enzymol.* **1980**, *65*, 499-560.

4. Pullman, A.; Pullman, B. *Q. Rev. Biophys.* **1981**, *14*, 289-380.

5. Kennard, O.; Hunter, W. N. *Quart. Rev. Biophys.* **1989**, *22*, 327-379.

RECEIVED January 26, 1994

Nitrosamine Occurrence:
Brief Discussions of Research

Chapter 38

Detection of Tobacco-Related Hemoglobin Adducts by Quadrupole Mass Spectrometry

E. Richter, B. Falter, C. Kutzer, and J. Schulze

Walther Straub Institute of Pharmacology and Toxicology,
Nussbaumstrasse 26, W–8000 Munich 2, Germany

Biomonitoring has been used in an attempt to quantitate the exposure of humans to foreign compounds. It has been successful in quantifying occupational exposure; the lower adduct levels to proteins or DNA in environmental exposure has been an obstacle in quantifying these expositions.

Several hemoglobin adducts have been proposed for the biomonitoring of individuals towards exposure to tobacco. Although an overall correlation between adduct levels of 4-aminobiphenyl (4-ABP) and the number of cigarettes smoked has been found, 4-ABP measurements cannot be interpreted with regard to individual exposure (1,2). An improvement in this situation may be achieved through simultaneous measurement of other parameters linked exclusively to tobacco exposure, e.g. tobacco-specific nitrosamines.

To improve interpretation of Hb adduct levels we additionally determined 4-hydroxy-1-(3'-pyridyl)-1-butanone (HPB) adducts stemming from the metabolism of tobacco-specific nitrosamines. For increased sensitivity we employed negative chemical ionization and quadrupole mass spectrometry (QMD 1000 mass spectrometer, Fisons Instruments, Mainz, FRG), with methane used as ionization gas.

When optimizing detection conditions for HPB this compound could be detected in standard samples in amounts as low as 0.5 pg. The slope of the standard curve was linear at least to 1 ng.

Attempts to calibrate this detection method for 4-ABP resulted in different curves obtained at different time points. To solve this inconsistency we repeated the measurement of 4-ABP. The signal obtained for an identical amount of 4-ABP dropped during these repetitions, as is shown in Figure 1.

Two different parts can be separated. An initial drop during the first 5-10 repetitions is caused by traces of the tuning compound heptacosa remaining in the ionization chamber. This can be demonstrated by an increase in signal intensity after tuning.

During the later repetitions the signal obtained dropped further by 50% in the run shown in Figure 1. In different test runs over 24 h the last signal was 5 to 65% of

0097–6156/94/0553–0352$08.00/0

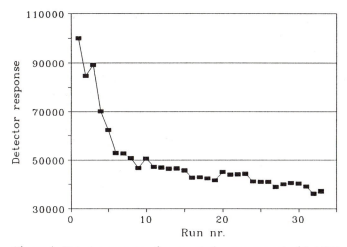

Figure 1: Detector response for repeated measurement of 4-ABP

the initial value. Original sensitivity was restored after intense cleaning of the ion source, especially the first reflector.

Similar effects were observed after repeated GC-MS-quantification for 4'-F-4-ABP, which is used as a recovery standard, as well as for anthraquinone. Both compounds exerted an initial decrease which has been attributed to the presence of heptacosa, as well as a continuing decrease which has not yet been explained.

The observed decrease is a substance specific effect rather than an instrumental problem. This is illustrated by the simultaneous measurement of both anthraquinone and HCB (Figure 2). The signal obtained for HCB is constant over 20 hours, whereas

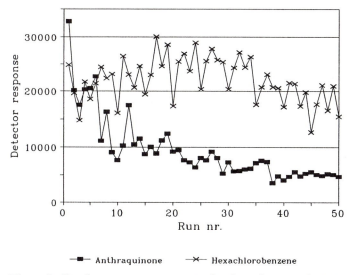

Figure 2: Simultaneous measurement of anthraquinone and HCB

the signal obtained for anthraquinone decreases. HPB like HCB did not show any decrease after repeated measurements.

Under optimal conditions, 4-ABP can be quantified from standard solutions as low as 50-100 fg. However for quantitation it is necessary to introduce a "sensitivity marker" indicating the actual sensitivity at the time of the measurement.

Results obtained from repeated quantification of 4-ABP in hemoglobin, as well as measurements from blood samples drawn at different time points, indicate a good reproducibility. So far 4-ABP adduct levels (pg/g HB; mean ± std err) were determined in 17 smokers (118 ± 26) and 9 nonsmokers (33 ± 8). These values are in accordance with data published previously (2). HPB adduct levels (fmol/g Hb) in 10 smokers (range 21 - 138) and 7 nonsmokers (range 10 - 61) were much more variable. The sample size is still to small to draw any conclusions on possible correlations between the two smoking markers.

REFERENCES

1. Wilson VL Weston A Manchester DK Trivers GE Roberts DW Kadlubar FF Wild CP Montesano R Willey JC Mann DL Harris CC *Carcinogenesis* (1989) *10*, 2149-2153
2. Skipper PL Tannenbaum SR *Carcinogenesis* (1990) *11*, 507-518

RECEIVED January 26, 1994

Chapter 39

Nonvolatile *N*-Nitrosamides in Dried Squid
Analysis by High-Performance Liquid Chromatography—Photolysis—Chemiluminescence

S. H. Kim[1] and J. H. Hotchkiss[2]

[1]Cheju National University, Cheju-do, Korea
[2]Institute of Food Science, Stocking Hall, Cornell University, Ithaca, NY 14853

The consumption of dried and salted fish products has been associated with nasopharyngeal and gastric cancers (*1*). N-nitroso compounds (NOC) have been suggested as one potential etiological factor (*2*). Relatively high amounts of volatile N-nitrosamines (VNA) are formed when dried squid is broiled (*3*). It is possible that dried squid also contains nonvolatile N-nitroso compounds (NVNOC) which might play a role in the risk associated with dried fish. However, analytical methods for NVNOC, and particularly N-nitrosamides in foods, have not been developed. Recently developed instrumentation, based on the photolysis of NOC to produce nitric oxide, makes detection of NVNOC possible by interfacing a chemiluminescence detector to an HPLC. We report here an analytical method for detecting nitrosamides in uncooked dried squid. As a model nitrosamide, we used N-nitrosotrimethylurea (NTMU), which is relatively stable, of low toxicity, and not reported to occur in nature. Our goal is to develop a method which we can apply to dried fish after cooking or reaction with nitrite. As a preliminary step, we undertook analysis of squid which was spiked with a model N-nitrosamide. It is likely that an analytical method capable of detecting NTMU would also detect other unknown NVNOC.

Experimental

Imported Korean dried squid was locally obtained. NTMU was synthesized by nitrosation of trimethylurea (TMU). HPLC-photolysis-chemiluminescence (HPLC-Nitrolite-TEA) conditions were as follows: HPLC, (4.6 x 250 mm Altex C-18 column with 5 μm particle size; injection volume of 20 μl); mobile phase, H_2O:ACN, 85:15; 1 mL/min. For UV-photolysis and interfacing to chemiluminescence detector, we used the Nitrolite (Thermedics Inc., Woburn, MA) which is based on a published design (*4*). Sample preparation is outlined in Figure 1. NTMU was extracted with MeOH, and the lipids were removed from the extract by a modification of the method of McLeod (*5*). This procedure removes lipids and waxes by precipitation at -85°C while the extract is agitated by bubbling N_2 gas to keep the particles small. The sample was filtered at -85°C in a cold-jacketed Buchner funnel. The filtrate was rotoevaporated to ca. 10ml at 35°C. The concentrate was then adjusted to pH 3-4 with 3N H_2SO_4. The extract was mixed with 8g Celite, packed over 15g Na_2SO_4 in a glass column (2 x 29cm) and eluted

0097–6156/94/0553–0355$08.00/0
© 1994 American Chemical Society

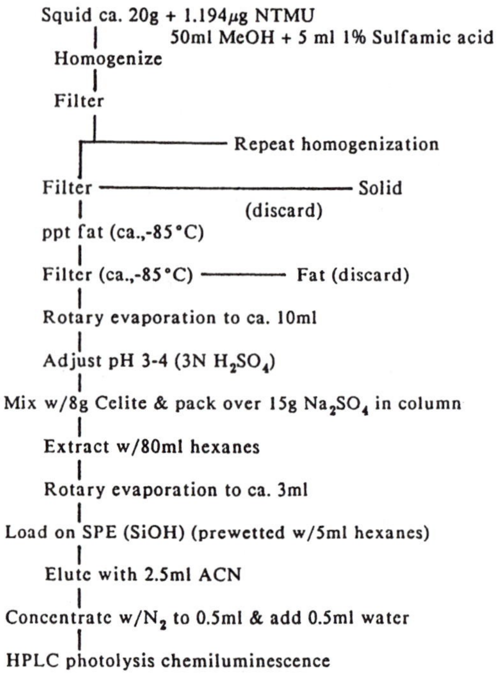

Squid ca. 20g + 1.194µg NTMU
  | 50ml MeOH + 5 ml 1% Sulfamic acid
Homogenize
  |
Filter
  |
    ┌─────────────── Repeat homogenization
  |
Filter ──────────────── Solid
  | (discard)
ppt fat (ca.,-85°C)
  |
Filter (ca.,-85°C) ──────── Fat (discard)
  |
Rotary evaporation to ca. 10ml
  |
Adjust pH 3-4 (3N H_2SO_4)
  |
Mix w/8g Celite & pack over 15g Na_2SO_4 in column
  |
Extract w/80ml hexanes
  |
Rotary evaporation to ca. 3ml
  |
Load on SPE (SiOH) (prewetted w/5ml hexanes)
  |
Elute with 2.5ml ACN
  |
Concentrate w/N_2 to 0.5ml & add 0.5ml water
  |
HPLC photolysis chemiluminescence

Fig 1. Analytical Scheme for NTMU
in Dried Squid

with 80ml hexane. The hexane was rotoevaporated to ca 3ml. The hexane extract was then loaded on a pre-wetted (5 mL hexanes) SPE(SiOH) column. NTMU was eluted from the column with 2.5 ml acetonitrile (ACN). The ACN was concentrated under a stream of N_2 gas to 0.5ml and 0.5ml H_2O added. This solution (20 ml) was analyzed by HPLC-Nitrolite-TEA.

Results

The HPLC-Nitrolite-TEA response to NTMU was linear (r=0.99) over a range of 2 to 20ng injected. The MeOH extract was viscous, cloudy and yellow after rotoevaporation to 10ml. Celite column extraction with hexanes and concentration to 3ml resulted in a fine precipitate which was removed by solid phase extraction (SPE). Moderately polar SPE packings were tested using ACN and dichloromethane (DCM) as eluting solvents to further clean the extract. Silica gel and florisil SPE with ACN as eluent, and cyano and amino SPE with DCM as eluent, produced satisfactory recoveries and baseline. Silica gel was chosen because ACN was compatible with the mobile phase. NTMU was most stable in the pH range of 3 to 5. The temperature of the water bath was important in achieving acceptable recoveries. With MeOH, the highest recovery was obtained of 30°C to 40°C. With hexanes, maximum recovery was obtained at 25°C and 30°C. The detection limit (3x noise) was 10µg/kg and recovery of NTMU from 60 µg/kg spiked squid was 79% (s.d.= 13%; c.v.= 16%; n=12).

Figure 2 is a representative chromatogram of a NTMU standard, spiked sample and unspiked sample. NTMU was not found in any unspiked samples but

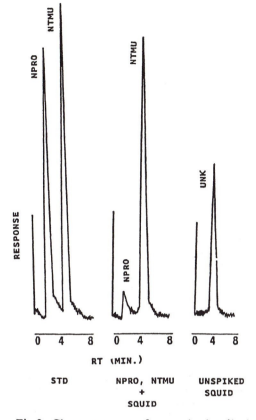

Fig 2. Chromatograms of a standard, spiked
sample and unspiked sample

several samples produced unknown peaks with different retention times than
NTMU. This indicates that dried squid contains unknown NOC. Because of the
selectivity of the photolysis-chemiluminescence reaction and this analytical
procedure, it is likely that these responses are due to unknown nonvolatile N-
nitroso compounds such as nitrosamides.

Literature Cited

1. Yu, M.C.; Ho, J.H.C.; Lai, S-H.; Henderson, B.E. *Cancer Res.*, **1986**, *46*, 956-961.
2. Mirvish, S. *J. Nat. Cancer Inst.* **1983**, *71(3)*, 631-647.
3. Matsui,M.; Ishibashi, T; Kawabata, T. *Bull. Japan. Soc. Fish.* **1984**, *50(1)*, 151-154.
4. Conboy, J.J.; Hotchkiss, J.H. *Analyst* **1989**, *114*, 155-177.
5. McLeod, H. *Anal. Chem.* **1972**, *44*, 1328-1330.

RECEIVED January 26, 1994

Chapter 40

N-Nitrosodiphenylamine in Diphenylamine-Treated Apples

Analysis by High-Performance Liquid Chromatography–UV Photolysis–Chemiluminescence

T. J. Lillard and J. H. Hotchkiss

Institute of Food Science, Stocking Hall, Cornell University,
Ithaca, NY 14583

Virtually all apples stored for later consumption are treated with diphenylamine (DphA) for scald, a respiratory deterioration at the apple surface. The level of DphA found in treated apples is on the order of 2 mg/kg and is concentrated in the skin (1). Due to the ease with which secondary amines are nitrosated, the possibility of finding N-nitrosodiphenylamine (NDphA) in apples arose. The carcinogenicity of NDphA is still an open question since animal assays have yielded both negative and positive results (2).

Method

Peels (10g) or 25g of chopped flesh of apples were frozen in liquid nitrogen and blended to a fine powder which was then mixed 5:4 with filter-grade Celite and packed over Na_2SO_4 and glass wool in a 1.9 cm diameter column. Hexane was used to extract the NDphA from the Celite column, 40 ml being collected. The hexane was passed through a 3 ml, 0.5 g silica gel SPE column (J.T. Baker). After a 5 ml hexane rinse, the NDphA was eluted with 2.5 ml 1:1 dichloromethane:hexane. 1.0 ml acetonitrile was added and the hexane evaporated under a stream of N_2.

Analyses were performed using HPLC-UV photolysis-chemiluminescence (3). HPLC conditions were as follows: Beckman 110B pumps and a 421 controller; Brownlee C-18 guard column; 4.6 x 250 mm Altex C-18 column with 5 μm particle size; mobile phase of 60:40 acetonitrile:water at 1 ml/min; and injection volume of 20 μl. For UV-photolysis and interfacing to the chemiluminescence detector, we used the Nitrolite (Thermedics Inc., Waltham, MA) which is based on a published design (3). The column output feeds through a microbore tube and then sprays into a helium-swept 1mm x 6m glass capillary column spiraled around a 200 W UV lamp. The nitrosyl group is cleaved by the UV light and then swept through cold traps at 0°C and -75°C prior to the -160°C cold trap attached to the TEA. For chemiluminescence detection, a TEA 543 by Thermedics was used. Peak area was computed by a Hewlett-Packard 3390A Integrator.

Results

The baseline of untreated apples was acceptable and steady using the extraction procedure outlined above and the limit of detection was approx. 3 μg/kg

0097–6156/94/0553–0358$08.00/0

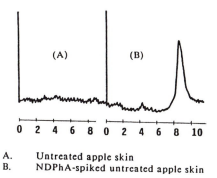

A. Untreated apple skin
B. NDPhA-spiked untreated apple skin

Figure 1. HPLC-photolysis-TEA chromatogram of apple skin known not to be treated with DPhA (A) and untreated apple skin spiked with NDPhA (B).

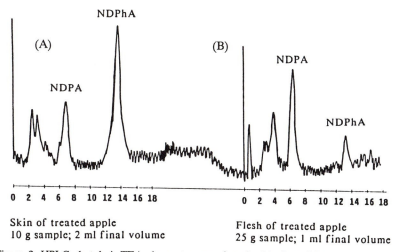

Skin of treated apple
10 g sample; 2 ml final volume

Flesh of treated apple
25 g sample; 1 ml final volume

Figure 2. HPLC-photolysis-TEA chromatogram of apple skin from DPhA treated apples (A) and the flesh from the same treated apple (B) (internal standard in N-nitrosodipropylamine, NDPA).

(3 x noise) based on spiked samples (Figure 1). NDPhA-treated or -spiked apples yielded a peak at 8.7 min., as did a standard solution of NDPhA. Mean percent recovery of apple skins spiked at 20 μg/kg, for 6 trials, was 71%, with a coefficient of variation of 13%. Untreated apples spiked with NDPhA-free DPhA at 10 mg/kg had no detectable NDPhA; thus NDPhA is not an artifact of the method.

Apples known or reported to be untreated, which were purchased at Cornell Orchards or at a local food cooperative, had no detectable NDPhA. Apples treated by Cornell Orchards with DPhA had an average level of 12 μg/kg NDPhA in the skins; but approx. 2 μg/kg, or just under the detection limit, in the skinless and coreless flesh (Figure 2). An Apple with an unknown history purchased at a local supermarket had 6 μg/kg NDPhA.

Conclusions

A method for the analysis of NDPhA has been developed. DPhA-treated apples show detectable levels of NDPhA; untreated apples do not. The concentration of NDPhA in skins of treated apples is at least 5 times greater than in the flesh.

Literature Cited

1. Gutenmann, W.H.; Lisk, D.J.; Blanpied, G.D. *J. Fd Safety,* 1990, *10*, 277-279.
2. Cardy, R.H.; Lijinsky, W.; Hildebrandt, P.K. *Ecotoxicol. Environ. Safety*, 1979, *3*, 29-35.
3. Conboy, J.J.; Hotchkiss, J.H. *Analyst* 1989, *114*, 155-177.

RECEIVED January 26, 1994

Chapter 41

Accumulation of Tobacco-Specific Nitrosamines during Curing and Aging of Tobacco

H. R. Burton and L. P. Bush

Department of Agronomy, University of Kentucky, Lexington, KY 40546-0091

Review of the literature on the tobacco-specific nitrosamines (TSNA) content of cured and processed tobacco has led to some conflicting conclusions. Brunnemann et al. (*1*) and Fischer et al. (*6*) reported there was a direct correlation between nitrate and TSNAs. Djordjevic et al. (*5*) reported no relationship between nitrate and TSNA's but there was a significant, positive correlation between secondary amine alkaloid content in cured tobacco and individual TSNA accumulation. It would seem there should be a better relationship between TSNAs and the level of nitrite in cured tobacco than either nitrate or alkaloid concentrations. This is based on the knowledge that nitrate concentration is 10^3 greater and alkaloid levels are 5 x 10^3 greater than TSNAs in cured tobacco (*3*). We propose that nitrite levels are rate limiting because of low levels in tobacco. Therefore, studies were initiated to determine if there was a correlation between nitrite and TSNAs in tobacco.

In a study on accumulation of TSNAs during curing, Burton et al. (*3*) reported there was a direct relationship between nitrite and TSNAs when tobacco was cured at high relative humidity. The large increase in TSNAs and nitrite occurred between the sixteenth and nineteenth day after harvest of stalk-cut tobacco. This observed increase was documented when cured tobacco was exposed to 32°C and 90% RH over a three week interval (*3*). It should be noted the levels of TSNAs and nitrite were approximately 1,000 ppm in the tobacco. Because these high levels of nitrosamines and nitrite do not reflect levels found in conventionally cured tobaccos, one would not expect to observe elevated levels of nitrite and TSNAs in tobaccos cured under ambient conditions. A three year study was initiated in 1988 to determine if there was a direct correlation between nitrite and TSNAs in conventionally cured tobaccos. Three different tobacco varieties were grown and cured using cultural practices for the production of burley tobacco. A flue variety (G28) was selected because flue cured varieties contain low levels of TSNAs. A burley variety (Ky14) also was grown along with a dark variety (Ky171). TSNA values for cured burley midvein ranged from 1ppm to 10ppm while nitrite-N values ranged from 2ppm to 140ppm. A significant correlation coefficient (r = 0.77) was obtained between nitrite and TSNA. Comparison of nitrite and NNN from the lamina

0097–6156/94/0553–0361$08.00/0
© 1994 American Chemical Society

and midrib also resulted in significant correlation coefficient. In general lamina contained the lowest levels of nitrite and NNN in comparison to levels contained in midvein tissue. When data regarding nitrite and TSNA levels from the three tobacco types were combined (n = 162) after three years of study, a significant correlation coefficient (r = 0.59) was obtained which indicated nitrosamine accumulation was limited by the plant's ability to accumulate nitrite during curing.

The distribution of the alkaloids and nitrite in the cured leaf also could limit TSNA accumulation. Under normal curing conditions TSNAs should accumulate when the reactants are in proximity to each other. There was no information available on the distribution of the precursors or products within the cured tobacco leaf. Recently Burton et al. (*4*) reported the distribution of TSNAs, individual alkaloids, nitrite, and nitrate in air-cured tobacco leaf. This study was achieved by segmenting air-cured tobacco lamina and midvein. The combined segments from 20 leaves were analyzed for TSNAs, alkaloids, nitrate, and nitrite. Results of this study showed alkaloid concentration was greatest at the periphery of the lamina and lowest at the basal portion of the midvein. This profile was obtained for all secondary amine pyridine alkaloids. If alkaloids were limiting TSNA accumulation, the TSNA concentration should be greatest at the periphery of the leaf. Nitrate concentration was highest along the center portion of the midvein. If nitrate were limiting one would predict TSNA and nitrite concentration be highest along the midvein portion of the leaf. The nitrite profile was quite different from the nitrate profile in the cured leaf. Nitrite was lowest at the tip of the leaf and highest at the base of the leaf including the midvein. The tip of leaf midvein contained the lowest concentration of nitrite. If nitrite was limiting nitrosamine accumulation, the TSNA distribution should be seemlier to the nitrite distribution in the cured leaf. The distribution profiles for TSNAs were almost identical to the nitrite distribution profile. This study indicated the accumulation of TSNAs in cured lamina is dependent on the ability of the senescing leaf to produce nitrite. The origin of nitrite most likely is due to microflora reduction of nitrate to nitrite.

Acknowledgments: The investigation reported in this paper was supported by the U.S. Department of Agriculture, Agricultural Research Service, under cooperative Agreement 58-6430-1-121 and is published with the approval of the Director of the Kentucky Agricultural Experiment Station (92-3-193).

REFERENCES

1. Brunnemann, K.D.; Masaryk, J.; Hoffmann, D. *J. Agric. Food Chem.* **1983**, *31*, 1221-1224.
2. Burton, H.R.; Bush, L.P.; Djordjevic, M.V. *J. Agric. Food Chem.* **1989**, *37*, 1372-1377.
3. Burton, H.R.; Childs, G.H., Jr.; Andersen, R.A.; Fleming, P.D. *J. Agric. Food Chem.* **1989**, *37*, 426-430.
4. Burton, H.R.; Dye, N.K.; Bush, L.P. *J. Agric. Food Chem.* **1992**, *40*, 1050-1055.
5. Djordjevic, M.V.; Gay, S.L.; Bush, L.P.; Chaplin, J.F. *J. Agric. Food Chem.* **1989**, *37*, 752-756.
6. Fischer, S.; Spiegelhalder, B.; Preussmann, R. *Carcinogenesis* **1989**, *10*, 1511-1517.

RECEIVED January 26, 1994

Chapter 42

Human Exposure to a Tobacco-Specific Nitrosamino Acid

A. R. Tricker[1,3], G. Scherer[1], F. Adlkofer[1], A. Pachinger[2], and H. Klus[2]

[1]Analytisch-biologisches Forschungslabor Professor Doktor F. Adlkofer, Goethestrasse 20, D–800 Munich, Germany
[2]Ökolab, Gesellschaft für Umweltanalytik, Hasnerstrasse 124a, A–1160 Vienna, Austria

Nicotine reacts slowly with nitrous acid in aqueous solution to produce the tobacco-specific N-nitrosamines 4-(N-methylnitrosamino)-4-(3-pyridyl)-1-butanal (NNA), N-nitrosonornicotine (NNN) and 4-(N-methylnitrosamino)-1-(3-pyridyl)-1-butanone (NNK) (*1*). It has been speculated that endogenous formation of Iso-NNAC may occur via direct oxidative nitrosation of nicotine, nitrosation of the nicotine metabolite, cotinine, and its hydrolysis product, 4-(methylamino)-4-(3-pyridyl)-butyric acid (*1*).

Iso-NNAC in Mainstream Cigarette Smoke

Iso-NNAC was found in cigarette tobacco (10-330 μg/g) and in the mainstream smoke (<5 ng/cigarette) of commercial cigarettes smoked under standard conditions. The transfer rate to mainstream cigarette smoke ranged between 1.1 and 2.5 % for filter and nonfilter cigarettes, respectively.

Biomonitoring Studies of Iso-NNAC Excretion

Following i.v. administration of 100 μg Iso-NNAC/rat, 24-h excretion in urine and feces was 67 and 6%, respectively. Iso-NNAC showed similar excretion characteristics to other N-nitrosamino acids normally found in human urine.

Urine samples (24 h) from adult smokers (n=20) and nonsmokers (n=12) were analysed for nitrate, nicotine, cotinine and Iso-NNAC. Iso-NNAC was only detected (limit of detection 20 ng/l) in 4 urine samples collected from smokers (44, 65, 74 and 163 ng/day). Three of these smoking volunteers and additional volunteers were requested to refrain from smoking for 5 days prior to administration of 4 mg nicotine capsules (n=8, total dose 12-40 mg) or cotinine (n=3, total dose 40-60 mg) with nitrate supplementation (150 mg NO_3^-). After administration of nicotine and cotinine, Iso-NNAC excretion was not found. Smoking status was validated

[3]Current address: Wissenschaftliche Abteilung, Verband der Cigarettenindustrie, Königswintererstrasse 550, 53227 Bonn, Germany

using urinary nicotine and cotinine determination and cessation was confirmed prior to administration of either nicotine or cotinine.

Conclusions

Since Iso-NNAC was not found after oral administration of nicotine or cotinine with nitrate, it is concluded that endogenous nitrosation of both nicotine and cotinine does not occur. This result is supported by (i) kinetic considerations that strongly argue against the endogenous nitrosation of tertiary amines such as nicotine (1), (ii) studies showing that chemical nitrosation of nicotine does not occur in human saliva and gastric juice ex vivo (3), and (iii) administration of nicotine to rats does not result in the formation of a globin adduct derived from endogenously formed NNN and NNK (4).

Since Iso-NNAC is rapidly excreted, shows no genotoxic properties in the hepatocyte primary culture (HPC)/DNA repair assay in the A/J mouse, and is not carcinogenic (5), the occasional occurrence of this compound in smokers urine cannot be considered to present a health risk.

Literature cited

1.   Caldwell, W.S., Greene, J.M., Plowchalk, D.R., deBethizy, J.D. *Chem. Res. Toxicol.*, **1991**, *4,* 513-516.
2.   Djordjevic, M.V.; Brunnemann, K.D.; Hoffmann, D. *Carcinogenesis (Lond.)*, **1989**, *10*, 1725-1731.
3.   Tricker, A.R.; Preussmann, R. In *Effect of nicotine on biological systems*; Adlkofer, F; Thurau, K. Eds.; Birkhäuser Verlag, Basel, **1991**, pp. 109-113.
4.   Hecht, S.S.; Kagan, S.S.; Kagan, M.; Carmella, S.G. In *Relevance to human cancer of N-nitroso compounds, tobacco smoke and mycotoxins*; O'Neill, I.K.; Chen, J.; Bartsch, H. Eds.; International Agency for Research on Cancer, Lyon, **1991**, Vol. 105, pp. 113-118.
5.   Rivenson, A.; Djordjevic, M.V.; Amin, S.; Hoffmann, D. *Cancer Lett.*, **1989**, *47*, 111-114.

RECEIVED October 5, 1993

Chapter 43

Significance of Nitrosamines in Betel Quid Carcinogenesis

Bogdan Prokopczyk, Jacek Krzeminski, and Dietrich Hoffmann

American Health Foundation, 1 Dana Road, Valhalla, NY 10595

Chewing betel quid has been estimated to be practiced by about 10% of the world's population (1). The high incidence of oral cancer in the Indian subcontinent has been causatively associated with this habit. The major carcinogens found in the saliva of betel quid chewers are three nitrosamines derived from Areca alkaloids (ADNA); 3-(methylnitrosamino)propionitrile (MNPN), nitrosoguvacoline (NG), and nitrosoguvacine (NGC). Capillary GC-TEA analyses have indicated the formation of additional nitrosamines in the saliva of betel quid chewers, namely, 3-(methylnitrosamino)propanal (MNPA), N-nitrosomethylethylamine, and 3-(methylnitrosamino)propionic acid. The confirmation of the identity of these compounds by mass spectrometry is now in progress.

MNPN is a strong carcinogen in F344 rats. Upon subcutaneous injection it elicits benign and malignant tumors of the esophagus, tongue, nasal cavity, and liver (2,3). MNPN was also tested for its tumor initiating activity on mouse skin and for its tumorigenic potential in the oral mucosa of rats (4). On mouse skin, MNPN showed only weak local tumor initiator activity. Twice daily swabbing of the oral cavity of rats with an aqueous solution of MNPN for up to 61 weeks led only to one oral tumor in 30 animals. Yet, regardless of the route of application, MNPN induced multiple distant tumors int he lungs, nasal cavity, liver and esophagus. MNPA is more cytotoxic to cultured human buccal epithelial cells than any other ADNA (5), and induces benign and malignant tumors in the lung of rats upon s.c. injection (6). Since MNPA-treated rats also developed tumors in the nasal cavity and in the liver, it appears that MNPA follows the pattern of other N-nitrosamines in terms of organ-specific carcinogenic activity. NG is weakly mutagenic to S.typhimurium TA1535 (7) and induces adenoma in the exocrine pancreas in rat (8).

MNPN is the most powerful Areca-derived carcinogen found in the saliva of betel quid chewers at levels of 0.50-11.39 ng/mL. It is activated in vivo by α-hydroxylation thereby forming two electrophiles. Methyldiazohydroxide, which results from the methylene-carbon oxidation, can methylate nucleophilic centers in DNA. When α-hydroxylation occurs on the methyl group of MNPN 2-cyanoethyldiazohydroxide if formed. This unstable

0097–6156/94/0553–0365$08.00/0

metabolite also reacts with nucleophilic centers of DNA. The formation of 7-methylguanine, 7-(2-cyanoethyl)guanine, O^6-methylguanine, and O^6-(2-cyanoethyl)guanine was detected in DNA of rats treated with MNPN *(9)*. In rat liver DNA, the levels of 7-methylguanine were 3.3-7.5 times higher than those of 7-(2-cyanoethyl)guanine. In contrast, the extent of formation of)^6-(2-cyanoethyl)guanine was similar to that of O^6-methylguanine. However, this ratio does not directly reflect the rate of metabolism of MNPN to a methylating or cyanoethylating intermediate since both the extent of alkylation and the profile of alkylation products can differ considerably. In DNA isolated from the nasal mucosa, the ratios of 7-methylguanine:7-(2-cyanoethyl)guanine were considerably lower (1.5:2.2) and the extent of formation of cyanoethylated guanines was considerably higher, suggesting that in this organ MNPN is more extensively metabolized to the cyanoethylating intermediate. Interestingly, the nasal mucosa was found to be a primary target tissue in the carcinogenicity assay. The involvement of cyanoethylated adducts in MNPN carcinogenicity requires further elucidation.

ACKNOWLEDGEMENTS

Supported by Grant CA-29580 from the National Cancer Institute.

REFERENCES

1. Fendell, L.D. and Smith, J.R. 1980, J. Oral Surg., 28: 455-456.
2. Wenke, G., Rivenson, S., and Hoffmann, D. 1984, Carcinogenesis, 5: 1137-1140.
3. Prokopczyk, B., Rivenson, A., Bertinato, P., Brunnemann, K.D., and Hoffmann, D. 1987, Cancer Res., 6: 467-471.
4. Prokopczyk, B., Rivenson, A., and Hoffmann, D. 1991, Cancer Lett., 60: 153-157.
5. Sudqvist, K., Liu, Y., Nair, J., Bartsch, H., Arvidson, K., and Grafstrom, R.C. 1989, Cancer Res., 49: 5294-5298.
6. Nishikawa, A., Prokopczyk, B., Rivenson, A., Zang, E., and Hoffmann, D. 1992, Carcinogenesis, 13: 369-372.
7. Rao, T.K., Hardigree, A.A., Young, J.A., Lijinsky, W., and Epler, J.L. 1977, Mutat. Res., 56: 131-145.
8. Rivenson, A., Hoffmann, D., Prokopczyk, B., Amin, S., and Hecht, S.S. 1988, Cancer Res., 48: 6912-6917.
9. Prokopczyk, B., Bertinato, P., and Hoffmann, D. 1988, Cancer Res., 48: 6780-6784.

RECEIVED January 31, 1994

Chapter 44

Characterization of N-Nitrosamino Acids in Tobacco Products and Assessment of Their Carcinogenic Potential

Mirjana V. Djordjevic, Jingrun Fan, Jacek Krzeminski,
Klaus D. Brunnemann, and Dietrich Hoffmann

American Health Foundation, 1 Dana Road, Valhalla, NY 10595

Ninety percent of the more than 300 nitrosamines bioassayed are organ-specific carcinogens in animals; some of these compounds are carcinogenic at doses as low as 0.1 mmol/kg. Several N-nitrosamines are suspected human carcinogens. Recently, our attention has been focused on N-nitrosamino acids (NAA) which occur widely in food products and in other consumer products. N-Nitrosamino acids are also known to be formed by endogenous nitrosation of amino acids. So far 16 nitrosamino acids have been identified in consumer products, 7 have been bioassayed and 3 have been found to be carcinogenic in mice and/or rats: N-nitrososarcosine (NSAR) induces liver carcinoma in male newborn mice and esophageal tumors in BD rats; 3-(methylnitrosamino)propionic acid (MNPA) induces lung tumors in female A/J mice and, 4-(methylnitrosamino)butyric acid (MNBA) causes bladder cancer in F344 rats.

The highest levels of carcinogenic N-nitrosamino acids were found in smokeless tobacco products (Table I), MNPA amounting to 70 μg/g dry weight of the tobacco of a recently introduced U.S. moist snuff brand. In addition to the known NAA, the aforementioned moist snuff brand contains at least five unknowns amounting to 50 ppm. To date, only the structure of a phenylalanine derivative, 2-(methylnitrosamino)-3-phenylpropionic acid, was confirmed.

Table I. N-Nitrosamino Acid Content in
U.S. Commercial Moist Snuff Brands
(μg/g dry weight)

	A	B	C	D	E[1]	Sweden[2]
NSAR	0.06	0.06	ND[3]	0.06	0.4-6.3	0.27
MNPA	5.13	3.62	2.72	2.20	6.0-70.0	3.17
MNBA	0.47	0.26	0.09	0.20	0.8-17.5	0.62
Total[4]	9.4	7.1	3.5	4.2	13-167.0	10.2

[1] Range of 9 samples purchased in different states in 1990 and 1991;
[2] Snuff in sachets imported from Sweden (average of 3 brands); [3] ND, not detected; [4] Includes NSAR, MNPA, MNBA, NPRO and iso-NNAC.
(Reproduced with permission from reference 3. Copyright 1991 British Industrial Biological Research Association.)

In a lung adenoma bioassay in A/J mice, in which MNPA was administered by i.p. injection, it proved to be about 1/10th as carcinogenic as the highly potent 4-(methylnitrosamino)-1-(3-pyridyl)-1-butanone [NNK; (*1*)]. The results of DNA alkylation studies with MNPA and NNK are well in line with the bioassay data (Table II). When MNPA is applied at a 10-fold higher dose than NNK, about the same levels of N^7- and O^6-methylguanine adducts are obtained in both mouse lung and liver. Considering that the levels of MNPA in smokeless tobacco exceed those of NNK about 10-fold, it is clear that this NAA makes a significant contribution to the overall carcinogenic burden of the tobacco user.

Table II. DNA Methylation by 3-(Methylnitrosamino)propionic Acid (MNPA) in Female A/J Mice

COMPOUND	DOSE μmol/mice	LUNG		LIVER	
		N^7-MeG	O^6-MeG	N^7-MeG	O^6-MeG
		pmol/μmol G		μmol/mol G	
MNPA	100	96	8	1,594	87
NNK	10	92	10	1,212	83
Untreated		ND	ND	ND	ND

Values are the average of 2 samples; ND, not detected

The nicotine-derived 4-(methylnitrosamino)-4-(3-pyridyl)butyric acid (iso-NNAC) is not tumorigenic in female A/J mice (1). Its lack of carcinogenicity, its low concentrations in cigarette tobacco [ND-28 ng/g dry weight; (*2*)], and its low transfer rate (1%) into the mainstream smoke make it a suitable biomarker of endogenous formation of carcinogenic tobacco-specific nitrosamines. Dosimetry of endogenous nitrosamine formation is an important tool for cancer risk assessment of tobacco users in view of the fact that *in vitro* nitrosation of nicotine under physiological conditions (37°C; pH 1.5, 5.5, 8.0; catalysis by thiocyanate) yielded up to 0.8 and 0.2% of N-nitrosonornicotine and NNK, respectively.

ACKNOWLEDGEMENT

Supported by Grant CA-29580 from the National Cancer Institute.

REFERENCES

1. Rivenson, A., Djordjevic, M.V., Hoffmann, D. 1989, Cancer Letters 47: 111–114.
2. Djordjevic, M.V., Sigountos, C.W., Brunnemann, K.D., Hoffmann, D. 1990, CORESTA Symposium Proceedings, SO4, pp. 54, Kallithea, Greece.
3. Hoffmann, D., Djordjevic, M.V., Brunnemann, K.D. 1991, *Fd. Chem. Toxic.* 29: 65-68.

RECEIVED January 26, 1994

Chapter 45

Analysis of Tobacco-Specific Nitrosamines in Tobacco and Tobacco Smoke

Klaus D. Brunnemann and Dietrich Hoffmann

American Health Foundation, 1 Dana Road, Valhalla, NY 10595

Several groups of nitrosamines have been identified in tobacco and to-bacco smoke; these include volatile nitrosamines (mainly nitroso-dimethylamine and nitrosopyrrolidine), non-volatile nitrosamines (such as nitrosodiethanolamine), nitrosamino acids (see abstract by Djordjevic et al. in this volume) and tobacco-specific nitrosamines (TSNA). The latter derive from the *Nicotiana* alkaloids. The major TSNA found in tobacco and tobacco smoke are N'-nitrosonornicotine (NNN), N'-nitrosoanatabine (NAT), N'-nitrosoanabasine (NAB), and 4-(methylnitrosamino)-1-(3-pyridyl)-1-buta-none (NNK). Of these, NNN and NNK are strong carcinogens in mice, rats and hamsters where they induce benign and malignant tumors of the lung, nasal cavity, mouth, esophagus, and/or pancreas. For the analysis of TSNA, 5-10 g of tobacco were extracted with citrate buffer pH 4.5 containing 20 mM ascorbic acid; N-nitrosopentylpicolylamine (NPePicA) served as internal standard. The buffer was then partitioned with dichloromethane (DCM) and the DCM fraction was chromatographed on 65 g Al_2O_3 using 250 ml DCM (to yield the VNA fraction) followed by 250 ml DCM-acetone (4:1 to yield the TSNA fraction). The latter was analyzed by GC-TEA. We analyzed 16 samples of moist snuff and found 0.8-64 μg NNN/g, 0.2-215 μg NAT/g, 0.01-6.7 μg NAB/g, and 0.08-8.3 μg NNK/g (expressed per dry tobacco weight). In six samples of dry snuff, we found 9.4-55 μg NNN/g, 11-40 μg NAT/g, 0.5-1.2 μg NAB/g, and 0.88-14 μg NNK/g. These levels of TSNA are at least three orders of magnitude higher than those N-nitrosamines found in any other consumer product.

A newly developed method for the extraction of TSNA in tobacco is based on the extraction with methanol-modified supercritical carbon dioxide (1). This method yielded up to 7 times more NNK than did conventional solvent extraction, while the yields of other TSNA were similar among the different procedures.

For the analysis of TSNA in cigarette smoke, 10-40 cigarettes (for mainstream smoke collection) or 5-10 cigarettes (for sidestream smoke collection) were smoked into two gas wash bottles containing citrate buffer pH 4.5 with 20 mM ascorbic acid as scavenger and NPePicA as internal standard, followed by a 90-mm Cambridge filter. Others claimed that this method may lead to artefact formation (2); however, in our hands there was no difference

0097–6156/94/0553–0369$08.00/0

whether the smoke was collected this way or onto a Cambridge filter treated with ascorbic acid or onto an untreated Cambridge filter directly *(3,4)*. The buffer and the filter were extracted with DCM and analyzed analogously to tobacco. Table I lists the major TSNA found in the mainstream smoke of some domestic cigarettes. Some foreign brands (Russian and French) yielded much higher TSNA data than domestic brands, while Japanese brands clearly had lower TSNA levels.

It is noteworthy that the brand yielding the highest TSNA levels in mainstream smoke yielded the lowest TSNA levels in the sidestream smoke and *vice versa*. Moreover, the sidestream/mainstream ratio for NNN ranged from 0.6-8.3 while that for NNK ranged from 1.5-21, a two to three fold increase. This result means that during smoldering of the cigarette more NNK is pyrosynthesized in the reducing zone of the burning cone. We observed a similar trend earlier *(5)*.

Table I. TSNA in the Mainstream Smoke of Domestic Cigarettes (ng/cig.)

Cigarette	Type	NNN	NAT	NAB	NNK	Total
A	NF	278	236	30	156	700
D	F	209	172	21	156	558
G	F, M	250	192	20	173	635
I	F, L	138	114	14	87	353
J	F, UL	40	37	4	17	98

NF=non-filter; F=filter; M=menthol; L=light; UL=ultra light.

We also analyzed the TSNA in the sidestream smoke of domestic cigarettes (Table II).

Table II. TSNA in the Sidestream Smoke of Domestic Cigarettes (ng/cig.)

Cigarette	Type	NNN	NAT	NAB	NNK	Total
A	NF	170	105	20	241	536
D	F	191	119	19	312	641
G	F, M	238	132	24	303	697
I	F, L	214	132	18	216	580
J	F, UL	330	221	41	352	944

For abbreviations see Table I.

We also applied supercritical fluid extraction (SFE) to the analysis of TSNA. The mainstream smoke of only one cigarette was collected on a 44-mm Cambridge filter and then extracted with supercritical CO_2 containing 5% methanol. 2-(2-Methylnitrosaminoethyl)pyridine served as internal standard. The recoveries of TSNA standards ranged from 60-68%. In contrast to tobacco, the SFE method yielded only 60-90% of the TSNA obtained with conventional solvent extraction. Efforts are underway to improve the efficiency by using different modifiers.

ACKNOWLEDGEMENTS

Supported by Grant CA-29580 from the National Cancer Institute.

REFERENCES

1.   Prokopczyk, B., Hoffmann, D., Cox, J.E., Djordjevic, M.V., and Brunnemann, K.D. Chem. Res. Toxicol. 5: 336-340, 1992.
2.   Caldwell, W.S. and Conner, J.M. 43rd Tobacco Chemists' Research Conf., Richmond, VA, Oct. 2-5, 1989; abstr. 45.
3.   Djordjevic, M.V., Sigountos, C.W., Brunnemann, K.D., and Hoffmann, D. J. Agric. Food Chem. 39: 209-213, 1991.
4.   Brunnemann, K.D., Cox, J.E., Liu, Y. and Hoffmann, D. 46th Tobacco Chemists' Research Conf., Montreal, Canada, Sept. 27-30, 1992; abstr. 22.
5.   Adams, J.D., O'Mara-Adams, K.J. and Hoffmann, D. Carcinogenesis 8: 729-731, 1987.

RECEIVED January 26, 1994

Author Index

Affiliation Index

Subject Index

Production: Charlotte McNaughton
Indexing: Deborah H. Steiner
Acquisition: Rhonda Bitterli
Cover design: Cesar Caminero

Printed and bound by Maple Press, York, PA

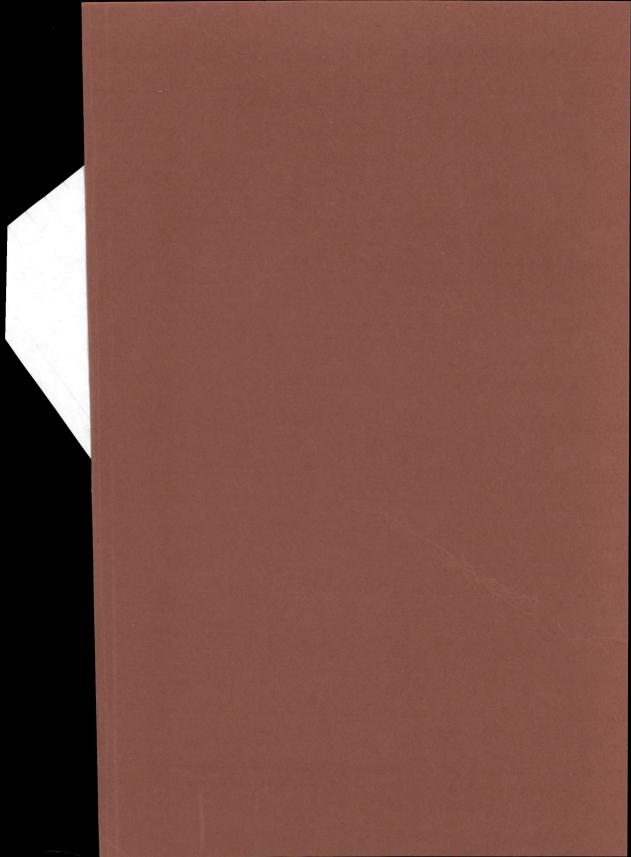

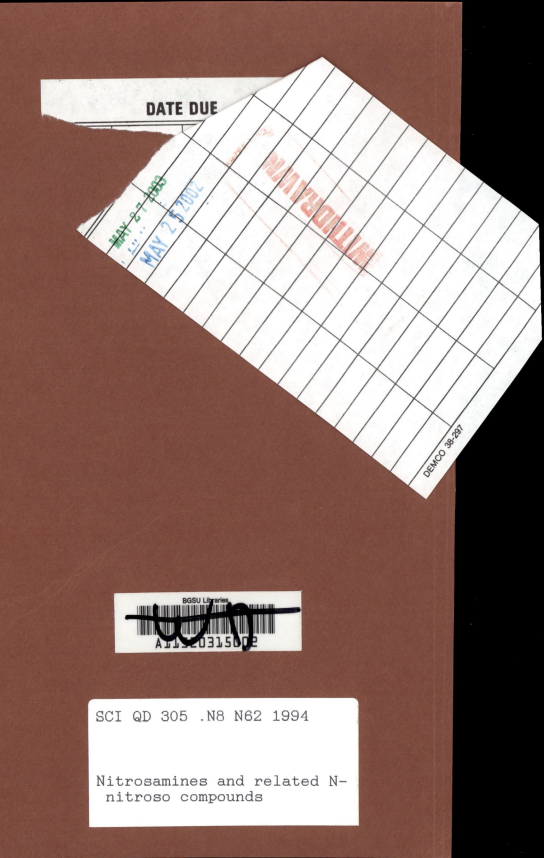